Handbook of
Neuroprosthetic
Methods

Biomedical Engineering Series

Edited by Michael R. Neuman

Published Titles

Electromagnetic Analysis and Design in Magnetic Resonance Imaging, Jianming Jin

Endogenous and Exogenous Regulation and Control of Physiological Systems, Robert B. Northrop

Artificial Neural Networks in Cancer Diagnosis, Prognosis, and Treatment, Raouf N.G. Naguib and Gajanan V. Sherbet

Medical Image Registration, Joseph V. Hajnal, Derek Hill, and David J. Hawkes

Introduction to Dynamic Modeling of Neuro-Sensory Systems, Robert B. Northrop

Noninvasive Instrumentation and Measurement in Medical Diagnosis, Robert B. Northrop

Handbook of Neuroprosthetic Methods, Warren E. Finn and Peter G. LoPresti

The BIOMEDICAL ENGINEERING Series
Series Editor Michael Neuman

Handbook of
Neuroprosthetic
Methods

Edited by
Warren E. Finn
Peter G. LoPresti

CRC Press
Taylor & Francis Group
Boca Raton London New York

CRC Press is an imprint of the
Taylor & Francis Group, an **informa** business

CRC Press
Taylor & Francis Group
6000 Broken Sound Parkway NW, Suite 300
Boca Raton, FL 33487-2742

First issued in paperback 2019

ISBN-13: 978-0-8493-1100-0 (hbk)
ISBN-13: 978-0-367-39559-9 (pbk)

Library of Congress Card Number 2002031437

Library of Congress Cataloging-in-Publication Data

Handbook of neuroprosthetic methods / edited by Warren E. Finn, Peter G. LoPresti.
    p. ; cm. -- (Biomedical engineering series)
  Includes bibliographical references and index.
  ISBN 0-8493-1100-4 (alk. paper)
  1. Neural stimulation--Handbooks, manuals, etc. 2. Prosthetics--Handbooks, manuals, etc. 3. Cochlear implants--Handbooks, manuals, etc. 4. Neurons--Handbooks, manuals, etc. I. Finn, Warren E. II. LoPresti, Peter G. III. Biomedical engineering series (Boca Raton, Fla.)
  [DNLM: 1. Nervous System Diseases--rehabilitation. 2. Prosthesis and Implants. 3. Electric Stimulation Therapy--methods. 4. Nervous System Physiology. 5. Prosthesis Design. WL 140 H2362 2002]
  RC350.N48 H36 2002
  616.8′046--dc21
                                        2002031437

Preface

Purpose

The purpose of *The Handbook of Neuroprosthetic Methods* is threefold. First, the book combines the most commonly employed concepts, applications, and knowledge from the many disciplines associated with neuroprosthetic research in a clear and instructive way. Second, the book provides examples of neuroprosthetic systems at different stages of development, from the more mature cochlear implant to the maturing areas of upper-limb and motor control to the relatively fledgling area of visual prostheses. The book explores the varying developmental processes to give the reader guidance on issues that have yet to be solved, successful strategies for solving such problems, and the potential pitfalls encountered when developing neural prostheses. Third, the book introduces key topics at a level that is useful to both new and practicing professionals working directly or indirectly with neuroprosthesis projects. In this way, the book provides an accessible common ground and perhaps fosters a more effective and productive collaborative environment for multidisciplinary teams working on protheses.

Organization

The book is organized into six main sections, starting with basic neurophysiology and ending with some of the emerging technologies that will have significant impact on the next generation of prostheses. Section I provides an overview of the significant events in the field of neuroprostheses and the broad problems that remain a challenge to the development of a functional and practical neuroprosthetic system. Section II addresses the main target of the neuroprostheses, the neuron. The main topics in this section are how neurons can become electrically excited to produce signals used to restore sensory or motor function and the ways this behavior can be modeled mathematically and predicted.

Section III addresses the important and difficult task of recording from and stimulating neurons, either in the laboratory or in devices implanted in humans. The book first addresses the design problem of finding an effective stimulus to adequately activate a population of neurons in a nerve pathway to restore function. Of importance here is the concept of not only sending the correct signals but also sending the signals in a way that does not injure or destroy the nerve cells with which one is communicating. Second, the book addresses the components and devices commonly used for communication in current systems. Third, the book addresses the task of listening to the chatter amongst neurons using many of the same devices utilized for signal transmission.

In Section IV, the book addresses some issues related to processing the recorded neural signals. The goal of signal processing is to understand how the neurons are communicating and how information is encoded within their communications. Section V provides examples of three neuroprosthetic systems at different stages of the development cycle. Section VI introduces some emerging technologies that promise to alter current approaches to neuroprosthetic design. Through this organization, the book accentuates the potential contribution of the biological and engineering fields brought together to solve the complex problems that are at the heart of the neuroprosthetic field.

Each chapter follows a basic format that we hope the reader finds useful. Each chapter begins with a brief history of the topic and then addresses the fundamental issues and concepts. It is hoped that the manner in which each topic is addressed provides understanding to the beginning practitioner as well as guidance to practitioners with more experience in the field. The last part of each chapter provides practical applications and examples that relate the topic to the actual design and implementation of a neuroprosthetic system or device. In this way, each chapter provides a connection between theory and practice that will help the reader better comprehend the material presented.

Warren Finn
Tulsa, Oklahoma

Peter LoPresti
Tulsa, Oklahoma

The editors

Warren E. Finn, Ph.D., is an Associate Professor of Medical Physiology in the Department of Pharmacology and Physiology in the Biomedical Sciences Program at the Oklahoma State University Center for Health Sciences. Dr. Finn earned his Baccalaureate and Masters of Science degrees in zoology at University of Wisconsin. In addition, he earned his Ph.D. in the biological sciences, physiology from Texas A&M University in College Station, TX. Dr. Finn teaches medical and graduate students in the fields of cellular, molecular and integrative neurophysiology. His research interests are in the areas of cell culturing and the electrophysiology of retinal neurons. He co-coordinates with Dr. LoPresti the Artificial Vision Project, a multidisciplinary research program studying vision neuroprosthetics. This project provides research training for students in medicine, biomedical sciences, and electrical engineering. Dr. Finn has worked on various bioengineering projects, such as the biophysics of hypothermia as a treatment for cerebral ischemia, the activation of sensory neurons in myocardial ischemia, and the electrophysiology of amblyopic eye disease. With Dr. LoPresti, he has authored studies on animal models for the development of retinal prostheses reported at various IEEE Engineering in Medicine and Biology Society Annual International Conferences. Dr. Finn contributed three chapters to the Handbook of Endocrinology, published by CRC Press. He is an active participant in policy and strategic planning in biomedical engineering through his activities in the development of intellectual property policies and technology transfer.

Peter G. LoPresti, Ph.D., an associate professor of electrical engineering at the University of Tulsa, teaches graduate and undergraduate courses in electronics, signal processing, and optical communications. He earned a Ph.D. in electrical engineering from the Pennsylvania State University and a B.S. in electrical engineering from the University of Delaware. His current research interests include visual neuroprosthetics, fiber-optic sensors, and optical networking, and he serves as the director for the Williams Communications Fiber-Optic Networking Laboratory at the University of Tulsa. Dr. LoPresti is highly dedicated to improving the quality of engineering education, having won both university and departmental teaching awards and serving as coordinator for the electrical engineering component of the Tulsa Undergraduate Research Challenge program, which provides accelerated learning and research experiences for undergraduates.

Contributors

David J. Anderson
University of Michigan
Ann Arbor, Michigan

Danny Banks
Monisys, Ltd.
Birmingham, England

Steven Barnes
Dalhousie University
Halifax, Nova Scotia, Canada

Rizwan Bashirullah
North Carolina State University
Raleigh, North Carolina

Chris DeMarco
North Carolina State University
Raleigh, North Carolina

Kenneth J. Dormer
The University of Oklahoma Health
  Sciences Center
Oklahoma City, Oklahoma

Kevin Englehart
Institute of Biomedical Engineering,
  and Department of Electrical and
  Computer Engineering
University of New Brunswick
Fredericton, New Brunswick,
  Canada

Warren E. Finn
Oklahoma State University
Tulsa, Oklahoma

Robert J. Greenberg
Second Sight, LLC
Valencia, California

Steven Barnes
Dalhousie University
Halifax, Nova Scotia, Canada

Warren M. Grill
Case Western Reserve University
Cleveland, Ohio

Jamille F. Hetke
University of Michigan
Ann Arbor, Michigan

Bernard Hudgins
Institute of Biomedical Engineering,
  and Department of Electrical and
  Computer Engineering
University of New Brunswick
Fredericton, New Brunswick,
  Canada

Mark S. Humayun
Doheny Retina Institute
University of Southern California
Los Angeles, California

Richard T. Lauer
Shriners Hospitals for Children
Philadelphia, Pennsylvania

Dongchul C. Lee
Case Western Reserve University
Cleveland, Ohio

Wentai Liu
North Carolina State University
Raleigh, North Carolina

Peter G. LoPresti
University of Tulsa
Tulsa, Oklahoma

Cameron C. McIntrye
The Johns Hopkins University
Baltimore, Maryland

Richard A. Normann
University of Utah
Salt Lake City, Utah

Philip Parker
Institute of Biomedical Engineering,
  and Department of Electrical and
  Computer Engineering
University of New Brunswick
Fredericton, New Brunswick,
  Canada

C. Pearson
University of Durham
Durham, England

P. Hunter Peckham
Rehabilitation Engineering Center
MetroHealth Medical Center
Cleveland, Ohio

Michael C. Petty
University of Durham
Durham, England

Frank Rattay
TU-BioMed
Vienna University of Technology
Vienna, Austria

Susanne Resatz
TU-BioMed
Vienna University of Technology
Vienna, Austria

Donald L. Russell
Carleton University
Ottawa, Ontario, Canada

Praveen Singh
North Carolina State University
Raleigh, North Carolina

David J. Warren
University of Utah
Salt Lake City, Utah

James D. Weiland
Doheny Retina Institute
University of Southern California
Los Angeles, California

Acknowledgments

Dr. Finn would like to thank his wife, Judith, and daughters, Kirstin and Arikka, for their help, support, and encouragement during this writing project. They are all excellent writers in their own right, and he greatly valued their advice during these months. He would also like to thank his colleagues David John and George Brenner for their encouragement and support in undertaking this project and the sharing of their wisdom as authors.

Dr. LoPresti would like to acknowledge the unwavering support of his family, both natural and adopted, through the process of developing this book. In particular, he acknowledges the support of Carrie and Joshua, who keep his spirits up and his priorities straight.

Drs. Finn and LoPresti both wish to thank their many students in medicine and electrical engineering who have participated in the Artificial Vision Project over the years. They appreciate their many hours of helpful discussion and persistence in the laboratory. They would also like to thank David Mooney, Assistant Librarian at the Oklahoma State University Center for Health Sciences, for the generous sharing of his knowledge and search skills of the world's literature. They also thank Jeffrey Shipman for his assistance with the historical timeline. They also wish to acknowledge the excellent editorial assistance of Susan Farmer, Helena Redshaw, Robert Stern, Susan Fox, and many others at CRC Press in making this book a reality.

Table of Contents

section one

Introduction

chapter one

Introduction to neuroprosthetics

Warren E. Finn and Peter G. LoPresti

Contents

1.1 Purpose of the handbook

Since 1990, the field of neuroprosthetics has grown at a tremendous rate. But, what exactly is meant by the term *neuroprosthetics*, and why are neuroprostheses of such consuming interest? For the purposes of this handbook, a neuroprosthetic is a device or system that does one of the following:

0-8493-1100-4/03/$0.00+$1.50
© 2003 by CRC Press LLC

1. Replaces nerve function lost as a result of disease or injury. The neuroprosthetic commonly acts as a bridge between functional elements of the nervous system and nerves or muscles over which control has been lost. Examples include the peripheral nerve bridges implanted into the spinal cord, lumbar anterior-root stimulator implants to allow standing in paraplegics, and systems to restore hand and upper limb movement in tetraplegics. The neuroprosthetic may also act as a bridge between the nervous system and a physical prosthesis, as is the case in upper limb replacement.
2. Augments or replaces damaged and destroyed sensory input pathways. The neuroprosthetic records and processes inputs from outside the body and transmits information to the sensory nerves for interpretation by the brain. Examples include the cochlear implant for restoring hearing and an assortment of retinal and visual cortex prostheses for restoring vision.

A common component of all the systems in Figure 1.1 is the need to interact directly with nerves. The system must either collect signals from nerves or generate signals on nerves, or both. The interaction may be with individual nerve cells and fibers or with nerve trunks containing hundreds to millions of axons. Just as important is the need to understand and speak the language of the nervous system and understanding that the language changes as the signaling requirements change. For example, the auditory and optic nerve systems have very different organizations, levels of signaling, and processing complexity as dictated by the different nature of the auditory and visual inputs. A neuroprosthesis, therefore, is a device or system that communicates with nerves to restore as much of the functionality of the nervous system as possible.

1.1.1 Why a handbook on neuroprosthetics?

The rapidly expanding interest and research in neuroprosthetics over the last decade paralleled the rapid increase in resources and literature devoted to the larger fields of bioengineering and biomedical engineering. A quick search of the Internet finds over 50 academic institutions with departments of bioengineering, many of which are less than a decade old, and many more with a bioengineering or biomedical engineering "emphasis" within traditional departments such as chemistry, electrical engineering, mechanical engineering, and biology. Professional publications that present research in bioengineering continue to increase in number and in size. An excellent example of this is the *IEEE Transactions on Systems, Man and Cybernetics*, which was founded as a single entity in January of 1971, was split into two parts in 1996, and had to be split yet again into three parts just two years later in 1998. New titles, such as the *IEEE Transactions on NanoBioscience* (due to be published in late 2002 or early 2003), reflect the effect of emerging technologies on the practice of bioengineering and

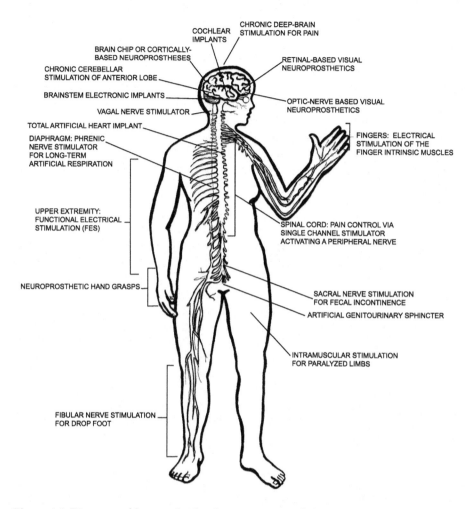

CHRONIC DEEP-BRAIN
STIMULATION FOR PAIN

COCHLEAR
IMPLANTS

BRAIN CHIP OR CORTICALLY-
BASED NEUROPROSTHESES

CHRONIC CEREBELLAR
STIMULATION OF ANTERIOR LOBE

BRAINSTEM ELECTRONIC IMPLANTS

VAGAL NERVE STIMULATOR

TOTAL ARTIFICIAL HEART IMPLANT

DIAPHRAGM: PHRENIC
NERVE STIMULATOR
FOR LONG-TERM
ARTIFICIAL RESPIRATION

UPPER EXTREMITY:
FUNCTIONAL ELECTRICAL
STIMULATION (FES)

NEUROPROSTHETIC HAND GRASPS

FIBULAR NERVE STIMULATION
FOR DROP FOOT

RETINAL-BASED VISUAL
NEUROPROSTHETICS

OPTIC-NERVE BASED VISUAL
NEUROPROSTHETICS

FINGERS: ELECTRICAL
STIMULATION OF THE
FINGER INTRINSIC MUSCLES

SPINAL CORD: PAIN CONTROL VIA
SINGLE CHANNEL STIMULATOR
ACTIVATING A PERIPHERAL NERVE

SACRAL NERVE STIMULATION
FOR FECAL INCONTINENCE

ARTIFICIAL GENITOURINARY SPHINCTER

INTRAMUSCULAR STIMULATION
FOR PARALYZED LIMBS

Figure 1.1 Diagram of human body showing many of the neuroprosthetic systems currently employed or in development.

biomedical engineering. Government spending and support, always important in the sciences, has also been increasing, as evidenced by such initiatives as the German Federal Ministry of Education and Science and establishment of the National Institute of Biomedical Imaging and Bioengineering in 2000 at the National Institutes of Health in the United States. While the wealth of support, research, and literature is a good thing, it can present even an experienced practitioner with the daunting task of assembling the basic knowledge required to design and implement an effective neuroprosthesis from widely spread resources. One reason for the book, then, is to provide a point of consolidation for key information that aids practitioners in more effectively finding the techniques and information they need.

Another reason for developing this book is the inherent interdisciplinary nature of the neuroprosthetics field. Practitioners from such disparate fields as electrical engineering, mechanical engineering, mathematics, physics, computer science, physiology, neurology, pharmacology, and cellular biology, to name just a few, must work together in a cooperative environment, without the benefit of a common vocabulary, a common set of methods for approaching problems, or a common set of analytical and experimental tools. Yet, methods and concepts from all of these areas are necessary to build an effective neural prosthesis. We give some examples of this marriage of engineering and biology later in this chapter. The second reason for the book, therefore, is to provide a common point of reference to facilitate interactions and understanding among practitioners from different backgrounds.

The final reason for developing this book is the potential impact that the development of neural prosthetics may have on society as a whole. Neural prosthetics have the power to significantly extend the lifespan of a person and increase the portion of that lifespan during which a person is an active and productive member of society. Profound changes in workforce demographics, national healthcare systems, and the way in which people participate in society in their later years are quite likely to follow as more people live longer and lead more active lives. Neural prosthetics have the potential to reverse, in total or in part, the loss of function not related to aging, which may lessen the physical and psychological impact of injury or disease. In addition, neuroprosthetics have the potential to extend the capabilities of the human body beyond its current limitations. Visual prostheses using semiconductor-based photoreceptors, for example, could extend the visual experience into the infrared. The potential impact on social and political systems will likely be significant as well, though this is beyond the scope of this book.

1.1.2 What this book hopes to accomplish

The purpose of this book is threefold. First, the book intends to combine the most commonly employed concepts, applications, and knowledge from the many disciplines associated with neuroprosthetics in a clear and instructive way. Mathematical and modeling theories combine with examples of their use in existing and future neuroprostheses to clarify their usefulness and demonstrate their limitations. Second, the book intends to provide examples of neuroprosthetic systems at different stages in their development, from the more mature cochlear implant to the maturing area of upper-limb control to the still unsettled world of visual prostheses. The book hopes that exploring the varying developmental processes will provide guidance for those developing other prostheses in regard to potential pitfalls, looming issues that must be solved, and successful strategies for solving difficult problems. Third, the book hopes to introduce key topics at a level that is useful to both new and practicing professionals working directly or indirectly with neuroprosthesis projects. By providing an accessible common ground, the book hopes to foster a more effective and productive collaborative environment.

1.2 Evolution of neuroprosthetics

Neuroprosthetics, while currently a field generating a lot of interest and scholarly work, has reached its current state through a traceable evolutionary path. Some of the key events and technologies that have driven this evolution are noted on the timeline in Figure 1.2. As one can see from the timeline, a long period of steady improvements in technology and groundbreaking experiments has led to an explosion of new and more functional systems in the last few years. In this section, we briefly discuss some key aspects of the evolution of neuroprosthetics.

1.2.1 Early experimentation and technologies

The field of neuroprosthetics has a long history, tracing its roots back to the 18th century. Luigi Galvani observed at that time that a frog's skeletal muscles contracted when in contact with both an anodic and a cathodic metal. Allesandro Volta, his contemporary, connected his newly discovered "battery" to his ear and discovered that an aural sensation could be induced electrically. These early experiments were severely limited by the available technologies, particularly in the equally fledgling area of electricity.

As technology and the fields of physiology and biology advanced, the experiments of Galvani, Volta, and other pioneers were periodically revisited. It was not until 1934, however, that the first attempt at developing something like a modern prosthesis was successful. At that time, the first true electronic hearing aid was developed, based on the work of Wever and Bray.[1] While admittedly crude, it was the first real indicator that meaningful improvement in sensory function by an electronic device was possible. However, real breakthroughs would not be achieved until engineering and biological methods improved to the point that communication on the cellular level was possible.

1.2.2 Key tools arrive and first successes reported

1.2.2.1 Key tools

In the middle and late 20th century, researchers received several key tools that would facilitate the development of more successful neuroprostheses. The first of these tools, the transistor and its platform, the integrated circuit, evolved over a period from 1947 through 1963. The transistor provided a host of capabilities in a package that would rapidly decrease in size. Currents could be more precisely controlled and switched using small voltage across a metallurgical junction between positively doped (p-type) and negatively doped (n-type) semiconductors, and this current could be made independent of the controlling circuit. The transistor was also capable of significantly amplifying weak signals without the need for bulky and radiative transformers, which eventually reduced the size and cost of signal amplification circuitry required to "hear" neurons and nerves talking. The technology

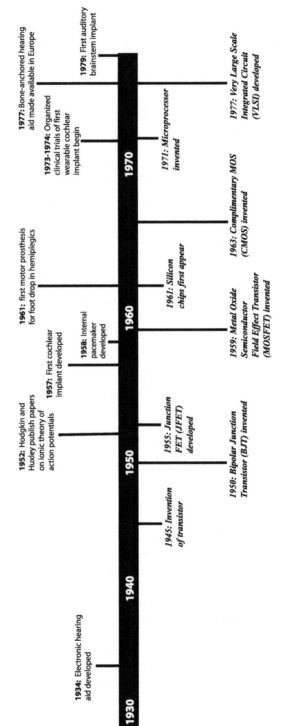

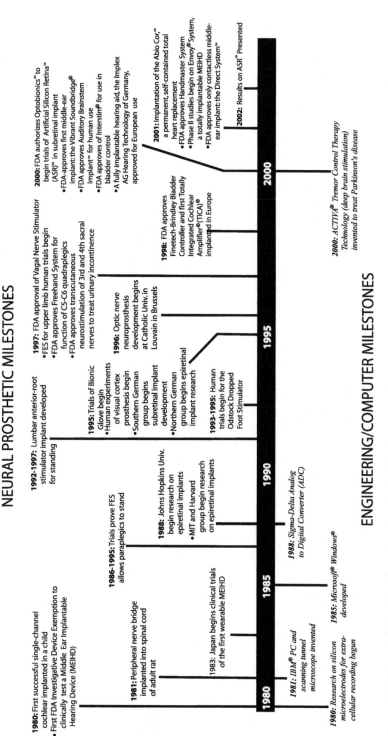

Figure 1.2 Historical timeline showing the relationship between engineering and computer milestones and progress in the development of neuroprosthetic devices.

required to produce the MOSFET (metal-oxide semiconductor field effect transistor) is a particularly important milestone, as this led to the development of charge-coupled device (CCD) cameras and variations capable of incorporating neurons into the structure for either recording or stimulation.[2,3] The development of integrated circuits (ICs) and silicon chips made the technology more manageable and simpler to use.

The seminal work of Hodgkin and Huxley appeared in 1952 and continues to influence the way researchers envision the communication between electronics and tissue to this very day (see Chapter three).[4] Based on a series of experiments on squid axons, Hodgkin and Huxley developed a physiological model of neuron behavior at an unprecedented level of detail. In particular, the model illuminated the process of action-potential generation, detailing the roles of membrane potential and ionic currents. The work also described unique methods for controlling and recording neural signaling. Present researchers use the information contained in the model to develop electrical models of neuron activity, select the proper methods for eliciting a desired response, and deciphering the origin of measured signals (see Chapters three, four, and five for basics; also see discussions in Section five).

As the ability to communicate with neurons grew and the models describing the process became more sophisticated, a tool was required to allow scientists to control more complex experiments, to process electrical signals more rapidly, and to isolate the contributions of individual neurons. The development of the microprocessor in 1971 and very-large-scale integration (VLSI) in 1977 provided the necessary tools. The techniques developed to construct VLSI circuits formed the foundation from which circuit miniaturization and micromachining of microelectromechanical systems (MEMS) and electrode arrays were developed (see Chapters six, seven, and thirteen, for example). Transistors and other electronic structures the size of a neuron were now possible. The microprocessors provided researchers with a means to program and automate experimental procedures, data collection, and data processing. While specialized processor chips (such as digital signal processing [DSP] and analog-to-digital converter [ADC] chips) would not come until later, the microprocessor still facilitated a dramatic increase in experimental complexity and control with a correspondingly dramatic decrease in execution time. As microprocessor technologies and architectures matured, experiments requiring simultaneous monitoring of multiple signals became feasible. This led to new discoveries such as the concerted signaling in the retina that offers clues to the processing of visual images.[5]

The combination of affordable microprocessors and VLSI led to one of the most important tools at the disposal of modern researchers, the affordable personal computer. The first IBM personal computer appeared in 1981, and the computer has since become a staple of every laboratory and research center. Not only has the computer increased the degree of automation in experimental work, but it has also made the processing and analysis of data easier. Complicated control algorithms for artificial limbs can be tested and modeled easily. Images from microscopes and analog signals from recording electrodes are captured directly into the computer, reducing the need for such time-consuming activities such as scanning pictures, developing film, and data entry. Many modern analysis and processing tools for extracting information from images and data, such as the wavelet transform in Chapter eight, would be laborious or impossible without the capabilities of the modern computer. Most importantly, many experiments and tests can be performed completely on the computer, without the need for biological subjects, in order to perfect designs and minimize hazards before live testing begins. More powerful programming languages and user-friendly interfaces have helped to ensure the continued utility of this most useful tool.

Finally, the development of the scanning tunnel microscope has allowed researchers to visually explore the world in which they operate. Combined with cell staining and other cell marking techniques, one is now able to observe the growth of artificial neuronal networks, explore the impact of implants on cell pathology, and actually view the tiny microelectrodes and similar devices, just for starters. Assays such as these provide invaluable information on the behavior of nerves and neurons and their interaction with foreign implants that improve the design of the next generation of devices and systems.

1.2.2.2 First successes in neuroprosthetics

As the tools noted above became available and more widespread, a number of successful prosthetic systems were developed. The development of just two such systems are detailed below and in Figure 1.2. Auditory prosthetics, building on the early success with the hearing aid, were among the first to benefit from the new technologies. The first cochlear implant was developed in 1957 by Djourno and Eyries[1] and consisted of electrodes placed on the auditory nerve and stimulated at different pulse rates. By the mid-1970s, the cochlear implant had been refined to the point that clinical trials were begun in the United States, and a bone-anchored hearing aid was made available in Europe. By 1980, the cochlear implant became important enough to warrant a U.S. Food and Drug Administration (US-FDA) Investigative Device Exemption to clinically test a middle ear implantable hearing device (MEIHD), and by 1983 clinical trials had begun in Japan as well. By the early 1990s, the cochlear implant began to gain widespread commercial and public acceptance.

The other prosthetic systems developed at this time addressed losses in motor function. The first motor prosthesis, targeted at foot-drop in hemiplegics, was developed in 1961. Despite this early success, it was not until the mid-1980s that clinical trials definitively proved that functional electrical stimulation (FES) of motor nerves and muscles was a valid approach. These trials showed that FES could allow paraplegics to perform the actions required to stand. By the mid-1990s, several versions of a neural prosthesis for standing had been developed and approved for human trials, and systems for biotic hands, upper-limb prostheses, and systems to treat urinary incontinence had begun in earnest.

1.2.3   Rapid expansion

In the last five years alone, the field of neuroprosthetics has seen a rapid growth, both in the enabling technologies and in the number of biological systems targeted by neuroprostheses. Again, Figure 1.1 provides a summary of the more prominent prosthetic systems under development.

In 1998, the US-FDA approved the Finetech–Brindley bladder controller for commercial use, and the first totally integrated cochlear amplifier (TICA) was implanted in Europe. In 2000 alone, the US-FDA approved the first middle ear implant, the auditory brainstem implant, and the Interstim implant for bladder control for use in humans. Also, a fully implantable hearing aid, the Implex AG Hearing Technology from Germany, was approved for European use. Prostheses for restoration of vision began to make significant progress with large-scale human trials of prostheses located in the visual cortex (1995), epiretinal space (1998), and subretinal space (2000). The results of the US-FDA-authorized subretinal trial were presented in 2002 (see Chapter 11 for details). Implantation of the Abio Cor, a permanent, self-contained heart replacement proceeded in 2001, along with Phase II studies on a totally implantable MEIHD. Also in 2001, the US-FDA approved the first contactless middle-ear implant and the Handmaster system for restoring hand functionality.

A host of technologies and extensive research have built upon the earlier developed tools to fuel this explosion in neuroprosthetics. The continued miniaturization of all forms of electronics, from cameras to processors to the electrodes themselves, has made it more feasible to communicate with larger numbers of nerves and neurons, therefore providing finer control over motor functions and finer sampling of sensory inputs to the ears and eyes. Communications technologies, both optical and electronic, have reduced the need for control wires and transcutaneous electrical connections and are key to liberating the implant recipient from excessive external apparatus and preventing infection. Emerging technologies such as MEMS (see Chapter thirteen), biomolecular electronics (Chapter fourteen), and artificially grown neuronal networks[6,7] have provided the means to better understand neural behavior and engineer effective and long-lasting prostheses. Materials and methods for reducing the rejection of the foreign prostheses by biological tissue have matured and are well known (see Section 12.3 of Chapter twelve for a basic discussion). The impacts of fields such as nanotechnology and genetic engineering have yet to be felt.

1.3   A marriage of biology and engineering

As evidenced by the timeline in Figure 1.2 and discussions of the previous section, the development and realization of neuroprosthetic devices require the talents and knowledge of both the biologist and the engineer. Even basic experimentation, whether in the laboratory or in a theoretical framework, cannot be performed without appropriate knowledge of biological and engineering techniques and technologies. In this section we discuss briefly some of the key issues that are central to the successful development of a neuroprosthesis and that require a combined biological/engineering solution.

1.3.1   How do neurons communicate with each other?

Before one can begin talking to nerves and neurons, one must first determine how they communicate with one another and determine how information is encoded within the communication. The challenge, then, is to monitor the signals transmitted by one or more neurons to a tightly controlled natural stimulus and correlate features of the signals with information contained in the stimulus.

Several methods have been devised to record from nerves and neurons, based on biological knowledge of how nerves conduct signals. Most nerves communicate via action potentials, a complex signal generated by an intricate coordination of ion movements across neuronal membranes (see Chapters two and three) and controlled by voltage potentials across the cell membrane. Recording devices must therefore tap or intercept voltages and ionic currents, and transform them into electrical signals suitable for processing. While the concept is relatively simple, implementing such a device is complicated by the millimeter to micrometer scale of most neurons and the small changes (millivolts or lower) in membrane potentials typically encountered. Material scientists are required to develop devices small and reliable enough to interact with a neuron. Devices such as the cuff electrode and suction electrode[8] measure a compound signal from the entire nerve, while single-wire electrodes and electrode arrays (see Chapter seven) aim to record from one or a small population of neurons, respectively. Electrical engineering techniques are required to extract the neural signal from biological and external noise sources and amplify them to manageable levels for processing. Solutions are found in physical differential amplifier-based head stages and in analog filters and amplifiers, in addition to the software or microprocessor-based solutions that use sampled representations of the recorded signal.

Correlating features of the neural signal with the original stimulus requires knowledge of the biological system under study and powerful analysis tools to examine the myriad of possibilities. In the ear, for example, the axons that extend from the cochlea to make up the auditory nerve are known to respond to specific sonic frequency ranges distributed along the length of the cochlea (see Chapter ten). This knowledge of how the neural system functions provides important clues to interpreting signals from different sections of the auditory nerve. It is also generally agreed upon that the frequency, timing, and duration of action potentials generated by a given neuron carry a significant amount of information. Analysis tools must therefore be able to track amplitude and frequency as a function of time. Wavelet theory has proven to be a useful tool, though not the only one, in addressing this challenge. Artificial neural networks and other sophisticated statistical analysis programs provide the computational power necessary to sift through a variety of possibilities and to arrive at the most likely relationships between stimulus and response (see Chapter twelve for example).

1.3.2   How does one communicate with neurons?

While listening in on the conversation between neurons is a challenge, learning how to inject our thoughts into the conversation is also difficult. We must manipulate voltages or inject currents to make ourselves heard without damaging the cells or their surroundings and ensuring that our message reaches the intended cell or group of cells. The challenge then is to find the most effective and safest way to communicate with neurons.

Extensive research continues to focus on how to best communicate with cells. While impaling a cell with an electrode is the most direct approach, the cell inevitably dies from the wound, and the approach is not practical for a functional neuroprosthesis. Many of the methods used to listen in on cells also function well as signal transmitters. Regardless of the method employed, the key issues that must be addressed are the amplitude of the stimulating signal (voltage or current), the duration and polarity of the signal, and the spatial selectivity. To be successful, the biologist must investigate how the natural processes of a cell are altered by a foreign stimulus and must determine the limits of this response before damage occurs. For example, a biphasic (two-polarity), charge-balanced signal best replicates the natural ebb and flow of ionic currents when a current stimulus is used (see Chapter four). Engineers, material scientists, and physicists must then find the best way to generate such a signal and deliver it to the cell. Sophisticated modeling techniques, such as those described in Chapters five and seven, estimate the voltage or current generated by competing electrode designs as a function of time and space within adjacent tissue. By adjusting the properties (surface area, geometry, and conductivity), researchers attempt to target specific cell groups with a sufficiently large stimulus. Different implant materials and architectures influence the electrical power required to deliver a desired density of charge to the cell, which in turn affects the electrical efficiency and heat generation of the implant. As the available technology continues to evolve, researchers continue to refine their techniques.

In addition to creating a safe and effective electrical connection with the cell, the implant must not physically endanger the cell and its surroundings. Many implant materials, such as semiconductors and most metals, are poisonous to the human body. Insulating materials such as silicone prevent any interaction between the poisons and the tissue without acting as a barrier to the electrical signals. An implant must not cause tearing or other physical damage to the tissue at the point of connection. The human body experiences significant amounts of movement and jarring impacts in even a normal day, and the implant must move in concert with the surrounding tissue to avoid injury. Implants must also allow the exchange of nutrients and waste to proceed naturally so that the surrounding tissue remains healthy. If the cells that communicate with the implant perish, the implant becomes ineffective. These issues are treated with greater detail in Chapter eleven.

1.3.3   How does one make an implant last?

Interrelated with the first two issues is the problem of making a device palatable, or at least invisible, to the natural defense mechanisms of the human body. In addition, the natural environment within the body is detrimental to the long-term integrity and survivability of implant materials. The challenge here is to design a system that will operate successfully for extended periods in a hostile environment.

One of the keys to a successful chronic implantation is a judicious choice of materials. Fortunately for researchers, a number of materials have been identified as safe for use in the human body and most likely to survive the biological environment (see Chapter eleven). Safe materials exist to perform all the basic functions required within a neural prosthesis, from carrying electrical charge to electrical and physical insulation. Engineered biological materials may add to this list and provide greater functionality and survivability of implants.[9,10] A design that incorporates these materials, therefore, is more likely to survive chronic implantation.

Another key issue is that of wound healing. All implants, regardless of design, create a wound of one sort or another upon placement in the body. Small wounds can continually occur as the implant sight moves and flexes with movement of the body. How the body reacts to these wounds determines the long-term effectiveness of the implant. For example, prostheses implanted on the surface of the retina are encapsulated by fibrous growths drawn from the vitreous humor if care is not taken during the implanting procedure. This encapsulation increases the physical distance between the implant and the targeted cells. Because the emitted electric field strength decreases inversely with distance, a point is rapidly reached where the field amplitude is insufficient to stimulate the targeted cell, and the prosthesis become ineffective. Engineers must work with biologists and surgeons to ensure that the implanting procedure does not adversely affect the lifetime of the implant.

1.4   Organization and contents of the book

The book is organized into five main sections beyond this introductory section. In Section II, the basic elements of neural behavior are described and modeled. Chapter two addresses the fundamental processes of neuron activity, including neuronal excitability and its regulation by membrane ion channels. These processes are modeled mathematically in Chapter three. The models presented in Chapter three are some of the basic tools for choosing the placement and type of stimulating electrodes and estimating the required stimulus amplitude to elicit a desired neural response.

Section III provides the bridge between the biological and engineering worlds through discussion of the interface between electronics and neural tissue. Chapter four describes the basic methods and findings related to stimulating neural activity with electrical signals. Here, one will find what methods are best for targeting specific cells or cell components (axon, dendrite, or soma) for activation and how to minimize cell damage during chronic stimulation. Chapter five details specific models for better predicting the behavior of electrodes within a biological medium, complementing the discussion in Chapter two. In Chapters six and seven, semiconductor-based systems for communicating with implants, integration of signal processing and power supplies with the implant, and procedures for recording activity from neurons are presented.

Section IV discusses some of the more common techniques required to interpret, process, and utilize neural signals for neuroprosthetic systems. Chapter eight details the basics and proper use of the powerful wavelet transform for extracting key informational components from neural signals. In Chapter nine, the problem of using neural signals in neuroprosthetic device design is addressed.

A description of several different neuroprosthetic systems and their development is the subject of Section V. Chapter ten details the state of the relatively more mature field of otological implants for hearing rehabilitation, while Chapter eleven details the ongoing development and challenges of visual prostheses. Chapter twelve provides an excellent overview of the motor prosthesis field, which has challenges very different from those encountered with sensory prostheses.

Finally, in Section VI, two technologies are discussed that will have a significant impact on the future of neural prosthetics. In Chapter thirteen, the continuing progress in the field of microelectronics is discussed, including the application of MEMS devices. Chapter fourteen summarizes the exciting new area of biomolecular electronics, where biological entities provide the bridge between electronics and neurons. Further information useful to the neuroprosthetics practitioner is included in the appendices in Section VII.

1.5 Summary

The field of neuroprosthetics is an exciting and challenging one. Practitioners from diverse backgrounds must communicate and innovate together to develop effective and practical systems and devices. It is our hope that this book will serve as a common ground on which these practitioners can come together, forge understandings, and acquire the basic knowledge they require. From better communication, better interaction, and enhanced awareness of concepts and methods can spring innovative and more sophisticated ideas and implementations that will not only benefit the prosthesis recipient, but will also help to further unlock the mysterious and wondrous workings of the human body and mind.

References

1. Miyamoto, R., Cochlear implants, *Otolaryngol. Clin. North Am.*, 28, 87–294, 1995.
2. Fromherz, P. and Stett, A., Silicon-neuron junction: capacitive stimulation of an individual neuron on a silicon chip, *Phys. Rev. Lett.*, 75, 1670–1673, 1995.
3. Meister, M., Pine, J., and Baylor, D.A., Multi-neuronal signals from the retina: acquisition and analysis, *J. Neurosci. Meth.*, 51, 95–106, 1994.
4. Hodgkin, A.L. and Huxley, A.F., A quantitative description of membrane current and its application to conduction and excitation in nerve, *J. Physiol.*, 117, 500–544, 1952.
5. Meister, M., Lagrado, L., and Baylor, D.A., Concerted signaling by retinal ganglion cells, *Science*, 35, 1535, 1995.
6. Corey, J.M., Wheeler, B.C., Brewer, G.J., Compliance of hippocampal neurons to patterned substrate networks, *J. Neurosci. Res.*, 30, 300–307, 1991.
7. Branch, D.W., Wheeler, B.C., Brewer, G.J., and Leckhand, D., Long-term maintenance of patterns of hippocampal pyramidal cells on substrates of polyethylene glycol and microstamped polylysine, *IEEE Trans. Biomed. Eng.*, 47, 290–300, 2000.
8. Stys, P.K., Ransom, B.R., and Waxman, S.G., Compound action potential of nerve recorded by suction electrode: a theoretical and experimental analysis, *Brain Res.*, 546, 18–32, 1991.
9. Zhong, Y., Yu, X., Gilbert, R., and Bellamkonda, R.V., Stabilizing electrode–host interfaces: a tissue engineering approach, *J. Rehabil. Res. Develop.*, 38(6), 2001.
10. Haipeng, G., Yinghui, Z., Jianchun, L., Yandao, G., Nanming, Z., and Xiufang, Z., Studies on nerve cell affinity of chitosan-derived materials, *J. Biomed. Mater. Res.*, 52(2), 285–295, 2000.

Neurons and neuron modeling

chapter two

Neuronal excitability: membrane ion channels

Steven Barnes

Contents

2.1 Introduction

The purpose of this chapter is to provide an understanding of how ion channels produce membrane excitability. The neuroprosthetic theories and devices described in this book depend to varying degrees on the excitability of neural membranes and on the function of many different types of voltage-gated ion channels. The overall problem is that neuroprosthetic devices, such as retinal implants, cochlear implants, deep brain stimulators, and motor prostheses, achieve their effect by selectively exciting a neuron or nerve. In the retina, it can be the ganglion cells or the bipolar cells; in the ear, the spiral ganglion neurons; or for motor prostheses, near the nerve–muscle junction.

In each example, an interface exists between a stimulating electrode and a neuronal membrane, and once stimulation of the neural membrane occurs the cell physiology itself must be relied upon to carry the signal to its termination points.

This chapter provides explanations of the voltage-dependent gating and the permeation properties of several types of ion channels in neuronal membranes. The emphasis here is on voltage-gated ion channels, but the involvement of other channel types in neuronal excitability should not be excluded. The first focus is on the *passive* electrical properties of the typical neuronal somatic membrane and axon to recognize the important role that *active* electrical properties play to overcome the restrictions on electrical signaling that the passive properties confer. The ion channels featured most prominently in this balance are voltage-gated sodium channels (Na channels) and numerous types of voltage-gated potassium channels (K channels). An additional class of voltage-gated channel types that are considered are calcium channels (Ca channels), which are active in virtually every cell type, neuronal or otherwise, and play important roles in cell signaling via Ca^{2+} as an intracellular messenger.

2.2   *The balance of passive and active properties of the neuronal membrane*

Signaling via membrane potential (voltage) changes in cells lacking active conductances is dramatically limited by the passive properties of the cells. Passive properties arise for the most part from the membrane capacitance afforded by the lipid membrane and the resistance of the intra- and extracellular fluids, as well as the resistance of the membrane itself. Lipid bilayers are of extremely high resistance, so this component of the passive repertoire is not considered a restricting factor. However, the combination of capacitance and resistance gives rise to spatial and temporal filtering of voltage signals. That is, the distance that a signal can propagate without serious decrement is limited, and the fidelity of a signal in time is markedly reduced, typical of low pass (RC) filter characteristics.

The extremely fine neuritic process that a spinal motor neuron employs to connect to its target muscles, or the axon projecting from the ganglion cells of the retina to the thalamus of the brain, must in a reliable manner carry high-frequency information of up to 1000 Hz over distances ranging from a few centimeters to a meter. Cable theory adequately approximates the distance that a sustained membrane potential change can be detected along an axon: $V(x) = V(0)\exp(-x/\lambda)$ in one dimension, with the length constant given by $\lambda = (r_m/(r_i + r_o))^{1/2}$. For an axon of 1 μm diameter, the length constant is on the order of a few hundred micrometers. A single exponential equation also approximates the time dependence of the rise and fall of the signal as $V(t) = V(0)\exp(-t/\tau)$, with the time constant given by $\tau = r_m c_m$. Changes in voltage are thus slowed dramatically by cells, where typical values for the time constant are on the order of 1 to 100 msec. (See Figure 2.1.)

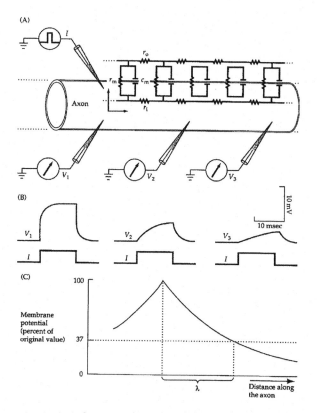

Figure 2.1 Decay in time and space of an electrical signal in an axon due to passive properties of the membrane. (A) The segmental resistances (r_o, resistance of outside medium; r_m, resistance of membrane; r_i, internal resistance within the axon) and capacitance (c_m, membrane capacitance) involved in the passive propagation of a depolarization. In the diagram, current (I) is injected at one end with a microelectrode, and membrane voltage at three positions (V_1, V_2, and V_3) is measured along the axon. (B) Voltage signal at the three positions. Close to the site of current injection, voltage V_1 is the largest and is only rounded slightly by the RC filtering properties of the passive membrane. With increasing distance from the site of current injection, voltage signals V_2 and V_3 are smaller in amplitude and reflect a marked slowing in rise time. (C) Plot of membrane voltage against distance along the axon. The decay of the amplitude of the signal follows a single exponential function characterized by the length constant, λ, the distance taken to reduce the signal to 37% of its original value. (From Hall, Z.W., *An Introduction to Molecular Neurobiology*, Sinauer Associates, Sunderland, MA, 1992. With permission.)

Considering the spatial and temporal limitations imposed by a passive membrane, propagation of the electrical signals is poor. The failure of an axonal process to carry information over a distance of more than a few hundred micrometers necessitated the appearance of ion channels, which, together with ionic gradient-generating and energy-requiring ion pumps and exchangers, provide the means for rapid and reliable regenerative electrical signaling.

2.3 *Active membranes overcome temporal and spatial degradations caused by passive properties*

Several different types of ion pumps and exchangers produce ionic gradients between the inside and outside of cells. Energy-requiring pumps, such as the sodium–potassium ATPase, which uses one ATP molecule to drive three Na^+ ions out and two K^+ ions in, are the primary source of the ionic gradient almost always found in neurons. The high sodium concentration outside the cell (for mammals, this is about 145 mM) and low concentration inside the cell (below 10 mM) produces an inward-driving force on Na^+ ions. The high K^+ concentration inside the cell (around 150 mM) and the low concentration outside (usually about 3 mM) produces an outward-driving force on K^+ ions. A calcium pump called the Ca^{2+}-ATPase uses ATP to extrude Ca^{2+} ions from the inside of cells and produces extremely large concentration gradients for this ion with external concentrations of 2 to 3 mM and internal concentrations around 10 nM. The Ca^{2+}-ATPase is assisted by a class of exchangers that utilize the inward Na^+ gradient to move Ca^{2+} out. Additional pumps and exchangers move Cl^-, bicarbonate (HCO_3^-), and other ions across the cell membrane.

With the gradients established, gating of ion channels can produce a remarkable diversity of electrical signals. The most frequently encountered electrical signal in nerve cells and muscles is the action potential, which represents a brief, transient, self-regenerating depolarization. The stereotypical action potential has rising, repolarizing, and after-hyperpolarizing phases that were described nearly a century ago. From a resting potential of near −60 mV (this value varies considerably between cells), an external stimulus (depolarizing influence) may bring the membrane to threshold (near −40 mV; again, the value will vary between cells). Once threshold is reached, the cell fires an all-or-none, regenerative action potential whose peak amplitude may reach +40 mV. The repolarization phase returns membrane potential to a value near the original resting potential, often with a brief period of "after hyperpolarization" during which the membrane potential is made more negative than the resting value. The time course of an action potential varies, with most mammalian neurons firing spikes having a half-width of well under 1 msec.

Physiologists studying the problem of cellular excitability were aware of the different ionic compositions of the intra- and extracellular media and concluded that ionic conductances must vary over time to account for the membrane polarizations seen during the action potential. Hodgkin and Huxley[6] voltage clamped the giant axon in squid and showed that an increased sodium conductance was responsible for the rise of the action potential and that increased potassium conductance was responsible for the repolarizing phase. Figure 2.2 shows the time courses of the transient sodium conductance increase (g_{Na}) and, following with a short but essential delay, the transient potassium conductance increase (g_K).

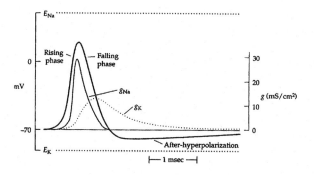

Figure 2.2 Time course of ionic conductances underlying the action potential. The Nobel-Prize-winning work of Hodgkin and Huxley identified transient increases in membrane conductance to Na^+ and K^+ ions that followed very specific temporal patterns. The initial increase in Na^+ conductance allows the cell to depolarize toward the Na^+ equilibrium potential (E_{Na}), an action that activates an additional increase in Na^+ conductance in a positive feedback loop. The delayed activation of an increase in K^+ conductance, together with inactivation of the Na^+ conductance, causes the cell to hyperpolarize toward the K^+ equilibrium potential (E_K), a value negative to the normal resting potential of the cell. The long after-hyperpolarization can be seen to arise from the prolonged increase in K^+ conductance. (From Hall, Z.W., *An Introduction to Molecular Neurobiology,* Sinauer Associates, Sunderland, MA, 1992. With permission.)

Hodgkin and Huxley[6] modeled total membrane current recorded under voltage clamp. Their recording of the membrane currents activated over the same voltages at which action potentials occur showed a fast, transient depolarizing current followed by a delayed and sustained hyperpolarizing current (Figure 2.3). Their observations gave rise to concept of ion channels long before channels could be identified as individual membrane proteins. Before the single-channel recording period was ushered in with the advent of patch clamping, physiologists and biophysicists could only record "macroscopic" or "whole cell" currents, which reflect the ensemble of thousands of individual channels undergoing their gating and permeation activities together.

Thousands of individual ion channels are responsible for the membrane conductance changes shown in these classical experiments. Ion channel activity was inferred from biophysical experiments analyzing the noise present in macroscopic ionic currents, and it was clear that highly selective toxins could block separable components of the ionic currents in cells, indicating that channels selective for different types of ions were present, but it was only with the patch clamp technique that single channel currents across the membrane of cells could be definitively identified.

The electrical signals in many neurons expand on the theme of action potential production, with variations in the rate, temporal pattern, individual spike width, magnitude, and after-potentials. Other electrical signatures involve the hyperpolarizing responses of cells. Generally, hyperpolarization is opposed by inwardly rectifying K^+ channels, although in many cases this

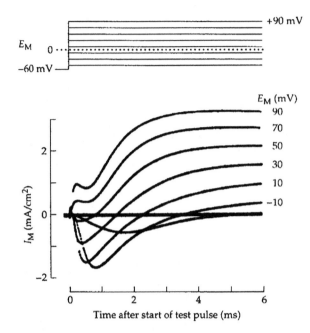

Figure 2.3 Voltage-clamped sodium and potassium currents in an axon. Under voltage clamp, membrane potential can be carefully and rapidly controlled to show the time dependence and degree of activation of voltage-dependent ionic conductances. In this recording, membrane potential is stepped from a holding potential of –60 mV (chosen to mimic the resting potential) to new values between –10 and +90 mV in steps of 10 mV. A downward (or "inward") current trajectory is recorded first at about –10 mV that represents activation of the Na^+ conductance, accounted for by the ensemble activity of thousands of Na channels. At more depolarized values of membrane potential, such as +90 mV, upward (or "outward") current is recorded. This is carried by K^+ ions through the K^+ conductance, again reflecting the activity of thousands of single K channels. (From Armstrong, C.M., *J. General Physiol.*, 54, 553–575, 1966, Rockefellar Press. With permission.)

opposition occurs over different time scales with nonselective cation channels activated by hyperpolarization or Cl⁻ channels that produce inward rectification at extreme membrane potentials.

2.4   Gating and permeation: the facts of life for ion channels

Biophysical descriptions of ion channel function generally fall within two categories: gating and permeation. Gating describes the temporal dependence of the opening and closing of a channel and the probability of finding a channel in an open or closed state as a function of membrane potential or the presence of a drug or neurotransmitter. Permeation describes the conductive properties of a channel in terms of its selectivity for specific ions, the rate at which ions can pass through the channel (which determines the maximum amplitude of the single channel current), and the effects and

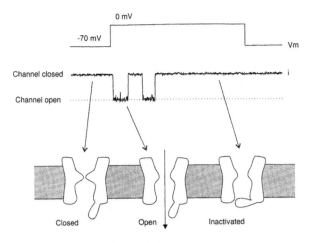

Figure 2.4 State diagram of a voltage-gated ion channel and its relation to channel activity. Most channels follow this scheme of closed–open–inactivated states, but with numerous variations. In many channels, the inactivated state is not present; in others, multiple open and closed states exist, resulting in complex kinetics in the transitions between opening and closing. In the classic view, the channel has at least two gates: (1) an activation gate that is sensitive to membrane potential and that allows the channel to make transitions from closed to open states, and (2) an inactivation gate on the cytoplasmic surface of the protein that renders the channel nonconducting in a voltage-independent manner following activation. The single-channel recording shows the channel in the closed state, then in the open state, and then after a step back to the closed state and a second opening, a transition to the inactivated state from which the channel does not return until sufficiently hyperpolarized (not shown). (From Ashcroft, F.M., *Ion Channels and Disease*, Academic Press, San Diego, CA, 2000. With permission.)

mechanisms of pore-blocking drugs and chemicals. Permeation and gating properties overlap in a few areas — for example, where an interaction of the permeating ion with the channel pore can affect the time spent in the open state of the channel.

Most ion channels have been described with gating schemes that follow the general pattern shown in Figure 2.4. The channel makes transitions between closed, open, and inactivated states. In the case of many channels (some are described later in this chapter), no inactivated state occurs or transitions to it are very slow or rarely made. Although the channel is nonconducting in both closed and inactivated states, inactivation is a specific, well-defined state from which transitions back to the open state are incredibly rare or nonexistent. The state transitions can be characterized with membrane-potential-dependent forward and reverse rate constants that ultimately describe the time dependence of channel activation (transitions from the closed to open states), inactivation (transitions to the inactivated state, which are generally voltage independent), and deactivation (transitions from the open back to the closed state). Smooth S-shaped curves plot the probability of a channel being open against membrane potential (activation curve).

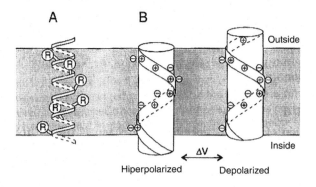

***Figure* 2.5** Catterall's twisting helix model for the translation of membrane depolarization into channel opening. Site-directed mutation studies identified the S4 transmembrane helix region of the protein as that responsible for voltage sensing. Positively charged side groups of residues match up with negative charges on other helical segments, but the strength of the electric field within the membrane exerts force on the S4 helix and can alter which pairs of charges match up. A ratcheting twist of S4 produces a conformational change in the protein, allowing the channel to open. Several other models have been proposed, related to this theme. (From Ashcroft, F.M., *Ion Channels and Disease*, Academic Press, San Diego, CA, 2000. With permission.)

While it was clear even in the early studies of ion channel gating that the channel must be capable of sensing the transmembrane potential, the mechanism responsible for this only became evident after ion channels were cloned and their molecular structure described. Hypotheses about which part of the protein had what function could then be tested on channels engineered to contain specifically substituted amino acid residues using the technique of site-directed mutation. It was shown that one of the putative transmembrane (TM)-spanning regions, one with a surprisingly large number of positively charged amino acids and called the S4, was responsible for the voltage dependence of the voltage-dependent steps in activation. The positively charged residues could be changed to residues with neutral side chains, with the result being that the channel no longer gated over the same range of membrane potentials. It has been hypothesized, but not proven, that the positive residues on the S4 TM segment enable the helix to couple membrane potential changes to conformational changes in the channel protein, so as to favor an open conducting state with more depolarized values of membrane voltage (Figure 2.5).

From the biophysicists' perspective, an ion channel required specific domains to carry out the various known functions that had been carefully described experimentally. Hille's depiction of an ion channel protein (Figure 2.6) carefully mapped out these domains in the era just before a thorough depiction based on molecular studies became available.

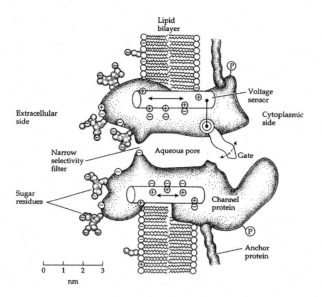

Figure 2.6 Hille's pre-molecular depiction of the working parts of a voltage-gated channel. The protein sits in the membrane so as to produce a ring structure to permit a central ion-permeable pore. The activation gate, shown here literally as a hinged structure, is coupled to the voltage sensor of the channel. To permit permeation by specific types of ions, the channel has a selectivity filter, a narrow region where the ions bind transiently on their way through the channel. In this depiction, a cytoplasmic region of the channel is poised to produce an inactivating block of the channel once the activation gate swings clear. Ion channels are, like other proteins, glycosylated, and they are tethered to the cells cytoskeleton via anchor proteins. (From Hille, B., *Ion Channels of Excitable Membranes*, Sinauer Associates, Sunderland, MA, 2001. With permission.)

2.5 Families of channels for Na^+, K^+, and Ca^{2+}

The explosion of molecular studies on ion channels that has occurred over the past decade has provided a rich description of the structural basis of these proteins. Not only have diverse, seemingly unrelated sequences been identified in a few cases as subserving similar functions between some classes of channels, but a provocative theme has also emerged strongly indicating the evolutionary roots of the channels within and between classes.

Figure 2.7 provides the current perspective on the three principle classes of voltage-gated channels. The Na channels are all built around the theme shown in the figure: an α subunit containing six transmembrane-spanning helices repeating in each of four motifs. Two subunits can accompany the channel, and these subunits in general may have roles in modulating the activity of the channel. Figure 2.7B shows the structure of the α_1 subunits of voltage-gated Ca channels; this principle subunit is very similar in overall form with the α subunit of the Na channel. The α_1 subunit can by itself form a complete channel, but accessory subunits (β_2, γ, and δ) bind to the channel

Figure 2.7 The principle subunits of voltage-gated ion channels. (A) Na channels have a single α subunit that is structured with four repeats of a six transmembrane helix motif. The four motifs are number I to IV, and the six helices in each motif are designated S1 to S6. Each S4 helix contains positively charged amino acid residues and comprises the channel's voltage sensor. The inactivation mechanism is located on the cytoplasmic linker between motifs III and IV. The Na channel pore is constructed from the linking regions found between S5 and S6 in each of the four motifs. Na channels are often composed of additional β₁ and β₂ subunits. (B) Voltage-gated Ca channels are structurally very similar to Na channels, with the same 6TM helix motif repeated four times. The S4 voltage sensors are similar, as are the P-loop linkers between the S5 and S6 helices of each motif. Inactivation of Ca channels is not the same, and several regions of the channels and quite different mechanisms are responsible for the varying degrees of inactivation seen in Ca channels. Ca channels almost always have a diversity of accessory subunits as shown. (C) The simpler K channel with its single-motif S1–S6 subunit. The same S4 voltage sensor region and P-loop pore domain between S5 and S6 are present. Inactivation of some voltage-gated K channels is mediated by the N-terminus region, where a ball of protein tethered to the S1 helix has been shown to literally plug the open channel. K channels are formed from four of these subunits, meaning that, with the diversity of subunits, heteromeric channels are found and channel properties are diverse. (From Hille, B., *Ion Channels of Excitable Membranes*, Sinauer Associates, Sunderland, MA, 2001. With permission.)

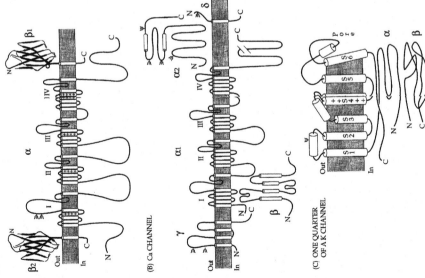

and are known to modify properties of the channel such as its interaction with neuromodulators and its inactivation kinetics. It is the K channel that appears to stand out because it is composed of four individual α subunits, each one resembling the 6TM motif of the Na and Ca channels.

2.6 Molecular properties of voltage gated Na channels

The Na channel of squid giant axon, whose kinetics were described by Hodgkin and Huxley,[6] certainly is archetypal for the subsequently discovered Na channels in a variety of other cell types. The hallmarks of Na channel gating are the rapid transition from closed to open states with membrane depolarization, and the subsequent rapid inactivation. Single-channel studies of Na channels brought forth a wealth of information about these transitions. With single-channel recordings, such as those shown in Figure 2.8, it was possible to show that channel activation occurred over a relatively broad temporal window and that in nearly every case inactivation was a voltage-independent, rapid, and essentially automatic process.

Inactivation of the Na channel is caused by a small structure composed of several amino acid residues on the cytosolic loop linking repeated motifs III and IV. Shortly after the voltage-dependent activation of the Na channel (at 37°C, the process occurs in approximately 100 μsec), the hinged lid occludes the open pore of the channel and essentially blocks the channel to further permeation.

Permeation of Na channels is determined by amino acid residues of the P-loop, a pore-forming domain present in each of the four motifs that links the TM5 and TM6 parts of each motif. Lining the pore of the Na channel are negatively charged residues in two of the four P-loops, and these are essential for binding Na^+ ions and conferring upon the channel selectivity for Na^+.[4]

Sequence analysis of the nine expressed human Na channels shows a very high degree of amino acid identity between the proteins. The dendrogram shown in Figure 2.9 emphasizes the similarities among human Na channels, reflecting the relatively recent evolutionary diversity in this class of channel, as amino acid substitutions are a measure of evolutionary distance.

2.7 The voltage-gated K channels show great structural diversity

Elucidation of the common structures of ion channels showed that K channels have the simplest subunit structure. Each subunit of a Shaker-type K channel has six transmembrane spanning helices. As described above, the S4 regions contain the voltage sensor, and the loops between S5 and S6, called the P-loops, fold inwardly and provide the pore of the channel. Site-directed mutagenesis has identified which amino acid res-

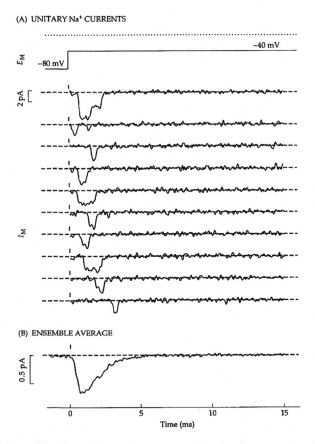

Figure 2.8 Unitary Na channel currents account for the fast transient Na⁺ conductance increase of the action potential. Panel (A) shows a single channel recording of a 19-pS Na channel presented with many depolarizations. Each time the membrane is depolarized, the Na channel undergoes its transition from closed to open and then to inactivated states. With each depolarization, this process occurs at a slightly different time due to the stochastic nature of channel gating. When the single openings are gathered in an ensemble average, as shown in panel (B), the current waveform resembles macroscopic recordings made from intact axons. This patch actually contains several Na channels, accounting for the double opening in the uppermost example. (Kindly provided by J.B. Patlak.)

idues of the P-loop are critical for permeation of K⁺ and block of the channel by chemicals such tetraethylammonium (TEA). At the N-terminus of the 6TM subunit is a cytoplasmic domain shown to be responsible for fast, so-called N-type, inactivation.

Potassium channels activate more slowly than Na channels, and from the beginning were termed "delayed rectifier" channels, owing to this delay in activation and their current-voltage behavior, which has the tendency to oppose strong depolarization (e.g., to give the membrane outwardly

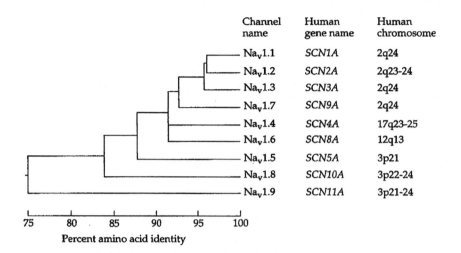

Channel name	Human gene name	Human chromosome
Na$_v$1.1	SCN1A	2q24
Na$_v$1.2	SCN2A	2q23-24
Na$_v$1.3	SCN3A	2q24
Na$_v$1.7	SCN9A	2q24
Na$_v$1.4	SCN4A	17q23-25
Na$_v$1.6	SCN8A	12q13
Na$_v$1.5	SCN5A	3p21
Na$_v$1.8	SCN10A	3p22-24
Na$_v$1.9	SCN11A	3p21-24

Figure 2.9 Relationships between nine different human Na channels and their chromosomal locations. The dendrogram shows the percent identity of amino acids on the various forms of the human Na channels. Channels having a high degree of amino acid identity are considered to be recent evolutionary ramifications. The names of the channel proteins are shown with the corresponding human gene. Additional Na channel genes are known to exist but have not been expressed. (From Goldin, A.L. et al., *Neuron*, 28, 365–368, 2000. With permission.)

rectifying properties). Figure 2.10 shows single K channel activity in response to a depolarization. The striking difference of these records, compared to those from the Na channel in Figure 2.8 (other than that the single-channel current is directed in the upward direction reflecting the outward flow of positive K$^+$ ions) is that the single channel opening are very long lived, for the most part lasting the duration of the depolarizing step. The ensemble current shown in panel B gives a clear perspective of the lack of inactivation in the case of the K channel.

The single channel and ensemble averaged currents shown for Na and K channels in Figures 2.8 and 2.10 readily account for the macroscopic currents underlying the action potential shown in Figure 2.3. During the action potential, the depolarizing phase is terminated by the combination of Na channel inactivation and K channel activation. The repolarizing phase of the action potential terminates, not because the K channels inactivate, but because just as the membrane becomes hyperpolarized the channels close, as shown in Figure 2.10.

Not all voltage-gated K channels follow the prototypical six-transmembrane-spanning helix structure. As Figure 2.11 shows, several classes of K channels utilize subunits with two or four transmembrane-spanning helices instead of the 6TM subunit. Some of these are inwardly rectifying K channels, channels that oppose hyperpolarization of the membrane and contribute in some cases to the resting membrane potential.

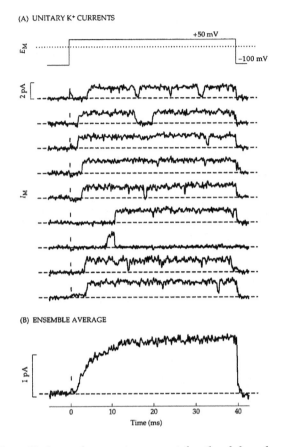

Figure 2.10 Unitary K channel currents account for the delayed and sustained K⁺ conductance increase of the action potential. (A) Single-channel recordings of K channels were recorded in a squid giant axon. A single channel with 20-pS conductance opened for most of the duration of each applied voltage step. The openings first occur with a much longer delay than the Na channel openings seen in Figure 2.9. The channel often makes transitions back to the closed state but then reopens. When membrane potential is returned to a negative value, where the channel has a low probability of being in the open state, the channel always closes rapidly. (B) The ensemble average shows a smooth activation time course and a non-inactivating steady-state value during the step to +50 mV. This trajectory agrees well with the macroscopic K⁺ currents shown in Figure 2.4. (With permission from Francisco Bezanilla [see Reference 5].)

The list of channels provided in Figure 2.11 includes many not identified in humans. The principle class of voltage-gated K channels is that designated by K_V1 to 9. $K_V1.1$, the Shaker channel, is the classic inactivating A-type K channel. It has roles in action potential repolarization, in some cases offering control of ramp depolarization of the membrane leading to threshold. Its name comes from its discovery in a fruit fly mutation that shakes its legs

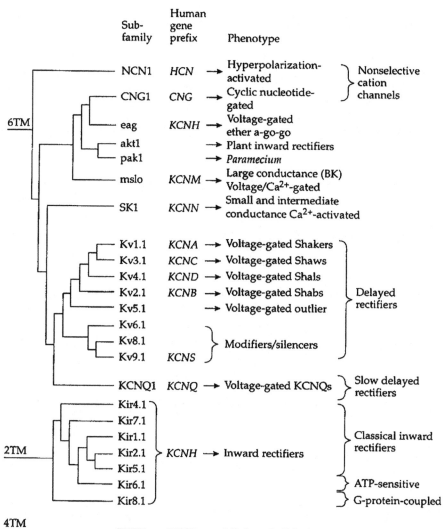

Figure 2.11 K channels exist in many forms. K channels come not only as 6TM subunits, but as 2TM and 4TM subunits as well. The 2TM subunits include inward rectifier K channels and ATP-sensitive channels. Of the 6TM subunits, the largest subfamily is the K_v group, including the sustained and rapidly inactivating K channels. Ca^{2+}-activated K channels include large and small conductance channels. Although not specifically K channels, the figure includes the nonselective cation channels HCN and CNG, which are gated by hyperpolarization and cyclic nucleotides, respectively. (From Hille, B., *Ion Channels of Excitable Membranes*, Sinauer Associates, Sunderland, MA, 2001. With permission.)

when anesthetized. The other K_V channels offer unique channel kinetics and single-channel conductances.

2.8   The family of voltage gated Ca channels

Although not involved directly in action potential generation or propagation in axons, Ca channels are critically important players in all cells including nerve cells. Some Ca channels help generate bursts of action potentials, some enable Ca^{2+} influx that then suppresses action potential firing for relatively long periods, and some are the coupling elements that transduce presynaptic membrane potential into intracellular Ca^{2+} signals that control neurotransmitter release. The roles of Ca channels are diverse and central in cell physiology.

Calcium channels have been divided and subdivided on biophysical, pharmacological, and, most recently, molecular bases. Low-voltage-activated (LVA) channels have been termed T-type, as they are transient and have tiny, single-channel conductances. Three have been cloned from humans and are designated $Ca_V3.1$, 3.2, and 3.3 (Figure 2.12). These channels activate and then inactivate, much like the Na channels.

High-voltage-activated (HVA) channels include all of the L-type Ca channels ($Ca_V1.1$, 1.2, 1.3, and 1.4) which are found in muscle (human gene CACNA1S), in heart and brain (CACNA1C and CACNA1D), and at synaptic terminals of photoreceptors and bipolar cells in the retina (CACNA1F). The $Ca_V2.1$, 2.2, and 2.3 group are responsible for synaptic transmission throughout the nervous system. Generally, HVA Ca channels show little inactivation.

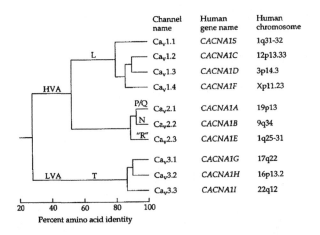

Figure 2.12 Classification of Ca channel proteins and human genes. The low-voltage-activated, T-type Ca channels are very similar kinetically to Na channels showing rapid inactivation and are involved in generating bursts of action potentials. The high-voltage-activated channels, further subdivided into L-type and P/Q-, N-, and R-type, have very slow inactivation or none at all. L-type channels are found in muscle (CACNA1S), as well as photoreceptors in the retina (CACNA1F), and in many cases allow the influx of Ca^{2+} that leads to important cellular responses such as the release of neurotransmitter. (From Ertel, E. et al., *Neuron*, 25, 533–535, 2000. With permission.)

Calcium channel structure is nearly the same as Na channel structure (Figure 2.7). As is the case for the Na channels, the four hydrophobic loops linking TM5 and TM6 of each repeating motif form the pore of the Ca channels. In the case of the L-type Ca channel, four glutamate residues, one from each motif, form a negatively charged ring that serves as the high-affinity binding site for Ca^{2+} ions.[7] Mutation of these glutamate residues reduced the selectivity of the channel for Ca^{2+}, in effect turning it into a Na^+ channel. Indeed, in the absence of divalent cations, most Ca channels conduct monovalent cations such as Na^+ or Li^+ very effectively. Thus, the ring of glutamates is essential for Ca^{2+} binding and it is this property that defines the Ca channels.

2.9 Conclusion

While the voltage-gated ion channels form families and the strong evolutionary links can be seen between these families, an enormous degree of diversity is still reflected in the structure of channels. The contribution to membrane excitability from other channel types should be considered. There are several classes of voltage-gated Cl channels that neurons express as well as voltage-gated nonselective cation channels. These are generally permeable to Na^+ and K^+ but retain some Ca^{2+} permeability as well, which may be important in intracellular Ca^{2+} signaling. In photoreceptors, which are considered to be nonspiking cells, nonselective cation channels activated by hyperpolarization play a prominent role in shaping the light response. Generally, these types of channels are not highly expressed in axons; therefore, their role in the transduction of neuroprosthetic signals may be extremely limited.

References

1. Ashcroft, F.M., *Ion Channels and Disease*, Academic Press, San Diego, CA, 2000.
2. Armstrong, C.M., Inactivation of the potassium conductance and related phenomena caused by quaternary ammonium ion injected in squid axons, *J. Gen. Physiol.*, 58, 553–575, 1969.
3. Hall, Z.W., *An Introduction to Molecular Neurobiology*, Sinauer Associates, Sunderland, MA, 1992.
4. Heinemann, S.H., Terlau, H., Stuhmer, W., Imoto, K., and Numa, S., Calcium channel characteristics conferred on the sodium channel by single mutations, *Nature*, 356, 441–443, 1992.
5. Hille, B., *Ion Channels of Excitable Membranes*, Sinauer Associates, Sunderland, MA, 2001.
6. Hodgkin, A.L. and Huxley, A.F., The components of membrane conductance in the giant axon of Loligo, *J. Physiol.*, 116, 473–496, 1952.
7. Yang, J., Ellinor, P.T., Sather, W.A., Zhang, J.F., and Tsien, R.W., Molecular determinants of Ca^{2+} selectivity and ion permeation in L-type Ca channels, *Nature*, 366, 158–161, 1993.

chapter three

Neuron modeling

Frank Rattay, Robert J. Greenberg, and Susanne Resatz

Contents

3.1 Introduction

A breakthrough in our understanding of the physics of neural signals, which propagate as membrane voltage along a nerve fiber (axon), was achieved by the ingenious work of Hodgkin and Huxley[1] on the nonmyelinated squid axon. This work helped explain the action potential or "spike" that conveys the all-or-none response of neurons. To explore the complicated gating mechanism of the ion channels, the stimulating electrode was a long uninsulated wire, thus every part of the neural membrane had to react in the same way; that is, propagation of signals was prevented (Figure 3.1A). Refinements of their method as well as the application of patch clamp techniques have supplied us with models for different neural cell membranes. Reliable prediction of membrane voltage V as a function of time is possible for arbitrary stimulating currents $I_{stimulus}$ with proper membrane models.

0-8493-1100-4/03/$0.00+$1.50

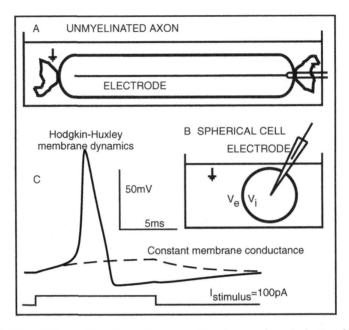

Figure 3.1 Stimulation without signal propagation (space clamp). A stimulating elec-trode in the form of an uninsulated wire inserted along an unmyelinated nerve fiber (A) or current injection in a spherical cell (B) causes the same inside voltage, V_i, for every part of the cell membrane. (C) In the first subthreshold phase, the voltage across the cell membrane follows the simple constant membrane conductance model (dashed line). Figure shows simulation of a 30-μm spherical cell sheltered by a membrane with squid axon ion channel distribution (i.e., by solving the Hodgkin–Huxley model with original data). Membrane voltage V is the difference between internal and external voltage: $V = V_i - V_e$; $V_e = 0$.

The main equation for internal stimulation of the soma, or any other compartment where current flow to other processes or neighbored compart-ments is prevented, always has the same form: One part of the stimulating current is used to load the capacity C_m (in Farads) of the cell membrane and the other part passes through the ion channels; that is,

$$I_{stimulus} = C_m \frac{dV}{dt} + I_{ion} \qquad (3.1)$$

The rate of membrane voltage change, dV/dt, follows as:

$$\frac{dV}{dt} = \left[-I_{ion} + I_{stimulus} \right] / C_m \qquad (3.2)$$

where the ion currents I_{ion} are calculated from appropriate membrane mod-els. Usually, the membrane models are formulated for 1 cm² of cell membrane and the currents in Eq. (3.1) become current densities.

A positive stimulating current applied at the inside of an axon or at any other part of a neuron will cause an increase of V according to Eq. (3.1), if the membrane has been in the resting state ($I_{stimulus} = 0$ and $dV/dt = 0$) before. In order to generate a spike, this positive stimulus current has to be strong enough for the membrane voltage to reach a threshold voltage, which causes many of the voltage-sensitive sodium channels to open. By sodium current influx, the transmembrane potential increases to an action potential without the need of further stimulating support. This means that as soon as the solid line in Figure 3.1C is some few millivolts above the dashed line we can switch off the stimulus without seeing any remarkable change in the shape of the action potential.

In general the excitation process along neural structures is more complicated than under space clamp conditions as shown in Figure 3.1. Current influx across the cell membrane in one region influences the neighboring sections and causes effects such as spike propagation (Figure 3.2). Besides modeling the natural signaling, the analysis of compartment models helps to explain the influences of applied electric or magnetic fields on representative target neurons. Typically, such a model neuron consists of functional subunits with different electrical membrane characteristics: dendrite, cell body (soma), initial segment, myelinated nerve fiber (axon), and nonmyelinated terminal. Plenty of literature exists on stimulated fibers, but little has been written about external stimulation of complete neurons.

In 1976, McNeal[2] presented a compartment model for a myelinated nerve fiber and its response to external point-source stimulation. He inspired many authors to expand his model for functional electrical stimulation of the peripheral nerve system, such as analysis of external fiber stimulation by the activating function,[3,4] unidirectional propagation of action potentials,[5] stimulation of a nerve fiber within a[6,7] selective axon stimulation,[8,9,10] and influence of fiber ending.[11] Of specific interest is simulation of the threshold and place of spike initiation generated with stimulating electrodes (e.g., by the near field of a point source or dipole) by finite element calculation for a specific implanted device or when the farfield influence from surface electrodes is approximated by a constant field.[12] Stronger electric stimuli or application of alternating currents generate new effects in neural tissue. All effects depend essentially on the electric properties of the neural cell membrane and can be studied with compartment modeling. The ion channel dynamics can be neglected during the first response of the resting cell, but the complicated nonlinear membrane conductance becomes dominant in the supra-threshold phase (Figure 3.1C). Consequently, the behavior can be analyzed with simple linear models or with more computational effort by systems of differential equations that describe the ion channel mechanisms in every compartment individually.

Modeling the sub-threshold neural membrane with constant conductances allows the analysis of the first phase of the excitation process by the activating function as a rough approach. Without inclusion of the complicated ion channel dynamics, the activating function concept explains the

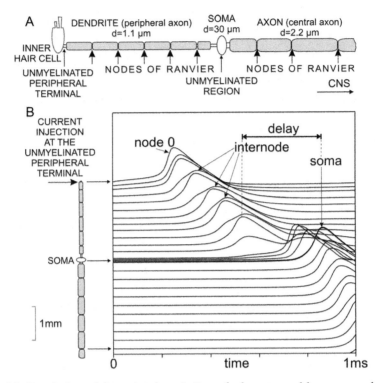

Figure 3.2 Simulation of the natural excitation of a human cochlear neuron by stimulation with a 50-pA, 250-μsec current pulse injected at the peripheral end. This bipolar cell is a non-typical neuron: (1) the dendrite is myelinated and often called peripheral axon; (2) in contrast to animal cochlear neurons, the soma and the pre- and post-somatic regions are unmyelinated; (3) a single synaptic input from the auditory receptor cell (inner hair cell) generates a spike that propagates with a remarkable delay across the current consuming somatic region. Note the decay of the action potential in every internode that again is amplified in the next node of Ranvier. Simulation uses internode with constant membrane conductance, a "warm" Hodgkin–Huxley model ($k = 12$) in the active membranes of the soma, and a 10-fold ion channel density in the peripheral terminal (node 0), all nodes, and pre- and post-somatic compartments. The lines are vertically shifted according to their real location of the rectified neuron. For details, see Rattay et al.[35]

basic mechanism of external stimulation, the essential differences between anodic and cathodic threshold values, its dependence on the geometric situation, the mechanism of one side firing, and the blocking of spike propagation by hyperpolarized regions, as well as several other phenomena.[13,14] Beside nerve fiber analysis, the activating function, which represents the direct influence of the electric field in every compartment, is useful for magnetic stimulation[15,16] and for direct stimulation of denervated muscle fibers[17] or, in generalized form, cardiac tissue[18,19] and arbitrary neurons.[20,12]

In previous work we have shown how the shape of a neuron affects the excitation characteristics and that several surprising phenomena may occur.

This is demonstrated, for example, by comparing the threshold current for neurons with small and large soma. Whereas thick axons are easier to stimulate than thin ones (known as the inverse recruitment order[21]), this relation does not hold for the size of the soma: a 100-μs pulse from a point electrode 430 μm above the soma requires −2.3 mA to stimulate a 30-μm-diameter soma neuron with a 2-μm axon but only −1.1 mA for a 10-μm soma. Increasing the electrode distance to 1000 μm results in the same −2.2-mA threshold for both cases; that is, the second surprise is that for the large soma the neuron excitation threshold increases slightly when the electrode is moved within a specific range toward the soma.[12] The explanation is that the axon, which is generally more excitable than the soma, loses more current to load the larger capacity of the large soma; this effect is more pronounced when the place of spike initiation within the axon is close to the soma.

Comparing a pyramidal cell with a cochlear neuron demonstrates the variety in architecture and signal processing principles in the dendrite and soma region. Some neuron types, such as the afferent bipolar cochlear neurons, are effective transmission lines, where nearly all of the spikes initiated at the synaptic contact with an inner hair cell arrive with some delay and small temporal variation (jitter) in the terminal region (Figure 3.2). In contrast to the cochlear neuron, a single synaptic input at the dendritic tree of a pyramidal cell produces only minimal change in membrane voltage at the soma and at the initial segment, and usually the collective effect of many synaptic activities is necessary to influence spike initiation significantly. Typically, the pyramidal cell response is dominated by the internal calcium concentration, which depends on two types of voltage-dependent calcium channels: (1) low-voltage-activated channels respond in the subthreshold range and may include generation of low-threshold spikes, and (2) high-voltage-activated channels in the dendrites respond, for example, to backpropagating sodium spikes.[22-24]

3.2   Models for the cell membrane

The cell membrane is not a perfect insulator for ion transport. When one Na channel opens for a short time, some thousand sodium ions may be driven into the cell by the high outside concentration. The second driving element is the voltage across the membrane, which requires a Na+ current according to Ohms law. The conductance of a patch of cell membrane for a specific ion type depends on the number of open channels and is defined by Ohm's law and a battery voltage according to the Nernst equation. A different internal concentration, c_i, and external ion concentration, c_e, cause the membrane voltage, E_m, when one type of ions is involved:

$$E_m = \frac{RT}{zF} \ln \frac{c_e}{c_i} \tag{3.3}$$

where the gas constant $R = 8.31441$ J/(mol.K); temperature, T, is in Kelvin; z is valence and the Faraday constant represents the charge of one mol of single-valenced ions $F = 96485$ c/mol. *Note:* At room temperature ($T = 20°C = 293.15$ K), the factor RT/F is about 25 mV.

In their experiments Hodgkin and Huxley[1] found the sodium current proportional to $E - E_{Na}$. Approximating the Nernst equation [Eq. (3.3)] with their data gives $E_{Na} = 45$ mV $= (25$ mV$)\ln(c_e/c_i)$ and results in $c_e/c_i = \exp(45/25) = 6.05$; that is, the outside sodium concentration is six times the inner one, and $E_K = -82$ mV $= (25$ mV$)\ln(c_e/c_i)$ requires that the potassium outside:inside concentration is 1:26.5.

The Goldman equation defines the steady state membrane voltage $E_m = E_{rest}$ when several ion types are involved. When K+, Na+, and Cl− ions are considered it reads as:

$$E_m = \frac{RT}{F} \ln \frac{P_K[K]_e + P_{Na}[Na]_e + P_{Cl}[Cl]_i}{P_K[K]_i + P_{Na}[Na]_i + P_{Cl}[Cl]_e} \tag{3.4}$$

where [K] is the potassium concentration and the suffixes i and e stand for inside and external, respectively. P_K, P_{Na}, and P_{Cl} are permeabilities measured in centimeters per second (cm/sec). Note that sodium and potassium are anions, but chloride is cathodic, therefore, $[Cl]_i$ appears in the numerator in contrast to the anionic concentrations. In the resting state, the membrane is most permeable to potassium ions, and the resting membrane voltage of about −70 mV (i.e., the inside is more negative compared to the extracellular fluid) is close to the potassium Nernst potential.

The nonlinear conductance of the neural cell membrane depends on different classes of ion channels[25] (1 through 3, below) and on the activity of ion pumps (4, below):

1. Voltage-dependent gating dominates excitation and neural signal propagation in the axon; open/close kinetics depend on the voltage across the cell membrane.
2. Calcium-dependent gating occurs mainly at the soma and dendrites; the calcium concentration regulates the opening of the channels as opening depends on intracellular calcium ion binding.
3. Transmitter gating and second messenger gating occurs at the pre- and postsynaptic membrane.
4. Ion pumps are membrane molecules that consume energy pump ions across the cell membrane in order to restore the high individual ion concentration on one side of the cell (e.g., high external sodium and high internal potassium concentration).

3.2.1 Axon models of the Hodgkin–Huxley type

The Hodgkin–Huxley model (HH model) was developed from a homogeneous nonmyelinated squid axon. It includes sodium, potassium, and leakage currents and has the form:

$$\frac{dV}{dt} = \left[-g_{Na}m^3h(V - V_{Na}) - g_K n^4(V - V_K) - g_L(V - V_L) + i_{stimulus}\right]/c \quad (3.5)$$

$$\frac{dm}{dt} = \left[-(\alpha_m + \beta_m)m + \alpha_m\right]k \quad (3.6)$$

$$\frac{dh}{dt} = \left[-(\alpha_h + \beta_h)h + \alpha_h\right]k \quad (3.7)$$

$$\frac{dn}{dt} = \left[-(\alpha_n + \beta_n)n + \alpha_n\right]k \quad (3.8)$$

$$k = 3^{0.1T - 0.63} \quad (3.9)$$

where V is the reduced membrane voltage, resulting from internal, external, and resting potential: $V = V_i - V_e - V_{rest}$; g_{Na}, g_K, and g_L are the maximum conductances for sodium, potassium, and leakage per cm², respectively; m, h, and n are probabilities (with values between 0 and 1) that reduce the maximum conductances of sodium and potassium according to experimental gating data; V_{Na}, V_K, and V_L are the sodium, potassium, and leakage battery voltages, respectively, according to the Nernst equation [Eq. (3.3)]; $i_{stimulus}$ is the stimulus current (μA/cm²); c is the membrane capacity per cm²; α and β are voltage-dependent variables fitted from experimental data to quantify the ion channel kinetics; k is a temperature coefficient that accelerates the gating process for temperatures higher than the original experimental temperature of 6.3°C; and temperature T is in °C. Although a temperature conversion is possible, it is interesting to note that, unlike real tissue, the Hodgkin–Huxley equations do not propagate action potentials above 31°C.[14] Parameter values, units, and further expressions of the HH model are listed in Table 3.1.

In 1964, Frankenhaeuser and Huxley[26] developed a model for the myelinated frog axon node (FH model) assuming HH-like gating mechanisms, but they derived the ion current formulation from the Nernst–Planck equation and added a nonspecific current density, i_p:

$$\frac{dV}{dt} = \left[-i_{Na} - i_K - i_P - i_L + i_{stimulus}\right]/c \quad (3.10)$$

$$i_{Na} = P_{Na}m^2h \frac{EF^2}{RT}\frac{[Na]_o - [Na]_i \exp(EF/RT)}{1 - \exp(EF/RT)} \quad (3.11)$$

$$i_K = P_K n^2 \frac{EF^2}{RT}\frac{[K]_o - [K]_i \exp(EF/RT)}{1 - \exp(EF/RT)} \quad (3.12)$$

Table 3.1 Expressions and Constants for Axon Membrane Models

	HH Model	FH Model	CRRSS Model	SE Model	SRB Model
α_m	$\dfrac{2.5 - 0.1V}{\exp(2.5 - 0.1V) - 1}$	$\dfrac{0.36(V - 22)}{1 - \exp\left(\dfrac{22 - V}{3}\right)}$	$\dfrac{97 + 0.363V}{1 + \exp\left(\dfrac{31 - V}{5.3}\right)}$	$\dfrac{1.87(V - 25.41)}{1 - \exp\left(\dfrac{25.41 - V}{6.06}\right)}$	$\dfrac{4.6(V - 65.6)}{1 - \exp\left(\dfrac{-V + 65.6}{10.3}\right)}$
β_m	$4.\exp\left(-\dfrac{V}{18}\right)$	$\dfrac{0.4(13 - V)}{1 - \exp\left(\dfrac{V - 13}{20}\right)}$	$\dfrac{\alpha_m}{\exp\left(\dfrac{V - 23.8}{4.17}\right)}$	$\dfrac{3.97(21 - V)}{1 - \exp\left(\dfrac{V - 21}{9.41}\right)}$	$\dfrac{0.33(61.3 - V)}{1 - \exp\left(\dfrac{V - 61.3}{9.16}\right)}$
α_n	$\dfrac{0.1 - 0.01V}{\exp(1 - 0.1V) - 1}$	$\dfrac{0.02(V - 35)}{1 - \exp\left(\dfrac{35 - V}{10}\right)}$		$\dfrac{0.13(V - 35)}{1 - \exp\left(\dfrac{35 - V}{10}\right)}$	$\dfrac{0.0517(V + 9.2)}{1 - \exp\left(\dfrac{-V - 9.2}{1.1}\right)}$
β_n	$0.125.\exp\left(-\dfrac{V}{80}\right)$	$\dfrac{0.05(10 - V)}{1 - \exp\left(\dfrac{V - 10}{10}\right)}$		$\dfrac{0.32(10 - V)}{1 - \exp\left(\dfrac{V - 10}{10}\right)}$	$\dfrac{0.092(8 - V)}{1 - \exp\left(\dfrac{V - 8}{10.5}\right)}$
α_h	$0.07.\exp\left(-\dfrac{V}{20}\right)$	$\dfrac{0.1(V + 10)}{1 - \exp\left(\dfrac{V + 10}{6}\right)}$	$\dfrac{\beta_h}{\exp\left(\dfrac{V - 5.5}{5}\right)}$	$-\dfrac{0.55(V + 27.74)}{1 - \exp\left(\dfrac{V + 27.74}{9.06}\right)}$	$\dfrac{0.21(V + 27)}{1 - \exp\left(\dfrac{V + 27}{11}\right)}$
β_h	$\dfrac{1}{\exp(3 - 0.1V) + 1}$	$\dfrac{4.5}{1 + \exp\left(\dfrac{45 - V}{10}\right)}$	$\dfrac{15.6}{1 + \exp\left(\dfrac{24 - V}{10}\right)}$	$\dfrac{22.6}{1 + \exp\left(\dfrac{56 - V}{12.5}\right)}$	$\dfrac{14.1}{1 + \exp\left(\dfrac{55.2 - V}{13.4}\right)}$
α_p		$\dfrac{0.006(V - 40)}{1 - \exp\left(\dfrac{40 - V}{10}\right)}$			$\dfrac{0.0079(V - 71.5)}{1 - \exp\left(\dfrac{71.5 - V}{23.6}\right)}$
β_p		$\dfrac{0.09(V + 25)}{1 - \exp\left(\dfrac{V + 25}{20}\right)}$			$\dfrac{0.00478(V - 3.9)}{1 - \exp\left(\dfrac{V - 3.9}{21.8}\right)}$
V_{rest} (mV)	-70	-70	-80	-78	-84

	HH	FH	CRRSS	SE	SRB
V_{Na} (mV)	115	—	115	—	—
V_K (mV)	-12	—	—	—	0
V_L (mV)	10.6	0.026	-0.01	0	0
g_{Na} (kΩ⁻¹/cm⁻²)	120	—	1445	—	—
$g_{K,fast}$ (kΩ⁻¹/cm⁻²)	36	—	—	—	30
$g_{K,slow}$ (kΩ⁻¹/cm⁻²)	—	—	—	—	60
g_L (kΩ⁻¹/cm⁻²)	0.3	30.3	128	86	60
c (kΩ⁻¹/cm⁻²)	1	2	2.5	2.8	2.8
$V(0)$	0	0	0	0	0
$m(0)$	0.05	0.0005	0.003	0.0077	0.0382
$n(0)$	0.32	0.0268	—	0.0267	0.2563
$h(0)$	0.6	0.8249	0.75	0.76	0.6986
$p(0)$	—	0.0049	—	—	0.0049
T_0	6.3°C	293.15 K = 20°C	37°C	310.15 K = 37°C	310.15 K = 37°C
$Q_{10}(\alpha_m)$	3	1.8	3	2.2	2.2
$Q_{10}(\beta_m)$	3	1.7	3	2.2	2.2
$Q_{10}(\alpha_n)$	3	3.2	3	3	3
$Q_{10}(\beta_n)$	3	2.8	3	3	3
$Q_{10}(\alpha_h)$	3	2.8	3	2.9	2.9
$Q_{10}(\beta_h)$	3	2.9	3	2.9	2.9
P_{Na} (cm/s)	—	0.008	—	0.00328	0.00704
P_K (cm/s)	—	0.0012	—	0.000134	—
P_P (cm/s)	—	0.00054	—	—	—
$[Na]_o$ (mmol/l)	—	114.5	—	154	154
$[Na]_i$ (mmol/l)	—	13.7	—	8.71	30
$[K]_o$ (mmol/l)	—	2.5	—	5.9	—
$[K]_i$ (mmol/l)	—	120	—	155	—

Note: HH model = Hodgkin–Huxley model; FH model = Frankenhaeuser–Huxley; CRRSS model = Chiu, Ritchie, Rogert, Stagg, and Sweeney model; SE model = Schwarz–Eikhof model; and SRB model = Schwarz, Reid, and Bostock model. Faraday constant $F = 96485$ C/mol; gas constant $R = 8314.4$ mJ/(mol.K).

$$i_p = P_p p^2 \frac{EF^2}{RT} \frac{[Na]_o - [Na]_i \exp(EF/RT)}{1 - \exp(EF/RT)} \tag{3.13}$$

$$i_L = g_L(V - V_L) \tag{3.14}$$

$$E = V + V_{rest} \tag{3.15}$$

$$\frac{dm}{dt} = -(\alpha_m + \beta_m)m + \alpha_m \tag{3.16}$$

$$\frac{dn}{dt} = -(\alpha_n + \beta_n)n + \alpha_n \tag{3.17}$$

$$\frac{dh}{dt} = -(\alpha_h + \beta_h)h + \alpha_h \tag{3.18}$$

$$\frac{dp}{dt} = -(\alpha_p + \beta_p)p + \alpha_p \tag{3.19}$$

Note that membrane voltage is denoted as E in Eqs. (3.11) to (3.13) and as V (reduced membrane voltage) in Eqs. (3.10) and (3.14). Temperature T is measured in degrees K; for temperatures other than 20°C = 293.15 K, the α and β values in Eqs. (3.16) to (3.19) must be modified. Parameter values and further expressions of the FH model are listed in Table 3.1. Sodium current plays the dominant role in the action potential of the mammalian node of Ranvier, and in contrast to the axon model of squid and frog there are almost no potassium currents.[27-29]

The CRRSS model, named after Chiu, Ritchie, Rogert, Stagg, and Sweeney, describes a myelinated rabbit nerve node extrapolated from orginal 14°C data to 37°C:[28,30]

$$\frac{dV}{dt} = \left[-g_{Na}m^2 h(V - V_{Na}) - g_L(V - V_L) + i_{stimulus} \right] / c \tag{3.20}$$

$$\frac{dm}{dt} = \left[-(\alpha_m + \beta_m)m + \alpha_m \right]k \tag{3.21}$$

$$\frac{dh}{dt} = \left[-(\alpha_h + \beta_h)h + \alpha_h \right]k \tag{3.22}$$

$$k = 3^{0.1T - 3.7} \tag{3.23}$$

Note that the temperature factor $k = 1$ for $T = 37°C$. Parameter values and further expressions of the CRRSS model are listed in Table 3.1.

Schwarz and Eikhof obtained a model of FH type from voltage clamp experiments on rat nodes.[29] From the original data, the Schwarz–Eikhof (SE) model results by assuming a nodal area of 50 μm².[14,31]

$$\frac{dV}{dt} = [-i_{Na} - i_K - i_L + i_{stimulus}] / c \tag{3.24}$$

$$i_{Na} = P_{Na} m^3 h \frac{EF^2}{RT} \frac{[Na]_o - [Na]_i \exp(EF / RT)}{1 - \exp(EF / RT)} \tag{3.25}$$

$$i_K = P_K n^2 \frac{EF^2}{RT} \frac{[K]_o - [K]_i \exp(EF / RT)}{1 - \exp(EF / RT)} \tag{3.26}$$

$$i_L = g_L (V - V_L) \tag{3.27}$$

$$E = V + V_{rest} \tag{3.28}$$

$$\frac{dm}{dt} = -(\alpha_m + \beta_m)m + \alpha_m \tag{3.29}$$

$$\frac{dn}{dt} = -(\alpha_n + \beta_n)n + \alpha_n \tag{3.30}$$

$$\frac{dh}{dt} = -(\alpha_h + \beta_h)h + \alpha_h \tag{3.31}$$

Parameter values and additional expressions of the SE model are listed in Table 3.1.

Schwarz et al.[33] derived the SRB model from human nerve fibers at room temperature. Single-action potentials are little effected by removing the fast or slow potassium currents, but a slow K conductance was required to limit the repetitive response of the model to prolonged stimulating currents. SRB model follows:

$$\frac{dV}{dt} = [-i_{Na} - i_{K,fast} - i_{K,slow} - i_L + i_{stimulus}] / c \tag{3.32}$$

$$i_{Na} = P_{Na} m^3 h \frac{EF^2}{RT} \frac{[Na]_o - [Na]_i \exp(EF / RT)}{1 - \exp(EF / RT)} \tag{3.33}$$

$$i_{K,fast} = g_K n^4 (V - V_K) \tag{3.34}$$

$$i_{K,slow} = g_{K,slow} p(V - V_K) \tag{3.35}$$

$$i_L = g_L (V - V_L) \tag{3.36}$$

$$E = V + V_{rest} \tag{3.37}$$

$$\frac{dm}{dt} = -(\alpha_m + \beta_m)m + \alpha_m \tag{3.38}$$

$$\frac{dn}{dt} = -(\alpha_n + \beta_n)n + \alpha_n \tag{3.39}$$

$$\frac{dh}{dt} = -(\alpha_h + \beta_h)h + \alpha_h \tag{3.40}$$

$$\frac{dp}{dt} = -(\alpha_p + \beta_p)p + \alpha_p \tag{3.41}$$

Additional SRB model data for 37°C are listed in Table 3.1. (Axon membrane models for biomedical applications are further discussed in References 14 and 31 through 36.)

Some authors neglect the weak K currents in the mammalian axon totally because of their small amounts, but up to five types of K ion channels are shown to have notable influences on axonal signaling, especially in the internode, an element often modeled as perfect insulator with capacity $C = 0$. Curious effects such as spontaneous switching between high and low threshold states in motor axons are suggested to be a consequence of K^+ loading within small internodal spaces.[37] Phenomena such as different excitability fluctuations after action potentials in sensory and motor neurons require far more sophisticated modeling. This means that for many applications the modeler's work is not finished by selecting the SRB model (as it is based on human data) or by fitting its parameters according to the shape and propagation velocity of a measured action potential.

The dendrite and especially the soma region is more difficult to simulate because a variety of ion channel types are involved.[37] Some membrane models are available,[39–45] but even in these models we cannot rely on a precise individual measurement of all the ion components that vary in channel density and open/close kinetics. Repetitive firing (bursting) may occur as response to stimulation or in pacemaker neurons without any input.

3.2.2 Influence of temperature

Most of the membrane model data were gathered at low temperatures, but for biomedical applications they have to be extrapolated to physiological temperatures. Usually, a specific constant, Q_{10}, is introduced to account for the acceleration in membrane-gating dynamics when the temperature is increased by 10°C. Hodgkin and Huxley used $Q_{10} = 3$ as a common coefficient for all gating variables m, n, and h [Eqs. (3.6)–(3.8)]:

$$k = Q_{10}^{(T-T_0)/10} \tag{3.42}$$

Raising the temperature causes shortening of the action potential and an increase of the spike propagation velocity in the unmyelinated squid axon. The spike duration in Figure 3.1 with original HH data (6.3°C) is about ten times longer than that of the cochlea neuron in Figure 3.2, which also was simulated with HH kinetics. The temporal membrane behavior in cat[46] was fitted with $k = 12$, corresponding to a temperature of 29°C (Eq. (3.42)); see Motz and Rattay[47] and Rattay,[14] p. 165). The HH spike amplitude becomes rather small for higher temperatures. An essentially reduced amplitude is not able to preserve the excitation process along the fiber, and for temperatures higher than 31 to 33°C action potentials will not propagate any more in squid axons (heat block).[14,48,49] Such amplitude reductions are not seen in the membrane models of myelinated axons (FH, CRRSS, SE, and SRB models), and the heat block phenomenon is absent both in myelinated and unmyelinated fibers of warm-blooded animals. In all axon models, spike duration shortens considerably when temperature is increased — for example, from 20°C (usual for data acquisition) to 37°C.[14,32] A temperature factor, $k = 12$, in the HH-model fits the temporal characteristics of action potentials in unmyelinated axons of humans and warm-blooded animals at 37°C.[14,35,50]

The original FH model[51] incorporates thermal molecular motion according to the laws of gas dynamics [Eqs. (3.11)–(3.13)], but the authors did not include the necessary Q_{10} values presented in Table 3.1 for the gating variables m, n, h, and p. The gating process becomes essentially accelerated for high temperatures and dominates spike duration.[14,32,48,52] Also, the threshold currents are influenced by the temperature: The warmer membrane is easier to excite (Figure 3.3). Therefore, the Q_{10} factors have to be involved in functional electrical nerve stimulation modeling. Note that many published results obtained with the original FH model must be corrected to be valid for 37°C.

3.2.3 Compartment models

Small pieces of a neuron can be treated as isopotential elements and a whole neuron is represented by an electric network (Figure 3.4). A current injected at the nth compartment has to cross the membrane as capacitive or ionic current and leave to the left or right side as longitudinal current; that is, application of Kirchhoff's law is an extension of Eq. (3.1):

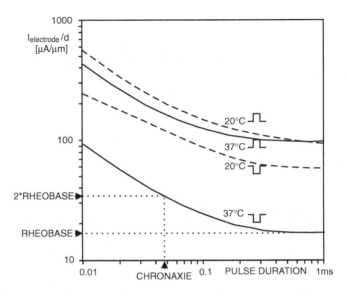

Figure 3.3 Threshold currents as a function of pulse duration (strength duration relation). A myelinated fiber of diameter *d* and internodal length $\Delta x = 100d$ is stimulated by a point source located at a distance Δx above a node. Excitation is easier for high temperatures (full lines) and cathodic pulse. The threshold current for long pulses becomes time-independent and is called *rheobase*. The pulse duration belonging to the doubled rheobase current is called *chronaxie*. Simulations use the FH model; internode membrane is assumed to be a perfect insulator, with $c = 0$.

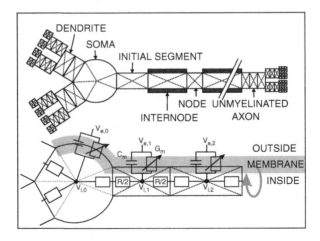

Figure 3.4 Scheme of a neuron with subunits and part of the simplified equivalent electrical network (batteries resulting from different ion concentrations on both sides of the membrane are not shown).

$$I_{injected,n} = C_n \frac{d(V_{i,n} - V_{e,n})}{dt} + I_{ion,n} + \frac{V_{i,n} - V_{i,n-1}}{R_n/2 + R_{n-1}/2} + \frac{V_{i,n} - V_{i,n+1}}{R_n/2 + R_{n+1}/2} \quad (3.43)$$

where n indicates the nth compartment, C is its membrane capacity, and $R/2$ is the internal resistance between the center and the border of the compartment. Introducing the reduced membrane voltage $V = V_i - V_e - V_{rest}$ leads to the following system of differential equations for calculating the time courses of V_n in every compartment:

$$\frac{dV_n}{dt} = \left[-I_{ion,\,n} + \frac{V_{n-1} - V_n}{R_{n-1}/2 + R_n/2} + \frac{V_{n+1} - V_n}{R_{n+1}/2 + R_n/2} + \ldots \right.$$
$$\left. + \frac{V_{e,n-1} - V_{e,n}}{R_{n-1}/2 + R_n/2} + \frac{V_{e,n+1} - V_{e,n}}{R_{n+1}/2 + R_n/2} + \ldots + I_{injected,\,n} \right] \bigg/ C_n \quad (3.44)$$

The dots in Eq. (3.44) stand for terms that have to be added in cases of more than two neighbor elements (e.g., at the cell body [soma] or at branching regions). For the first and last compartments, Eq. (3.44) has a reduced form. The membrane surface, A_n, of every compartment has to be calculated to find $C_n = A_n c_n$ (c_n is the specific membrane capacitance) and $I_{ion} = A_n i_{ion}$. The ionic membrane current density, i_{ion}, is computed with an appropriate membrane model; for cylinder elements (d, diameter; Δx, length), we obtain $A_n = d_n \pi \Delta x_n$, $R/2 = 2\rho_i \Delta x_n / (d_n^2 \pi)$, where the internal resistivity ρ_i is often assumed as $\rho_i = 0.1$ kΩcm.

In a spherical cell body with several processes (as in Figure 3.4), the internal resistances to the neighbor compartments depend on the compartment diameters; that is, $R_{soma \rightarrow dendrite1}/2 < R_{soma \rightarrow axon}/2$ if $d_{dendrite1} > d_{axon}$. In more detail, $A_{soma} = 4r_{soma}^2 \pi - \Sigma(2r_{soma}\pi h_j)$, with subscript j indicating the jth process) such that $h_j = r_{soma} - z_j$, where $z_j = \sqrt{r_{soma}^2 - (d_{process,\,j}/2)^2}$. The somatic resistance to the border of the jth process is:

$$\frac{R_{soma,\,j}}{2} = \frac{\rho_i}{2r\pi} \ln\left(\frac{r_{soma} + z_j}{r_{soma} - z_j} \right) \quad (3.45)$$

The electric and geometric properties change from compartment to compartment and the modeler has to decide about the degree of complexity that should be involved. For example, the surface area of a dendrite compartment and the soma can be simulated by a cylinder and a sphere or an additional factor is included to represent the enlargement by spines or windings. Halter and Clark[53] and Ritchie[54] presented detailed ion current data for the internode, whereas many authors neglect internodal membrane currents. A compromise is to simulate the internode as a single compartment with both membrane conductance G_m and membrane capacity C proportional to $1/N$, where N is the number of myelin layers.[12]

Synaptic activation of a neuron can be simulated by current injection at the soma or at dendrite compartments (Figure 3.2). In such cases, all external potentials, V_e, are assumed to be 0 in Eq. (3.44). On the other hand, neuroprosthetic devices generate neural activities by application of electric fields. Instead of current injection, the terms including the external potentials V_e become the stimulating elements in Eq. (3.44).

3.2.4 Activating function

The driving term of the external potential on compartment n (Eq. (3.44)) is called the activating function f_n:

$$f_n = \left[\frac{V_{e,n-1} - V_{e,n}}{R_{n-1}/2 + R_n/2} + \frac{V_{e,n+1} - V_{e,n}}{R_{n+1}/2 + R_n/2} + ... \right] \Big/ C_n \qquad (3.46)$$

The physical dimension of f_n is V/sec or mV/msec. If the neuron is in the resting state before a stimulating current impulse is applied, the activating function represents the rate of membrane voltage change in every compartment that is activated by the extracellular field. That is, f_n is the slope of membrane voltage V_n at the beginning of the stimulus. Regions with positive activating function are candidates for spike initiation, whereas negative activating function values cause hyperpolarization.

The temporal pattern generated in a target neuron by a single active electrode of a cochlear implant is analyzed in the following with the activating function concept. The electrode is assumed as an ideal point source in an infinite homogeneous extracellular medium, which results in the simple relation:

$$V_e = \frac{\rho_e \cdot I_{electrode}}{4\pi r} \qquad (3.47)$$

where V_e is measured at a point with distance r from the point source and ρ_e is the extracellular resistivity.

The insert of Figure 3.5A shows the values of V_e and f along a cochlear neuron when stimulated with 1 mA from a monopolar electrode. The alternating sequence of positive and negative activating function values indicates first spike initiation at peripheral node P2 (Figure 3.5B) and a second spike initiation region in the central axon for stronger positive monophasic stimuli (Figure 3.5C). Cathodic stimulation also generates two regions of spike initiation — first, excitation in P1 (Figure 3.5D), then excitation in P4 (Figure 3.5E). The activating function predicts easier excitation with cathodic currents with values of –3890 V/sec vs. 2860 V/sec in P1 and P2, respectively. Note, however, that f quantifies the very first response, which is flattened by the capacity (compare dashed line in Figure 3.1) and is further influenced by the neighbor elements.

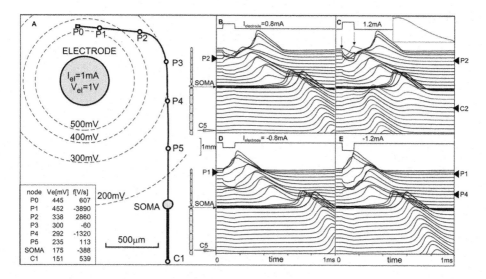

Figure 3.5 Human auditory nerve response evoked by a stimulating impulse from a cochlear implant. (A) Position of a cochlear neuron relative to a spherical electrode (same geometry as in Figure 3.2; unmyelinated terminal P0, nodes P1–P5 of peripheral myelinated thin process, unmyelinated region around the soma, and a thicker central axon). Extracellular resistivity $\rho_e = 0.3$ kΩcm, and calculation of potential from a 1-mA point source [Eq. (3.47)] results in 1 V at $r = 0.24$ mm, which is the electrode radius. The values of the activating function f shown in insert (A) are proportional to the slopes of the neural response curves at stimulus onset in (B) and (C). The largest f value of P2 predicts the place of spike initiation in (B). All slopes of membrane voltage in the central nodes are positive and strengthen the stimulus current to 1.2 mA, causing additional spike initiation at peripheral node C2 (C). The activating function value of P0 is also positive, but the much stronger negative activity in P1 compensates the excitation process quickly; insert (C) shows the part of the membrane voltage of the P0 compartment during the stimulus pulse as marked by arrows, with 5× magnification. In part (D), –800 µA per 100-µsec pulse initiates a spike at P1 which is in accordance with the activating function concept. In part (E), the smaller intensity of f at P4 allows spike initiation at –1200 µA. The P4 spike develops after the P1, but nevertheless the P4 spike activates the central axon. Note the different arrival times at the measuring electrode C5.

3.2.5 Activating function for long fibers

In 1976, McNeal introduced the first efficient model for computing peripheral nerve fiber responses in functional electrical nerve stimulation. Many authors followed his simplification of an ideal internode membrane, where both membrane conductance and membrane capacity are assumed to be 0. In this case, the main equation, Eq. (3.44), reduces to:[4,12]

$$\frac{dV_n}{dt} = \left[-i_{ion,\,n} + \frac{d\Delta x}{4\rho_i L} \left(\frac{V_{n-1} - 2V_n + V_{n+1}}{\Delta x^2} + \frac{V_{e,\,n-1} - 2V_{e,\,n} + V_{e,\,n+1}}{\Delta x^2} \right) \right] \Big/ c \quad (3.48)$$

with constant fiber diameter d, node-to-node distance Δx, node length L, and axoplasma resistivity ρ_i; both membrane capacity c and nodal ion current i_{ion} are per cm². The activating function:

$$f_n = \frac{d\Delta x}{4\rho_i Lc} \frac{V_{e,n-1} - 2V_{e,n} + V_{e,n+1}}{\Delta x^2} \tag{3.49}$$

becomes proportional to the second difference of the extracellular potential along the fiber.

Equations (3.48) and (3.49) model the unmyelinated fiber when the length of the active membrane is equal to the compartment length: $L = \Delta x$. With $\Delta x \to 0$, Eq. (3.49) reads as:

$$f = \frac{d}{4\rho_i c} \frac{\partial^2 V_e}{\partial x^2} \tag{3.50}$$

Equations (3.49) and (3.50) verify the inverse recruitment order; that is, thick fibers are easier to stimulate[3,21] (Figure 3.5) by the linear relation between d and f.

For a monopolar point source in an infinite homogeneous medium, the activating function is easy to evaluate. For example, for a straight axon (Eq. (3.47); Figure 3.6A,B), we obtain:

$$V_e = \frac{\rho_e \cdot I_{el}}{4\pi} \left((x - x_{el})^2 + z_{el}^2 \right)^{-0.5} \tag{3.51}$$

where *el* stands for electrode and z measures the center-to-center distance between a small spherical electrode and the axon, and (see Eq. (3.50)):

$$f = \frac{d\rho_e \cdot I_{el}}{16\rho_i c\pi} \left((x - x_{el})^2 + z_{el}^2 \right)^{-2.5} \left(2(x - x_{el})^2 - z_{el}^2 \right) \tag{3.52}$$

Equation (3.50) implies the relation between the curvature of the V_e graph and the regions of excitation in a target fiber; solving $f(x) = 0$ gives the points of contraflexure of V_e that separate the regions of depolarization ($f > 0$) from the hyperpolarized ones ($f < 0$). In our example, this results in the vertical dashed lines at $x = x_{el} \pm 0.707z_{el}$. In the excited ($f > 0$) regions, the V_e graph is always above its tangents (Figure 3.6B; regions I and III for anodic stimulation). Strong curvature of V_e means large f values; therefore, the central part of region II becomes extremely hyperpolarized at anodic stimulation (Figure 3.6C). In relation to the shape of the activating function, the threshold for cathodic stimulation is about four times weaker (Figure 3.6E and F) which is in accordance with experimental findings.[55]

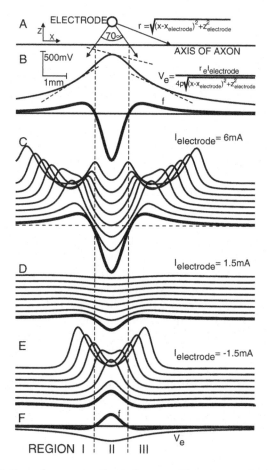

Figure 3.6 Stimulation of an unmyelinated axon with a monopolar electrode. During the 100-μsec stimulus pulse, the activating function defines three regions of responses, separated by vertical dashed lines. (B)–(D) Vertically shifted snapshots of membrane voltages in 100-μsec steps. Thick lines demonstrate the similarity between the activating function and the membrane voltage at the end of the 100-μsec stimulus signal. Simulation is of HH data at 29°C, $\rho_e = 0.3$ kΩ, electrode fiber distance $z = 1$ mm, and fiber diameter 10 μm.

Deviations between the thick lines for f and for V_e at the end of the 100-μsec pulse (just above threshold) are caused by (1) an inner-axonal current component resulting from membrane currents at the other compartments, (2) exponential loading process of the membrane capacity (in the activating function concept, the dashed line in Figure 3.1 is approximated by the tangent at stimulus onset), and (3) the beginning of the voltage-sensitive ion channel-gating processes. Compared to the prediction by f, there is a small enlargement of the hyperpolarized region II seen at the zero-crossings of the thick line in Figure 3.6C. This is mostly caused by the inner-axonal current, where the strong hyperpolarized central part has a negative influence at the borderlines to regions I and III. For the same reason the points of spike initiation (Figure 3.6C) are lateral to the maximum points of f (Figure 3.6B).

With some computational effort, the linear theory can be improved by considering the influences of the inner-axonal current and the exponential loading process of the membrane capacity.[56] Especially for close electrode fiber distances, the subthreshold steady-state membrane voltage will considerably deviate from the first response.[57,58] But, the activating function concept demonstrates the instructive relation between a simple formula and observed phenomena. For example, Eq. (3.49) indicates that a straight part of an axon in a constant field region is not stimulated directly; however, the bending part in a constant-field region will be de- or hyperpolarized, depending on the orientation of the axon and field. Such effects are studied in spinal root stimulation (see Figure 3.3 of Reference 59).

Note that the equations of this section require constant data along the fiber. At a fiber ending, widening or branching the treatment is according to the description of Eq. (3.44). The fiber-end influence may be modeled by a boundary field driving term in addition to the activating function.[11] In some applications (heart and skeletal muscle, nerve bundle), many excitable fibers are closely packed and the effects of current redistribution are handled by the bidomain model. The tissue is not assumed to be a discrete structure but rather two coupled, continuous domains: one for the intracellular space and the other for the interstitial space. This space-averaged model is described with a pair of coupled partial differential equations that have to be solved simultaneously for the intracellular and interstitial potentials.[7,18,19,60] Despite the complex situation, several stimulation phenomena of the heart muscle are already predicted by the activating function concept from single-fiber analysis.[60]

3.2.6   Membrane current fluctuations

Single-fiber recordings show variations in the firing delays even when stimulated with current pulses of constant intensity.[62,63] The main component of the stochastic arrival times (jitter) is explained by ion channel current fluctuations. Models of the Hodgkin–Huxley type define the gating variables m, n, and h as probabilities, but no variance is included. The addition of a simple noise term in single compartment modeling can fit several of measured cochlear nerve data.[47] Other phenomena — for example, a bimodal distribution of spike latencies, as discussed later (Figure 3.7B) — require full compartment analysis with a stochastic term in every active compartment.[35] Including membrane current fluctuations is significant in the following applications: (1) modeling temporal firing patterns (e.g., in the cochlear nerve), (2) detailed analysis of fiber recruitment (for example, constant stimulation conditions generate three types of axons within a nerve bundle: one group fires at every stimulus, the second fires stochastically, and the third shows solely subthreshold responses), (3) the transition from the myelinated to the unmyelinated branching axon is less secure in regard to signal transfer,

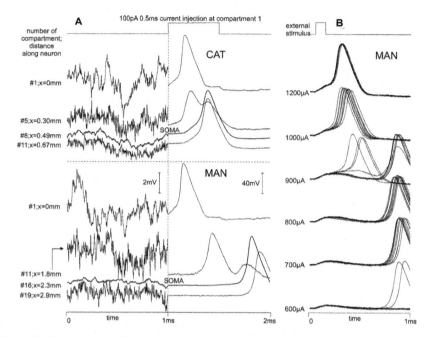

Figure 3.7 Simulation of ion current fluctuations in the cochlear neuron membrane. (A) Membrane voltage in the resting (left) and activated (right) neuron in a human and cat. Stimulation by current injection at the lateral end is close to natural spike initiation by synaptic activation. As a consequence of the RC circuit properties of the network, the voltage fluctuations become low pass filtered and neighboring compartment behavior is correlated. The unmyelinated terminals are assumed to have the same 1-μm diameter with same ion channel density as the peripheral nodes, but a length relation of 10 to 1.5 μm results in 2.58 times stronger membrane current fluctuation at the endings. The strong influence of the first compartment to the network is still seen in the cat soma region (left) but it is lost in the human case because of the longer peripheral axon (five nodes and a large unmyelinated soma region). Voltage fluctuations in the soma region are generally smaller than in the nodes. Loading the large capacity of the unmyelinated human soma requires the spike about 400 μsec to cross this region, which is three times the value of the myelinated cat case (right). (B) Responses of the fifth node in the central human axon (compartment 25 marked as C5 in Figure 3.5) to anodic external stimulation with 100 μsec pulses (10 runs for every stimulus intensity; situation shown in Figure 3.5). Weak stimuli fail sometimes (600 μA: 8 of 10) and have larger jitter, and their arrival is more delayed, as they are generated in the periphery. At 900 μA, spikes are also initiated in the central axon. Note that this experimentally known bimodal response is a consequence of ion channel current fluctuations. For details, see Rattay et al.[35] and Rattay.[70]

and spikes can stochastically be lost[50,64], (4) during the integration of neural inputs of an integrate-and-fire neuron.

The elaborate models of single-channel current influences are recommendable for detailed investigations.[65-67] Combining this work with a compartment model will show most of the noisy behavior in nerve fibers as reported by Verween and Derksen.[66] Membrane noise increases with deviation from the resting voltage and is greater for hyperpolarization than for depolarization. To reduce the enormous computational effort of simulating every channel activity we follow a suggestion of Rubinstein[69] and assume the noise current to be proportional to the square root of the number of sodium channels.[35,70] At every compartment, a Gaussian noise variable, *GAUSS* (mean = 0, standard deviation = 1), is introduced that changes its value every 2.5 μsec. The amplitude of the noise term is proportional to the square root of the number of sodium channels involved. Noisy membrane currents (in μA) are defined as:

$$I_{noise,n} = GAUSS \cdot k_{noise} \sqrt{A_n \cdot g_{Na}} \tag{3.53}$$

where A_n denotes the membrane area of compartment n (in cm^2), and g_{Na} (mS cm^{-2}) is the maximum sodium conductance. A standard factor k_{noise} = 0.05 μA.mS$^{-1/2}$ common to all compartments fits the observations of Verveen and Derksen.[66]

Simulated voltage fluctuations are compared for cochlear neurons in cats and humans. Both neurons are assumed to have the same peripheral (1 μm) and central (2 μm) axon diameters, but in the cat the peripheral axon is shorter (distance from soma to peripheral terminal in cats is 0.3 mm; in humans, 2.3 mm), and the cat somatic region is insulated by myelin sheets. Only the 10-μm-long peripheral terminal and the nodes are without myelin; additionally, in humans, the soma region is unmyelinated (Figure 3.2). The following effects are seen in Figure 3.7A:

1. Thick fibers show less membrane fluctuations. The peripheral compartments (1 and 5 in cats, 1 and 11 in humans) have smaller noise amplitudes than the central axon (11 in cats, 19 in humans). The phenomenon observed by Verveen and Derksen[66] can be explained in a simple model with corresponding states of homogeneous myelinated axons. During the excitation process, the nodal currents are proportional to axon diameter, but the noisy current increases with the square root of diameter only (nodal length is independent of diameter).

2. Membrane voltage fluctuations in one compartment have two components: the own membrane current and the noisy inner currents from the neighboring compartments. Note the correlated sinusoidal-like shapes of the three lower left cat curves. The cat myelinated soma is assumed to have no sodium channels; therefore, the displayed membrane voltage has lost the first component and becomes the averaged neighboring compartment noise filtered by the soma capacity effect.

And, as shown in Figure 3.7B:

3. Threshold voltage has to be redefined (e.g., as 50% probability of generating a spike); effectiveness increases from 600 µA (20%) to 700 µA (100%).
4. Weak stimuli have larger jitter because of the longer noise influence during the subthreshold part of excitation. Jitter of peripherally evoked spikes decreases from 600 to 900 µA; the same for the central ones — from 900 to 1200 µA.
5. Spike initiation regions can be sensitive to stochastic components (900 µA) .

3.3   The electrically stimulated retina

Photoreceptor loss due to retinal degenerative diseases such as age-related macular degeneration and retinitis pigmentosa is a leading cause of blindness. Although these patients are blind, they possess functioning bipolar and ganglion cells that relay retinal input to the brain. Retina implants are in development in order to evoke optical sensations in these patients. The success of such an approach to provide useful vision depends on elucidating the neuronal target of stimulation. A compartmental model for extracellular electrical stimulation of the retinal ganglion cell (RGC) has been developed.[71] In this model, a RGC is stimulated by extracellular electrical fields with active channels and realistic cell morphology derived directly from neuronal tracing of an amphibian retina cell.

The goal of the simulation was to clarify which part of the target cell is most excitable. It was previously shown that patients blind from retinitis pigmentosa resolve focal phosphenes when the ganglion cell side of the retinal surface is stimulated. None of the patients reported the perception of wedges; all reported spots of light or spots surrounded by dark rings. Where is the initiation point of the signal that causes these optical sensations? Is the threshold to propagate an action potential down the axon lowest for an electrode over the axon, the soma, or somewhere over the dendritic arbor, which may spread up to 500 µm in diameter and overlap the dendritic field of other ganglion cells?[72]

To explore the relative thresholds of ganglion cell axons, somas, and dendrites a mudpuppy (*Necturus maculosus*) RGC with a large dendritic field and a long axon was chosen. This cell has been mapped in three dimensions.[73] The entire traced cell, except the soma, is divided into cylindrical compartments. The soma can be modeled as a compartmentalized sphere[71] or as a single spherical compartment as described above. The number of compartments was increased until the thresholds did not change by more than 1%. Finally, more than 9000 compartments ensure fine numerical solutions for the RGC, as shown in Figure 3.8.

An extracellular field from an ideal monopolar point or disk electrode in a homogeneous medium were applied to these compartments. An electrical network as shown in Figure 3.4 represents the whole retinal ganglion cell. Several constants were specified based on whole-cell recording data

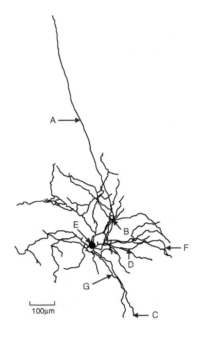

Figure 3.8 A topographical view of the target cell shown with its axon projecting upwards. The soma of the retinal ganglion cell is modeled as a 24-μm sphere. This size was chosen to approximate the diameter of the actual traced mudpuppy soma. The electrode is modeled as an ideal point source in an infinite homogeneous medium. For calculation of V_e, see Eq. (3.47); $\rho_e = 1/60$ Ωcm, the resistivity of normal (0.9%) saline, which is similar to the resistivity of the vitreous humor.[76] The height of the electrode was held fixed at 30 μm above the center of the soma, a distance comparable to that of the retinal ganglion cells to the surface of the retina in our region of interest. The electrode positions are marked by the letters A through G. The linear distances from the soma (in μm) are: (A) over the axon ~503, (B) over the axon ~130, (C) opposite the axon ~334, (D) perpendicular to the axon ~160, (E) directly above the soma, (F) perpendicular to the axon ~302, and (G) opposite the axon ~121. The vertical distance (not shown) from the electrode to the center of the nearest compartment is (in μm): (A) 33.0, (B) 30.0, (C) 34.0, (D) 30.0, (E) 30.0, (F) 29.5, and (G) 38.0.

at 22°C, including membrane capacitance ($c = 1$ μF/cm²) and cytoplasmic resistance (0.11 kΩcm).[75] These values are assumed to be uniform throughout the cell.

For comparison, the electric conductance of the cell membrane was evaluated with a linear passive model, a Hodgkin–Huxley model with passive dendrites (HH),[1] and a model composed of all active compartments with five nonlinear ion channels (FCM).[76] At the start of all simulations, the membrane potential is initialized to a resting potential of –70 mV. The linear passive model reduces each patch of membrane to a simple parallel RC circuit with a leak. The leak conductance was modeled as a battery at –70 mV in series with a conductance of 0.02 mS/cm².

The retinal ganglion cell in many species (including humans) has a nonmyelinated axon inside the eye. Some simulations were performed with the classic description for unmyelinated axons, the Hodgkin–Huxley model. The Hodgkin–Huxley channel kinetics are applied to the soma and the axon, with the constants g_{Na} = 120 mS/cm², E_{Na} = 50 mV, g_K = 36 mS/cm², E_K = –77 mV, g_l = 0.3 mS/cm², E_l = –54.3 mV. The dendrites are passive in this model.

The FCM model includes five nonlinear ion channels and one linear leakage channel.[44,45,76] The kinetic model was constructed on the basis of voltage clamp data in retinal ganglion cells of tiger salamanders and rats.[77,78] The model consists of I_{Na} and I_{Ca}; the delayed rectifier, I_K; $I_{K,A}$; and the Ca-activated K current, $I_{K,Ca}$. $I_{K,Ca}$ was modeled to respond to Ca influx, and a variable-rate, Ca-sequestering mechanism was implemented to remove cytoplasmic calcium. I_{Ca} and $I_{K,Ca}$ do play an important role in regulating firing frequency. The basic mathematical structure for voltage-gating is based on the Hodgkin–Huxley model:

$$C_m \frac{dE}{dt} = -\bar{g}_{Na} m^3 h (E - E_{Na}) - \bar{g}_{Ca} c^3 (E - E_{Ca}) - \bar{g}_K n^4 (E - E_K)$$
$$- \bar{g}_A a^3 h_A (E - E_K) - g_{K,Ca} (E - E_K) - \bar{g}_l (E - E_l) + I_{stimulus} \tag{3.54}$$

where

$$\bar{g}_{Na} = 50 \, mS/cm^2 , E_{Na} = 35 \, mV, \bar{g}_{Ca} = 2.2 \, mS/cm^2, \bar{g}_K = 12 \, mS/cm^2,$$

$$E_K = -75 \, mV, \bar{g}_A = 36 \, mS/cm^2, \bar{g}_l = 0.005 \, mS/cm^2, E_l = -60 \, to -65 \, mV.$$

The gating of $I_{K,Ca}$ is modeled as

$$g_{K,Ca} = \bar{g}_{K,Ca} \frac{\left([Ca^{2+}]_i / Ca^{2+}_{diss}\right)^2}{1 + \left([Ca^{2+}]_i / Ca^{2+}_{diss}\right)^2} \tag{3.55}$$

where $\bar{g}_{K,Ca}$ = 0.05 mS/cm² and the Ca²⁺-dissociation constant, Ca^{2+}_{diss}, is 10^{-3} mmole/l. $[Ca^{2+}]_i$ is allowed to vary in response to I_{Ca}. The inward flowing Ca²⁺ ions are assumed to be distributed uniformly throughout the cell; the free $[Ca^{2+}]$ above a residual level — Ca^{2+}_{res} = 10^{-4} mmole/l — are actively removed from the cell or otherwise sequestered with a time-constant (τ = 1.5 msec). Thus, $[Ca^{2+}]_i$ follows the equation:

$$\frac{d[Ca^{2+}]_i}{dt} = \frac{-sI_{Ca}}{2vF} - \frac{[Ca^{2+}]_i - [Ca^{2+}]_{res}}{\tau} \tag{3.56}$$

where F is the Faraday constant, s/v is the ratio of surface to volume of the compartment being studied, and the factor of 2 on v is the valency. For a spherical soma with $r = 12$ μm, Eq. (3.56) becomes:

$$\frac{d[Ca^{2+}]_i}{dt} = -0.000013\,I_{Ca} - 0.666667\,([Ca^{2+}]_i - 0.0001) \qquad (3.57)$$

E_{Ca} is modeled as variable according to the already mentioned Nernst equation:

$$E_{Ca} = \frac{RT}{2F}\ln\frac{[Ca^{2+}]_e}{[Ca^{2+}]_i} \qquad (3.58)$$

where $[Ca^{2+}]_e$ is the external calcium ion concentration = 1.8 mmole/l.

Similar to the Hodgkin–Huxley model, the rate constants for m, h, c, n, a, and h_A solve the first-order kinetic equation (Table 3.2):

$$\frac{dx}{dt} = -(\alpha_x + \beta_x)x + \alpha_x \qquad (3.59)$$

The multi-compartment model includes representations for dendritic trees, soma, axon hillock, a thin axonal segment, and an axon distal to the thin segment. On these parts, the five ion channels were distributed with varying densities, simulated by varying the value of g_{max} (mS/cm^2) for each channel (Table 3.3).[44,45,76] Using the multicompartmental simulation package NEURON[79], the action potential threshold values were computed for the electrode positions as marked in Figure 3.8.[71]

Table 3.2 Rate Constants for Voltage-Gated Ion Channels

α	β
$\alpha_m = \dfrac{-0.6(E+30)}{e^{-0.1(E+30)} - 1}$	$\beta_m = 20e^{-(E+55)/18}$
$\alpha_h = 0.4e^{-(E+50)/20}$	$\beta_h = \dfrac{6}{e^{-0.1(E+20)} + 1}$
$\alpha_c = \dfrac{-0.3(E+13)}{e^{-0.1(E+13)} - 1}$	$\beta_c = 10e^{-(E+38)/18}$
$\alpha_n = \dfrac{-0.02(E+40)}{e^{-0.1(E+40)} - 1}$	$\beta_n = 0.4e^{-(E+50)/80}$
$\alpha_A = \dfrac{-0.006(E+90)}{e^{-0.1(E+90)} - 1}$	$\beta_A = 0.1e^{-(E+30)/10}$
$\alpha_{hA} = 0.04e^{-(E+70)/20}$	$\beta_{hA} = \dfrac{0.6}{e^{-0.1(E+40)} + 1}$

Table 3.3 FCM Model Channel Densities at Soma, Dendrite, and Axon

g_{max}	Soma (mS/cm²)	Dendrite	Axon (0–3%)[a]	Axon (3–9%)[a]	Axon (9–100%)[a]
g_{Na}	70	40	150	100	50
g_{Ca}	1.5	3.6	1.5	0	0
g_K	18	12	18	12	15
g_{Ka}	54	36	54	0	0
$g_{K,Ca}$	0.065	0.065	0.065	0	0

[a] Percent total axon length from soma; total axon length approximately 1 mm.

A practical retinal prosthesis will perhaps have disk electrodes, because flat electrodes are easy to produce. To replace the point electrode source by a field from an equipotential metal disk in a semi-infinite medium, the external potential, V_e, under the electrode is calculated by:

$$V(r,z) = \frac{2V_0}{\pi} \cdot \arcsin \frac{2a}{\sqrt{(r-a)^2 + z^2} + \sqrt{(r+a)^2 + z^2}} \qquad (3.60)$$

where (r, z) is the radial and axial distance from the center of the disk in cylindrical coordinates $z \neq 0$.[80] V_0 is the potential, and α is the radius of the disk. In simulation, both 50- and 100-μm-diameter disks required the same cathodic threshold voltage. In human trials, the current required to reach threshold also did not vary when a monopolar stimulating electrode was changed from 50-μm-diameter to 100-μm.[71,81]

3.4 Concluding remarks

Calcium currents and the inside calcium concentration often have significant influence on the temporal spiking pattern of electrically stimulated neurons, because calcium is an important intracellular signaling molecule with rapid effects on many processes. A rather simple approach demonstrates the agreement between the FCM model and the firing behavior of retinal ganglion cells;[44,45,76] a similar model exists for pyramidal cells.[42] Detailed modeling includes the variation of calcium concentration with distance from the cell membrane, buffer-elements and pumps.[82]

Selective stimulation of neural tissue is a great challenge for biomedical engineering. A typical example in bladder control is activation of the detrusor muscle without activation of the urethral sphincter and afferent fibers when stimulating sacral roots with tripolar cuff electrodes. Analysis with the activating function is of help for finding the polarized and hyperpolarized regions but nonlinear membrane current modeling has to be included to quantify cathodic and anodic block or anodic break phenomena in selected target fibers and to determine the stimulus pulse parameters for an operating window.[14,83,84]

Table 3.4 Cathodic 100-μsec Current Thresholds for
Point-Source Stimulation[a]

Electrode	Passive Model	HH Model	FCM Model
A	0.900	1.58	1.73
B	1.00	1.79	1.97
C	0.954	75.1	55.2
D	0.699	14.4	2.72
E	1.00	1.00	1.00
F	0.812	65.1	2.80
G	1.05	9.58	2.63

[a] Normalized for position over the soma and calculated with the three membrane models. The absolute threshold current at the soma (electrode position E) for the passive model was −32.9 μA; for the Hodgkin–Huxley model, −43 μA; and for the FCM (five nonlinear ion channels) model, −71 μA. The threshold response for the passive model is defined as a 15-mV depolarization.[13]

Stimulation of neurons in a network is often difficult to explore because of the complex neural geometry. Every type of the involved neurons has to be analyzed in regard to the response and its sensitivity. Note that neurotransmitter release can be initiated or influenced by the applied field without the generation of a spike.[12] This may be of specific importance at the retina where several types of neurons never generate action potentials but their degree of neurotransmitter release heavily depends on transmembrane voltage.

The extracellular potential along a target cell is input data for the excitation process. Some case studies are based on the combination of ideal point sources in infinite homogeneous media or idealized surface electrodes,[71,80,85] but most clinical applications require more precise calculation of the field. Finite element software such as ANSYS is commonly used[70,86,87] but so, too, are finite differences[6] and the boundary element method.[88]

Simulation tools such as NEURON (http://www.neuron.yale.edu/)[79] and GENESIS (http://www.bbb.caltech.edu/GENESIS)[89] are recommended for solving neural compartment models. Compartment modeling (i.e., Eq. (3.44)), together with a set of equations for the ion currents, results in a system of ordinary differential equations that can be solved directly. This procedure (methods of lines) will cause numerical problems for fine spatial segmentation. Integration routines for stiff differential equations, implicit integration methods, or solving with the Crank Nicholson method are of help to reduce the computational effort.[90]

References

1. Hodgkin, A.L. and Huxley, A.F., A quantitative description of membrane current and its application to conduction and excitation in nerve, *J. Physiol.*, 117, 500–544, 1952.

2. McNeal, D.R., Analysis of a model for excitation of myelinated nerve, *IEEE Trans. Biomed. Eng.*, BME-23, 329–337, 1976.

3. Rattay, F., Analysis of models for external stimulation of axons, *IEEE Trans. Biomed. Eng.*, 33, 974–977, 1986.

4. Rattay, F., Analysis of models for extracellular fiber stimulation, *IEEE Trans. Biomed. Eng.*, 36, 676–682, 1989.

5. Sweeney, J.D. and Mortimer, J.T., An asymmetric two electrode cuff for generation of unidirectionally propagated action potentials, *IEEE Trans. Biomed. Eng.*, BME-33, 541–549, 1986.

6. Veltink, P.H., van Alste, J.A., and Boom, H.B.K., Simulation of intrafascicular and extraneural nerve stimulation, *IEEE Trans. Biomed. Eng.*, BME-35, 69–75, 1988.

7. Altman, K.W. and Plonsey, R., Point source nerve bundle stimulation: effects of fiber diameter and depth on simulated excitation, *IEEE Trans. Biomed. Eng.*, 37, 688–698, 1990.

8. Sweeney, J.D., Ksienski, D.A., and Mortimer, J.T., A nerve cuff technique for selective excitation of peripheral nerve trunk regions, *IEEE Trans. Biomed. Eng.*, BME-37, 706–715, 1990.

9. Veraart, C., Grill, W.M., and Mortimer, J.T., Selective control of muscle activation with a multipolar nerve cuff electrode, *IEEE-Trans. Biomed. Eng.*, 40, 640-653, 1990.

10. Tyler, D.J. and Durand, D.M., Intrafascicular electrical stimulation for selectively activating axons, *IEEE Eng. Med. Biol.* 13, 575-583, 1994.

11. Nagarajan, S.S., Durand, D., and Warman, E.N., Effects of induced electric fields on finite neuronal structures: a simulation study, *IEEE Trans. Biomed. Eng.*, BME-40, 1175–1188, 1993.

12. Rattay, F., The basic mechanism for the electrical stimulation of the nervous system, *Neuroscience*, 89, 335–346, 1999.

13. Coburn, B., Neural modeling in electrical stimulation review, *Crit. Rev. Biomed. Eng.*, 17, 133–178, 1989.

14. Rattay, F., *Electrical Nerve Stimulation: Theory, Experiments and Applications*, Springer-Verlag, New York, 1990.

15. Basser, P.J., Wijesinghe, R.S., and Roth, B.J., The activating function for magnetic stimulation derived from a 3-dimensional volume conductor model, *IEEE Trans. Biomed. Eng.*, 39, 1207–1210, 1992.

16. Garnham, C.W., Barker, A.T., and Freeston, I.L., Measurement of the activating function of magnetic stimulation using combined electrical and magnetic stimuli, *J. Med. Eng. Technol.*, 19, 57–61, 1995.

17. Reichel, M., Mayr, W., and Rattay, F., Computer simulation of field distribution and excitation of denervated muscle fibers caused by surface electrodes, *Artif. Organs*, 23, 453–456, 1999.

18. Sobie, E.A., Susil R.C., and Tung, L., A generalized activating function for predicting virtual electrodes in cardiac tissue, *Biophys. J.*, 73, 1410–1423, 1997.

19. Efimov, I.R., Aguel, F., Cheng, Y., Wollenzier, B., and Trayanova, N., Virtual electrode polarization in the far field: implications for external defibrillation, *Am. J. Physiol. Heart. Circ. Physiol.*, 279, H1055–H1070, 2000.

20. Rattay, F., Analysis of the electrical excitation of CNS neurons, *IEEE Trans. Biomed. Eng.*, 45, 766–772, 1998.

21. Blair, E.A. and Erlanger, J., A comparison of the characteristics of axons through their individual electrical responses, *Am. J. Physiol.*, 106, 524–564, 1933.

22. Huguenard, J.R., Low-threshold calcium currrents in central nervous system neurons, *Annu. Rev. Physiol.*, 58, 329–348, 1996.

23. Schiller, J., Helmchen, F., and Sakmann, B., Spatial profile of dendritic calcium transients evoked by action potentials in rat neocortical pyramidal neurones, *J. Physiol.*, 487, 583–600, 1995.

24. Mainen, Z.F. and Sejnowski, T.J., Modeling active dendritic processes in pyramidal neurons, in *Methods in Neuronal Modeling: From Ions to Networks*, 2nd ed., Koch, C. and Segev, I., Eds., MIT Press, Cambridge, MA, 1998, pp. 171–209.

25. Destexhe, A, Mainen, Z.F., and Sejnowski, T.J., Synthesis of models for excitable membranes, synaptic transmission, and neuromodulation using a common kinetic framework, *J. Comput. Neurosci.*, 1, 195–230, 1994.

26. Frankenhaeuser, B., Sodium permeability in toad nerve and in squid nerve, *J. Physiol.*, 152, 159–166, 1960.

27. Horáckova, M., Nonner, W., and Stämpfli, R., Action potentials and voltage clamp currents of single rat Ranvier nodes, *Proc. Int. Union Physiol. Sci.*, 7, 198, 1968.

28. Chiu, S.Y., Ritchie, J.M., Rogart, R.B., and Stagg, D., A quantitative description of membrane currents in rabbit myelinated nerve, *J. Physiol.*, 313, 149–166, 1979.

29. Schwarz, J.R. and Eikhof, G., Na currents and action potentials in rat myelinated nerve fibres at 20 and 37°C, *Pflügers Arch.*, 409, 569–577, 1987.

30. Sweeney, J.D., Mortimer, J.T., and Durand, D., Modeling of mammalian myelinated nerve for functional neuromuscular electrostimulation, in *IEEE 9th Annu. Conf. Eng. Med. Biol. Soc.*, Boston, MA, 1987, pp. 1577–1578.

31. Rattay, F., Simulation of artificial neural reactions produced with electric fields, *Simulation Practice Theory*, 1, 137–152, 1993.

32. Rattay, F. and Aberham, M., Modeling axon membranes for functional electrical stimulation, *IEEE Trans. Biomed. Eng.*, BME 40, 1201–1209, 1993.

33. Schwarz, J.R., Reid, G., and Bostock, H., Action potentials and membrane currents in the human node of Ranvier, *Eur. J. Physiol.*, 430, 283–292, 1995.

34. Wesselink, W.A., Holsheimer, J., and Boom, H.B., A model of the electrical behaviour of myelinated sensory nerve fibres based on human data, *Med. Biol. Eng. Comput.*, 37, 228–235, 1999.

35. Rattay, F., Lutter, P., and Felix, H., A model of the electrically excited human cochlear neuron. I. Contribution of neural substructures to the generation and propagation of spikes, *Hear. Res.*, 153, 43–63, 2001.

36. Burke, D., Kiernan, M.C., and Bostock, H., Excitability of human axons, *Clin. Neurophysiol.*, 112, 1575–1585, 2001.

37. Baker, M.D., Axonal flip-flops and oscillators, *Trends Neurosci.*, 23, 514–519, 2000.

38. Magee, J.C., Voltage-gated ion channels in dendrites, in *Dendrites*, Stuart, G., Spruston, N., and Häusser, M., Eds., Oxford University Press, London, 1999, pp. 139–160.

39. Belluzzi, O. and Sacchi, O., A five conductance model of the action potential in the rat sympathetic neurone, *Progr. Biophys. Molec. Biol.*, 55, 1–30, 1991.

40. Winslow, R.L. and Knapp, A.G., Dynamic models of the retinal horizontal cell network, *Prog. Biophys. Mol. Biol.*, 56, 107–133, 1991.

41. McCormick, D.A. and Huguenard, J.R., A model of the electrophysiological properties of thalamocortical relay neurons, *J. Neurophysiol.*, 68, 1384–1400, 1992.

42. Traub, R.D., Jefferys, J.G., Miles, R., Whittington, M.A., and Toth, K., A branching dendritic model of a rodent CA3 pyramidal neurone, *J. Physiol.*, 481, 79–95, 1994.

43. DeSchutter, E. and Smolen, P., Calcium dynamics in large neuronal models, in *Methods in Neuronal Modeling: From Ions to Networks*, 2nd ed., Koch, C. and Segev, I., Eds., MIT Press, Cambridge, MA, 1999, pp. 211–250.

44. Fohlmeister, J.F. and Miller, R.F., Impulse encoding mechanisms of ganglion cells in the tiger salamander retina, *J. Neurophysiol.*, 78, 1935–1947, 1997.

45. Fohlmeister, J.F. and Miller, R.F., Mechanisms by which cell geometry controls repetitive impulse firing in retinal ganglion cells, *J. Neurophysiol.*, 78, 1948–1964, 1997.

46. Hartmann, R., Topp, G., and Klinke, R., Discharge patterns of cat primary auditory fibers with electrical stimulation of the cochlea, *Hearing Res.*, 13, 47–62, 1984.

47. Motz, H. and Rattay, F., A study of the application of the Hodgkin–Huxley and the Frankenhaeuser–Huxley model for electrostimulation of the acoustic nerve, *Neuroscience*, 18, 699–712, 1986.

48. Hodgkin, A.L. and Katz, B., The effect of temperature on the electrical activity of the giant axon of the squid, *J. Physiol.*, 109, 240–249, 1949.

49. Huxley, A.F., Ion movements during nerve activity, *Ann. N.Y. Acad. Sci.*, 81, 221–246, 1959.

50. Rattay, F., Propagation and distribution of neural signals: a modeling study of axonal transport, *Physics Alive*, 3, 60–66, 1995.

51. Frankenhaeuser, B. and Huxley, A.L., The action potential in the myelinated nerve fibre of *Xenopus laevis* as computed on the basis of voltage clamp data, *J. Physiol.*, 171, 302–315, 1964.

52. Frankenhaeuser, B. and Moore, L.E., The effect of temperature on the sodium and potassium permeability changes in myelinated nerve fibers of *Xenopus laevis*, *J. Physiol.*, 169, 431–437, 1963.

53. Halter, J.A. and Clark, J.W., A distributed-parameter model of the myelinated nerve fiber, *J. Theor. Biol.*, 148, 345–382, 1991.

54. Ritchie, J.M., Physiology of axons, in *The Axon: Structure, Function and Pathophysiology*, Waxman, S.G., Kocsis, J.D., and Stys, P.K., Eds., Oxford University Press, Oxford, 1995, pp. 68–96.

55. Ranck, J.B., Jr., Which elements are excited in electrical stimulation of mammalian central nervous system: a review, *Brain Res.*, 98, 417–440, 1975.

56. Warman, E.N., Grill, W.M., and Durand, D., Modeling the effects of electric fields on nerve fibers: determination of excitation thresholds, *IEEE Trans. Biomed. Eng.*, 39, 1244–1254, 1992.

57. Plonsey, R. and Barr, R.C., Electric field stimulation of excitable tissue, *IEEE Trans. Biomed. Eng.*, 42, 329–336, 1995.

58. Zierhofer, C.M., Analysis of a linear model for electrical stimulation of axons: critical remarks on the "activating function concept," *IEEE Trans. Biomed. Eng.*, 48, 173–184, 2001.

59. Rattay, F., Minassian, K., and Dimitrijevic, M. R., Epidural electrical stimulation of posterior structures of the human lumbosacral cord: 2. Quantitative analysis by computer modeling, *Spinal Cord*, 38, 473–489, 2000.

60. Henriquez, C.S., Simulating the electrical behavior of cardiac tissue using the bidomain model, *Crit. Rev. Biomed. Eng.*, 21, 1–77, 1993.

61. Knisley, S.B., Trayanova, N., and Aguel, F., Roles of electric field and fiber structure in cardiac electric stimulation, *Biophys. J.*, 77, 1404–1417, 1999.

62. Van den Honert, C. and Stypulkowski, P.H., Temporal response patterns of single auditory nerve fibers elicited by periodic electrical stimuli, *Hear. Res.*, 29, 207–222, 1987.

63. Javel, E. and Shepherd, R.K., Electrical stimulation of the auditory nerve: II. Effect of stimulus waveshape on single fibre response properties, *Hear. Res.*, 130, 171–188, 1999.

64. Horikawa, Y., Simulation study on effects of channel noise on differential conduction at an axon branch, *Biophys. J.*, 65, 680–686, 1993.

65. Holden, A.V., *Models of the Stochastic Activity of Neurons*, Springer-Verlag, Berlin, 1976.

66. Sigworth, F.J., The variance of sodium current fluctuations at the node of Ranvier, *J. Physiol.*, 307, 97–129, 1980.

67. Verveen, A.A. and Derksen, H.E., Fluctuation phenomena in nerve membrane, *Proc. IEEE*, 56, 906–916, 1968.

68. DeFelice, L.J., *Introduction to Membrane Noise*, Plenum Press, New York, 1981.

69. Rubinstein, J.T., Threshold fluctuations in an N sodium channel model of the node of Ranvier, *Biophys. J.*, 68, 779–785, 1995.

70. Rattay, F., Basics of hearing theory and noise in cochlear implants, *Chaos, Solitons Fractals*, 11, 1875–1884, 2000.

71. Greenberg, R.J., Velte, T.J., Humayun, M.S., Scarlatis, G.N., de Juan, E., Jr., A computational model of electrical stimulation of the retinal ganglion cell, *IEEE Trans. Biomed. Eng.*, 46, 505–514, 1999.

72. Toris, C.B., Eiesland, J.L., and Miller, R.F., Morphology of ganglion cells in the neotenous tiger salamander retina, *J. Comp. Neurol.*, 352, 535–559, 1995.

73. Velte, T.J. and Miller, R.F., Dendritic integration in ganglion cells of the mudpuppy retina, *Visual. Neurosci.*, 12, 165–175, 1995.

74. Coleman, P.A. and Miller, R.F., Measurement of passive membrane parameters with whole-cell recording from neurons in the intact amphibian retina, *J. Neurophysiol.*, 61, 218–230, 1989.

75. Fohlmeister, J.F., Coleman, P.A., and Miller, R.F., Modeling the repetitive firing of retinal ganglion cells, *Brain Res.*, 510, 343–345, 1990.

76. Geddes, L.A. and Baker, L.E., The specific resistance of biological material-A compendium of data for the biomedical engineer and physiologist, *Med. Biol. Eng.*, 5, 271–293, 1967.

77. Lukasiewicz, P. and Werblin, F., A slowly inactivating potassium current trunkates spike activity in ganglion cells of the tiger salamander retina, *J. Neurosci.*, 8, 4470–4481, 1988.

78. Lipton, S.A. and Tauck, D.L., Voltage-dependent conductances of solitary ganglion cells dissociated from rat retina, *J. Physiol.*, 385, 361–391, 1987.

79. Hines, M.L. and Carnevale, N.T., NEURON: a tool for neuroscientists, *Neuroscientist*, 7, 123–135, 2001.

80. Wiley, J.D. and Webster, J.G., Analysis and control of the current distribution under circular dispersive electrodes, *IEEE Trans. Biomed. Eng.*, BME-29, 381–385, 1982.

81. Humayun, M.S., de Juan, E., Dagnelie, G., Greenberg, R.J., Propst, R., and Phillips, D.H., Visual perception elicited by electrical stimulation of retina in blind humans, *Arch. Ophthalmol.*, 114, 40–46, 1996.

82. DeSchutter, E. and Bower, J.M., An active membrane model of the cerebellar Purkinje cell: I. Simulation of current clamps in slice, *J. Neurophysiol.*, 71, 375–400, 1994.

83. Rijkhoff, N.J., Holsheimer, J., Koldewijn, E.L., Struijk, J.J., Van Kerrebroeck, P.E., Debruyne, F.M., and Wijkstra, H., Selective stimulation of sacral nerve roots for bladder control: a study by computer modeling, *IEEE Trans. Biomed. Eng.*, 41, 413–424, 1994.

84. Bugbee, M., Donaldson, N.N., Lickel, A., Rijkhoff, N.J., and Taylor, J., An implant for chronic selective stimulation of nerves, *Med. Eng. Phys.*, 23, 29–36, 2001.

85. Rattay, F., Modeling the excitation of fibers under surface electrodes, *IEEE-Trans. Biomed. Eng.*, BME-35, 199–202, 1988.

86. Coburn, B. and Sin, W.K., A theoretical study of epidural electrical stimulation of the spinal cord. Part I: Finite element analysis of stimulus fields, *IEEE Trans. Biomed. Eng.*, 32, 971–977, 1985.

87. Rattay, F., Leao, R.N., and Felix, H., A model of the electrically excited human cochlear neuron. II. Influence of the three-dimensional cochlear structure on neural excitability, *Hear. Res.*, 153, 64–79, 2001.

88. Frijns, J.H., de Snoo, S.L., and Schoonhoven, R., Improving the accuracy of the boundary element method by the use of second-order interpolation functions, *IEEE Trans. Biomed. Eng.*, 47, 1336–1346, 2000.

89. Bower, J. and Beeman, D., *The Book of GENESIS*, 2nd ed., TELOS, New York, 1997.

90. Mascagni, M.V. and Sherman, A., Numerical methods for neural modeling, in *Methods in Neuronal Modeling: From Ions to Networks*, 2nd ed., Koch, C. and Segev, I., Eds., MIT Press, Cambridge, MA, 1999, pp. 569–606.

section three

Stimulating and recording
of nerves and neurons

chapter four

Stimulating neural activity

James D. Weiland, Mark S. Humayun, Wentai Liu,
and Robert J. Greenberg

Contents

4.1. Introduction

The study of interactions between electricity and biological tissue has been an active research area for over 100 years. The use of electrical stimulation as an effective therapy to treat human diseases has only recently been realized, thanks in equal parts to advances in electronics, neuroscience, and medicine. In the 18th century, Allesandro Volta and Luigi Galvani both investigated sensory and muscular responses to electrical stimulation. Galvani stimulated the muscles of frogs, while the intrepid Volta was his own test subject, placing electrodes in either ear and reporting a "crackling sound" when power was applied. Further study of nerve and muscle identified these tissues as excitable (using electrical signals to communicate) and resulted in more careful investigations of electrical stimulation as a potential therapy.

0-8493-1100-4/03/$0.00+$1.50
© 2003 by CRC Press LLC

Among the pioneers in the field of neural prostheses were Djourno and Eyries,[1] who reported the first cochlear implants, and Brindley,[2] who implanted an electrical stimulation device on the cortex of a blind test subject. When Djourno and Eyres implanted four patients with a simple device, a single wire coupled to a sound transducer, they were able to distinguish several distinct sounds but could not understand speech. House advanced the cochlear implant by implanting stimulation devices in the cochlea and cochlear nucleus.[3] Brindley implanted a blind test subject with a device capable of stimulating the brain at 80 locations, and Dobelle has implanted a similar device in humans. One of Dobelle's patients has been implanted for over 20 years and has demonstrated the ability to identify 6-inch letters at a distance of 5 feet, corresponding to 20/120 vision.[4,5] These early results in humans demonstrated the potential of electrical stimulation as a means to replace lost sensory function and motivated a growing area of research into stimulation devices and physiology of electrically stimulated muscle and nerve.

This chapter describes methods for measuring the neural response to electrical stimulation. A brief review of membrane biophysics is presented to establish the mechanism by which electric current can mimic the natural neural response. Several analysis methods are described for biopotential recording, and alternatives to biopotential recording are discussed. Also presented is a review of experiments that demonstrate manipulation of the stimulus site by control of the waveform. Limits of stimulus waveforms are presented, as well. The chapter concludes with examples of implementation of the principles in commercial neural prostheses.

4.2   Basic theory and methods

Although sometimes considered an elemental building block, the neuron itself is a sophisticated biological system with a variety of subcellular structures designed to maintain a chemical and electrical equilibrium across the cell membrane; respond to chemical, electric, mechanical, or light stimuli; and signal messages to other neurons. The excitable membrane of the neuron has molecular sensors that respond to a variety of stimuli. If the response changes the electrical characteristics of the cell membrane, then the electrical potential across the cell (the membrane potential) will change. Signals are transmitted through nerve cells by temporary, controlled fluctuations in the membrane potential.

4.2.1   Membrane biophysics of neurons

Membrane biophyiscs is a broad area of research to which entire textbooks have been devoted. The discussion below is derived mainly from two sources: *Bioelectricity: A Quantitative Approach*, by Plonsey and Barr,[6] and *Principles of Neural Science*, 3rd ed., by Kandel et al.[7] Nerve cells have a measurable electrical potential that exists across the cell membrane. Typical

Table 4.1 Concentration, Permeability, and Nernst Potential for Various Ions in Squid Axon

Ion	Concentration (mM/l) Intracellular	Extracellular	Relative Permeability At Rest	For Active Membrane	Nernst Potential
K+	280	10	1.0	1.0	−83.9
Na+	61	485	0.04	20.0	52.2
Cl−	51	485	0.45	0.45	−56.7

resting potential is −60 mV measured inside with reference to outside. The electrical potential exists due to active and passive cell processes that maintain a difference in the chemical concentration of various ions that exist in the intra- and extracellular fluids.

The ions most responsible for maintenance of membrane potential are Na+, Cl−, and K+, although in some cells (cardiac and photoreceptor) Ca^{2+} plays a major role in signaling. The concentrations of these ions are different inside and outside the cell. Potassium (K+) has a higher concentration inside the cell, and sodium (Na+) and chloride (Cl−) have higher extracellular concentrations. Table 4.1 shows the concentrations in the squid axon.[6]

In human nerve cells, the absolute numbers are two to three times lower, but the ratio of intracellular to extracellular is similar. The cell membrane at rest is not equally permeable to all ions; that is, some ions diffuse more readily than others. The diffusion force will drive the ions toward equal concentration inside and outside the cell. However, taking an extreme example, if only positively charged ions could cross a membrane and negatively charged ions were completely blocked from diffusing, the diffusion of only positive charge would create an electrical imbalance that would slow and then halt the diffusion. The potential at which the diffusion force is equal and opposite to the electrical force is called the Nernst potential and can be calculated from Eq. (4.1),

$$V_{mp} = 58 / [Z_p \log_{10}([C_p]_e / [C_p]_i)] \; mV \qquad (4.1)$$

where V_{mp} is the membrane voltage for the *p*th ion; Z_p is the charge of the *p*th ion, $[C_p]_e$ is the extracellular concentration of the *p*th ion, and $[C_p]_i$ is the intracellular concentration of the *p*th ion. The Nernst potential is for a single ion. The various ions in the intra- and extracellular fluid each have Nernst potentials. The resting permeabilities and Nernst potentials for each ion are listed in Table 4.1. By weighting the Nernst potential of each ion with the membrane permeability for that ion, it is possible to calculate the membrane potential at steady state using the Goldman equation:

$$V_m = (RT / F)\ln((P_K[K^+]_e + P_{Na}[Na^+]_e$$

$$+P_{Cl}[Cl^-]_i) / (P_K[K^+]_i + P_{Na}[Na^+]_i + P_{Cl}[Cl^-]_e)) \qquad (4.2)$$

where P_x is the permeability of the membrane to ion x. Though this equation is only valid if the membrane is at rest (net current is zero), qualitatively it is clear that a change in permeability for one or more ions will alter the membrane potential.

Electrical charge moves across the cell membrane through ion channels and ion pumps. The rest of the cell membrane is impermeable to ion flow. The number and nature of ion channels in the membrane determine the permeability of the membrane. Ion channels can always be open or can be opened and closed in response to a stimulus, but once they are open they act as pores that allow ions to flow to equilibrium. Ion channels can be specific to particular ions but are otherwise passive; that is, the chemical or electrical driving forces will determine the direction of the net ion flow through the channel. In contrast, ion pumps move ions against the driving force gradient, consuming energy in the process.

Signals are initiated by changes in the membrane permeability, which change the membrane potential (see Eq. (4.2)). Membrane permeability can be changed in a variety of ways depending on the type of cell. If the cell is a sensory transducer, such as a photoreceptor in the retina, then the presence of the excitation signal results in a change of membrane potential. The photoreceptors are unusual in that they are depolarized (excited) in the absence of light. Other sensory receptors, such as the hair cells of the cochlea and olfactory cells of the nose, are depolarized in response to their excitation, either sound or odor. The excitation signal might trigger a chemical cascade that opens ion channels or the ion channels may be mechanically opened. The second way that membrane potential can be modified is by synaptic transmission, that is, through a signal transmitted from a connecting neuron. Chemical neurotransmitters released by a nearby neuron bind to a specific protein on the membrane of the target neuron, leading to the opening of chemically gated ion channels.

The stimulus that creates a change in the membrane permeability affects the equilibrium equation and thus alters the cell membrane potential, which triggers the opening of a special type of ion channel that is sensitive to transmembrane potential. These ion channels are called *voltage-gated ion channels*, and depolarization changes membrane permeability by opening voltage-gated channels. Initially, the stimulus will result in a proportional response from the cell. In this case, the cell membrane will respond with what is called a graded potential. This means that there is a proportional relationship between the amount of depolarization and the strength of stimulus. Graded potential cells, such as the bipolar cells in the retina, transmit information over short distances; the space constant for signal propogation is small (the signal will decay over distance). Some cells have the ability to "spike," to produce an action potential, a self-propogating signal that transmits over great lengths. The anatomical structure that serves as the transmission line for the action potential is the cell axon, a relatively long projection from the cell body (soma). The action potential is generated when the cell depolarization exceeds a threshold. Once a cell is depolarized beyond that point, a process begins that will generate an action potential, even if the stimulus that produces the initial depolarization is removed.

Hodgkin and Katz determined the permeability of the squid axon membrane to sodium, potassium, and chloride ions, both at rest and at the peak of the action potential. They obtained the data by altering the extracellular ion concentrations and monitoring the current. The results suggested that the action potential generation is due to a dramatic change in the permeability of the membrane to sodium. The permeabilities from this work are shown in Table 4.1.

This sequence of events that occurs during an action potential was described in detail by Hodgkin and Huxley,[1] who won the Nobel Prize for their seminal work in membrane biophysics. The initial depolarization (from synaptic transmission or sensory stimuli) will cause voltage-gated ion channels to activate (open). Voltage-gated ion channels exist for both sodium and potassium, but the sodium channels have a much shorter time constant. The quick response from the sodium channels results in an increase in the outward current, which depolarizes the membrane even further. The ion channels cannot open or close instantaneously, so, once opened, they will stay open for a finite period of time. This allows the depolarization to continue even after the stimulus is removed. An action potential is generated when the depolarization exceeds a threshold, at which point the outward current exceeds the inward current, creating a positive feedback situation (i.e., more outward current, more depolarization, more open sodium channels). The depolarization is quickly (within 1 to 2 msec) reversed by two mechanisms: inactivation (closing) of voltage-gated sodium channels and activation of voltage-gated potassium channels. Sodium activation and sodium inactivation are accomplished by two distinct, voltage-sensitive mechanisms. The sodium inactivation mechanism and the potassium activation mechanism have longer time constants than sodium activation, but they work together to restore membrane potential to resting potential.

After the action potential, it takes several milliseconds for the voltage-gated ion channels to return to their resting state. The voltage-gated sodium channels will change from inactive (cannot be opened) to resting (closed, but will open in response to voltage change). The potassium channels will change from active (open, thereby making the membrane highly permeable to inward potassium current) to resting (closed, but will open in response to depolarization). This transitional period is called the *refractory period* and has two stages. The absolute refractory period is the time during which the increased potassium permeability prevents an action potential generation regardless of the stimulus. The relative refractory period follows, during which time an increased stimulus is needed to generate an action potential. Hodgkin and Huxley quantified the behavior of sodium and potassium currents with mathematical equations. This work, and that of Frankenhauser and Huxley, remains the basis for much of the neuron modeling that is performed today.

The action potential has been described as an all-or-nothing phenomenon, implying that information on stimulus strength cannot be communicated by these cells. However, the firing rate of the neuron, the number of spikes per second, can be used to code information about stimulus strength. In addition, it has been demonstrated in the retina that cells may work in concert to transmit more information than if they worked independently

In summary, the basic steps of transient ion channel response to depolarization are: (1) initial depolarization; (2) voltage-gated sodium channel activation, leading to more depolarization; (3) more voltage-gated sodium channels being open due to continued depolarization and net outward current; (4) inactivation of sodium channels and activation of potassium channels; (5) repolarization of cell membrane; and (6) ion channels return to the resting state. Ion channels are sophisticated structures built from transmembrane proteins. The study of ion channel membrane biophysics is a field unto itself and a detailed discussion is beyond the scope of this chapter. While the description above is a generalization of membrane behavior to demonstrate how the neuron is capable of signaling by transiently altering its membrane properties, the basic mechanism of action potential generation is shared by most neurons.

4.2.2   *Electrical activation of voltage-gated ion channels*

Electrical energy applied to the neuron has the ability to elicit a graded potential and action potential from the neuron, thereby allowing electrical stimulation to replace lost function in the sensory or motor systems. Applying a negative current via an extracellular stimulating electrode has the effect of depolarizing the membrane.[9] The membrane depolarization from the external, artificial stimulus initiates the natural processes of transient membrane depolarization. A sufficient electrical stimulus can lead to action potential generation.

The stimulating electrode typically serves as the cathode because negative current leads to membrane depolarization. However, because a current generator has a source and a sink, a return electrode is required to complete the current loop through the tissue. Monopolar stimulation occurs when the return electrode is large and relatively remote from the stimulating electrode. Because the return electrode is large, the density of the current at this site is small so neuron activation only occurs at a single site, the *monopole*. If the return electrode is similar in size to the stimulating electrode, then the current density at the return is similar to the stimulation, allowing for neuron activation to occur at both electrodes in the pair, although such activation is less likely at the anode as anodic current is less effective at producing a response. This configuration is referred to as a *dipole* and has the advantage of reducing the spread of electrical current.

Mathematical models of electrical stimulation suggest that current density, not current, is the important parameter for calculating membrane potential, but experimentally this principle does not always hold. Current density is directly proportional to the electric field created by a stimulating electrode, and the electric field is frequently used as the input to models of stimulation. It follows that a reduction in the electrode size should be accompanied by a proportional reduction in stimulus current, but experimentally this is not true over all ranges of electrode sizes. A study using skin electrodes of varying diameter over several orders of magnitude found that response threshold was independent of diameter below a certain diameter but does become dependent on diameter as diameter increases.[10] The skin stimulation study hypothesized that current was conducted through a number of discrete channels through the skin, and the effective current density was dependent on the number of channels activated. Hence, the effective current density could be different from the calculated current density, the latter assuming uniform conduction through an isotropic media. Mathematical models of neural activation with extracellular electrodes use the stimulating current or current density as input, but these models typically assume an ideal electrode geometry (point source or sphere) in an isotropic conducting media.[11,12] The more complex situation of current flow between two electrodes in an anisotropic media will require numerical techniques to corroborate experimental data.

Electrical activation of neurons can also be achieved through intracellular stimulation. The technique used to inject current inside the cell is the patch clamp.[13,14] Glass pipette electrodes are used to penetrate the cell membrane. A high-impedance seal that forms between the membrane and pipette prevents stimulus current from leaking out the hole formed by the pipette. Small currents (nA) significantly affect membrane potential. Although this is an extremely efficient means of stimulating the cell in terms of how much energy is needed to produce an action potential, it would be difficult to realize an implantable device with intracellular electrodes, so this approach will not be considered further.

Electrical stimulus thresholds (i.e., the amount of electrical stimulus required to produce an action potential) are sometimes defined in coulombs (C), a basic unit of electrical charge (1 C equals the charge of 6×10^{18} electrons).[15] The more familiar electrical current unit, amperes (A), is charge per unit time, or C/sec. The amount of charge in a rectangular stimulus pulse is simply the product of the stimulus strength in amperes and the stimulus duration in seconds. For more complex pulse shapes, the stimulus charge can be determined from the area under the pulse. Because a stimulus pulse is typically less than 1 msec and stimulus current less than 1 mA, microcoulombs (μC) or nanocoulombs (nC) are used to quantify the stimulus charge threshold. One reason for using stimulus charge to define threshold, as opposed to stimulus current, is that current will change dramatically with the duration of the pulse, but stimulus charge, though not constant, will change less. Therefore, using charge allows the stimulus threshold to be stated as a single parameter. If current is used to define the threshold, the pulse duration must also be included. The best practice may be to present results with all of the pertinent parameters.

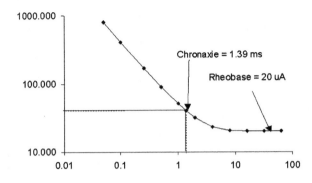

Figure 4.1 Simulated strength–duration curve. The rheobase current is the minimum current required to stimulate the neuron regardless of pulse duration. Chronaxie is the pulse duration corresponding to twice the rheobase current. At shorter pulse duration, the curve is nearly linear.

The relationship between stimulus current threshold and stimulus duration is often expressed graphically with a strength–duration curve.[6] Stimulus threshold current is plotted vs. stimulus pulse duration over at least one order of magnitude. Parametric models can be created from the empirically determined strength–duration curves to characterize the response of a cell to electrical stimulus over a range of pulse duration. These curves can also be used to quantify the rheobase current and the chronaxie of the cell. Rheobase is the minimum current required to stimulate a cell, regardless of the duration of the stimulus pulse. Chronaxie is the duration of the stimulus pulse required when twice the rheobase current is used. A shorter chronaxie implies that the cell will respond with less stimulus charge at that current level; that is, the cell is more sensitive. Figure 4.1 shows a simulated strength–duration curve. As stimulus pulse duration increases, threshold current decreases but threshold charge increases. Also, note the region of near-linear stimulus charge with shorter pulses. This demonstrates the value of threshold charge as a parameter to define the properties of the cell. For stimulus pulse durations from 0.05 to 0.5 msec, the threshold charge changes only 10% even though the pulse duration changes one order of magnitude. Shorter pulses are preferred, as they require less total energy.

A second method of investigating electrical response is the amplitude–intensity function. This plots the stimulus strength at a fixed-pulse duration vs. the amplitude of the electrically elicited response.[16] The response in this case is an evoked potential or summed response of many neurons. This typically yields an increasing line that will saturate at some stimulus level (Figure 4.2). One advantage of the input–output function is that it allows a predication of true threshold by extrapolation of the input–output function to the intersection of the x-axis. The strength–duration curve threshold is affected by noise and sometimes determined subjectively.

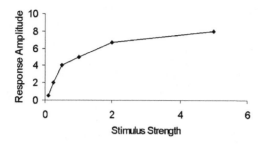

Figure 4.2 Simulated input–output function. The response amplitude saturates as the stimulus increases. It is possible to estimate the true threshold by extrapolation of the function to the *x*-axis intercept.

Evoked potentials are produced by averaging many responses to the same stimulus. Averaging N trials will improve the signal-to-noise ratio by $\sqrt{N}$ (100 averages will increase the signal-to-noise ratio by 10).[17] This noise reduction strategy cannot be applied to studies of single units (action potentials from a single cell) because the action potential is so short (<1 msec) and, due to the stochastic nature of action potential firing, can be offset by 1 msec between trials. An offset of 1 msec will result in diminishing by half the signal if two trials are averaged. To overcome this, single units are treated as discrete events occurring at a specific time. Over many trials, it is possible to accumulate spike event times after an electrical stimulus (or any stimulus). These spike times are then grouped into time windows called *bins*; for example, a 1-sec recording can be divided into 100 10-msec bins. The results of this sorting are displayed in a post-stimulus time histogram (PSTH). This is a straightforward procedure if only one cell is responding with the same action potential each time and the action potential signal is well above the noise, allowing a simple threshold rule for event detection. The PSTH allows us to demonstrate that the single unit firing is correlated to the stimulus if a particular bin rises above the spontaneous firing rate. However, in most instances the analysis becomes more difficult. The action potential may not be significantly above the noise, or multiple action potentials from multiple cells may be recorded simultaneously. Signal processing strategies have been developed to attack these problems, but a thorough review is beyond the scope of this chapter.

Strength–duration curves, amplitude–intensity functions, and PSTH are time-honored methods of biopotential analysis. These have been and will continue to be the first and best measures of the response of neural tissues to electrical stimulation. They are easily interpreted and can be acquired with relatively simple, inexpensive equipment; however, biopotential analysis is not without its limitations, particularly when used to assess electrically elicited neural response. First and foremost is the problem of the electrical stimulus artifact.

The electric field created in the tissue by the electrical stimulus can saturate amplifiers and mask an electrical response. This is especially problematic when the recording electrodes are positioned close to the stimulus electrodes. Potential solutions for artifact reduction have been discussed by Parsa et al.[18] Simple methods to reduce the artifact include changing the bandpass of the amplifier to allow for a quicker recovery, using a sample and hold amplifier to prevent saturation, and using differential recording with the reference electrode on an equipotential line with the recording electrode. Methods for avoiding artifact include recording "downstream" or at the proximal neural areas. An example would be recording in the inferior colliculus and stimulating in the auditory nerve. A delay of several milliseconds between stimulus and response would allow the artifact to subside before the response is present.

A second shortcoming of biopotential recording is the limited amount of tissue that is observed. The response is only recorded at a single point, or multiple points if several electrodes are used. This requires that the recording electrode be positioned such that it will record the response from the cell being activated by the stimulating electrode. The presence of interneurons can confound results.

To augment data from biopotential recording, neural responses can be recorded with a variety of optical and imaging methods. Voltage-sensitive fluorescent dyes reflect the actual membrane potential, while optical intrinsic signaling relies on changes in optical properties of tissue related to increased metabolic activity. Voltage-sensitive dyes (VSDs) are frequently used in brain slice recording.[19] This preparation lends itself to voltage-sensitive dyes, as the dissected tissue can easily be processed to load the dye in the cells. Further, the thickness of the tissue, which will affect the dye penetration, is controllable. It is more difficult to use dyes to penetrate intact tissue, such as an isolated retina, which may have some barriers to diffusion (note that a retinal slice is different from an isolated retina). It has been demonstrated that VSDs reflect the activity in the dendritic postsynatpic area of the neuron; that is, VSDs are markers of subthreshold activity. It is possible to combine VSD imaging with electrical recording to map both subthreshold and action potential activity.[20]

Optical intrinsic signaling has been used to study tissue activity for some time. In general, optical imaging does not require the use of fluorescent dyes or other extrinsic markers. The basic premise is that neuronal activity causes a transient increase in cell volume or blood flow that results in a change in light scatter or absorption. Optical imaging can be used to study activity on the surface of the brain or in a dissected section of brain tissue, such as the retina. Optical imaging has also been used to study electrically evoked changes in activity in isolated brain tissue slices. Kohn et al.[21] used three trains of electrical pulses of frequencies 10, 2, and 1 Hz. They found that it was possible to observe an increase in the transmitted light resulting from stimulation. The increase in transmitted light is most likely due to transient cellular swelling. The electrical activity of the stimulated neurons causes Na^+

and Cl– to be drawn into the cell. The increase in cellular volume is due to water drawn into the cell to equilibrate the osmotic pressure across the cell membrane. As the cellular volume increases, the concentration of scattering particles within the cell decreases, causing the light scatter of the cell to decrease. The reduced light scatter allows more light to be transmitted by the brain tissue slice. If the imaging system was measuring reflected light instead of transmitted light, the image intensity would likely decrease, as the reflected light is directly related to the light scatter.

Optical imaging of tissue requires the use of sensitive equipment to obtain the image. The changes in optical properties are small and require expensive imaging chips to detect these changes above the noise of the camera. It must be possible to visualize the tissue that is being analyzed. If cortical tissue is being analyzed, then the skull must be removed over the area of interest.

Functional magnetic resonance imaging (fMRI) is used to detect increased brain activity in response to specific stimuli. It has the advantage of being truly noninvasive. No direct electrical interface is required nor is direct visualization of the neural tissue necessary. Preliminary experiments indicate that the electrical stimulus will result in a stimulus artifact that obscures any measurable response. To counter this, it is possible to record the scan after the electrical stimulus has ceased. Even without this limitation, fMRI would be limited to spatial information. Also, the spatial resolution is low, typically 1 mm or greater. Finally, fMRI is difficult to perform, requiring expensive equipment. This type of data may supplement electrical responses, but probably would be of limited value alone.

The biochemistry of the neuron is altered by continuous increased activity, and these changes can be detected with histological markers. These methods require that the experimental animal be sacrificed a short time after the stimulus and the tissue processed appropriately. One such marker is the protein Cfos, which is transiently expressed 1 to 4 hours after neural excitation.[22] This protein can be marked with antibodies and has been used to investigate electrical stimulation of the cochlea and light responses in the retina.[23–25] In general, it can be used to indicate the extent of excitation for a given stimulus. For example, in the Saito study,[24] the cochlea was electrically activated in an area known to respond to sound waves in a particular frequency range. Histological evaluation of the dorsal cochlear nucleus and inferior colliculus showed increased Cfos expresssion in a narrow band, indicating that only that frequency band of the auditory system was activated. However, Cfos cannot distinguish between primary and secondary excitation, so unless the neural structure is anatomically well defined, such as the tonotopic lamina of the inferior colliculus or the ocular dominance columns of the striate cortex, then Cfos marking will be diffuse. Also, only one experimental condition can be tested because the animal must be sacrificed shortly after testing to process the tissue. Finally, the antibody is species dependent, so while the method may work well in one animal, it may not work in another if the Cfos antibody in the second animal is not well characterized.

4.3   Achieving desired response with stimulus waveform manipulation

4.3.1   Electrode shape and position dependence

The shape and position of the stimulating electrode affect the electrical response of neural tissue because the electric field created by the stimulating current is affected by these same electrode parameters. The effects of shape are most pronounced when the electrode is in close proximity to the neurons. Planar electrodes have a nonuniform current density that is significantly higher at the edges than in the center of the electrode.[26] The current density at a circular electrode within a nonconducting surface has an analytical solution that predicts that the current density at the edge of the electrode will be infinite. In practice, such effects are seen as skin burns at the perimeter of circular *defribrillators*. This may affect the response by selectively stimulating neurons in these areas of higher current density where the desired outcome would be to equally stimulate all neurons near the electrode. The nonuniform electric field of the planar electrode can be moderated by recessing the electrode surface in the nonconducting substrate or by changing the shape of the electrode itself to a half sphere.[27]

The position of the electrode will affect how well a response can be targeted to a small area of tissue. Due to current spread, an electrode far from the desired area will stimulate a larger area then a closer electrode. However, the current density is decreased with current spread, so more current is required to stimulate when the electrode is positioned farther away from the target neurons. In spite of these facts, optimal electrode position must balance anatomical considerations against proximity to neurons to arrive at the best solution. In the case of positioning a stimulating electrode on the retina, an electrode on the surface of the retina is less disruptive to the anatomy than an electrode that would penetrate into the retina. The electrical properties of electrodes implanted in the cortex are known to be significantly affected by the foreign-body reaction that surrounds the electrode. The reactive tissue has an insulating effect that degrades the electrical contact between the tissue and the electrode.

4.4   Stimulus pulse dependence

4.4.1   Charge balancing and cell damage

The electrode position and shape are analogous to computer hardware; that is, they are physically defined parameters that are impossible to change once set in place. What can be changed is the software, or, in the case of an electrical stimulating device, the stimulus pulse. It is possible to target specific neural structures by varying the stimulus pulse parameters. However, some general rules must be followed when electrically stimulating neural tissue with metal electrodes. First is the concept of charge balanced stimulation, first reported by Lilly in 1961.[28] For a single stimulation waveform, the total net charge must be zero. This can be accomplished either by supplying equal cathodic and anodic current from the stimulator or by a blocking capacitor that will slowly discharge after a monopolar stimulus pulse. If charge balance is not obtained, then, over time, the net charge accumulation on the electrode may increase the electrode potential to the point where harmful quantities of gaseous oxygen or hydrogen are produced (bubbling). In practice, stimulators are designed to have blocking capacitors. The size of the capacitor must be chosen carefully, as a capacitor that is too small will severely limit the duration of the pulse that can be used.

A charge-balanced waveform is not necessarily a safe waveform. Within the course of a stimulus pulse that is overall charge balanced, it is possible to momentarily exceed empirically established safety limits for electrode potential, charge density, or total charge. Two safety limits should be considered: neural damage limits and electrochemical limits. Neural damage limits are dependent on the ability of biological tissue to withstand the electric current without degrading. Electrochemical limits are based on the ability of the electrode to store or dissipate electrical charge without exceeding the *water window*, which is the potential window outside of which significant bubble formation is evident at the interface.

Neural damage limits have been studied extensively by Pudenz, Agnew, and McCreery at the Huntington Medical Research Institute for long-term stimulation in the peripheral and central nervous systems. Three important findings from this extensive work include the relationship between total charge and charge density, the comparison between capacitive electrodes and Faradaic electrodes, and the transient decreased neural excitability in response to electrical stimulation.

With respect to the amount of electrical charge that can be applied before tissue damage is evident, there is a dependency between the charge per phase and the charge density. The charge per phase is the total charge regardless of the electrode size.[29] Charge density is obviously dependent on the size of the electrode for a given amount of charge. When smaller electrode areas are used, the amount of charge that can be used before neural damage occurs corresponds to a charge density of 1 mC/cm^2. However, with larger electrodes, the neural damage threshold on charge limits the charge density to 0.1 mC/cm^2. It is hypothesized that some minimum amount of charge is required to result in damage to neural tissue. An analogy can be drawn to the rheobase current, the minimum amount of current required to activate a neuron. If a very small amount of charge is used, the charge density can be larger. This effect favors the use of smaller electrodes, because they can support a higher charge density with a smaller amount of charge.

The mechanism of neural damage is based on the type of electrode used. Briefly, faradaic electrodes use both capacitive coupling and reduction oxidation reactions to electrically stimulate tissue, while capacitive electrodes use only capacitive current. A direct comparison of faradaic vs. capacitor electrodes revealed that neural damage was evident with similar stimulus conditions for both electrodes. This suggests that the damage resulted from the passage of current through tissue, not from harmful bioproducts produced by chemical reactions at the electrode.[30]

The last result of the Huntington group that will be discussed is that of stimulus-induced decreased neural excitability.[31] It was shown that, when stimulating the cochlear nucleus continuously, the response threshold (recorded at the inferior colliculus) was increased with continuous pulsing. More current was required to elicit an equivalent response after several hours of continuous pulsing. This decreased excitability could not be explained histologically. Also, the effect was transient; that is, with a night of rest the original threshold was regained. This effect was only evident with continuous pulsing at a relatively high rate (more than 200 pulses/sec).

Electrochemical charge limits, the amount of charge that can be delivered by the electrode prior to the formation of gas bubbles, are also empirically determined by measuring the current vs. potential characteristics of the electrode and by direct observation of the electrode surface during stimulation. This area has been studied extensively at EIC Labs in Newton, MA, by Brummer et al.,[32–34] who have investigated a variety of materials for electrical stimulation. The material most commonly used for electrical stimulation is platinum, which can safely supply 0.1 to 0.4 mC/cm^2 of charge depending on the pulse characteristics. Alternative materials that have higher limits such as iridium oxide (1 to 3 mC/cm^2) and titanium nitride (0.6 to 0.9 mC/cm^2) have been studied extensively in simulated and real conditions but have not been widely used in commercial devices.

4.4.2   Selective stimulation of cells

Within the limits outlined above, it is possible to use electrical stimulation to effectively activate nerve cells, as demonstrated by cochlear implants and deep brain stimulators. Short pulses of electrical current are used to activate neurons and replace some functionality previously lost to disease. The shape of the current pulse can be manipulated to target specific cells that are not necessarily the cells closest to the electrode. This is because different types of cells have different response characteristics to electrical stimuli. In the retina, this can be readily demonstrated because different cell types are grouped in well-organized layers. Greenberg[35,36] demonstrated that using longer pulses allows bipolar cells to be selectively activated over ganglion cells in the frog retina. This was determined by comparing the strength duration curves from retinal stimulation under a variety of experimental conditions. The three experimental conditions tested were normal retina, light-saturated retina (photoreceptors hyperpolarized), and retina with cadmium added (cadmium blocks synaptic transmission). In each case, a distinct strength–duration curve was obtained by recording retinal ganglion cell responses (Figure 4.3). While all cells were depolarized to some degree by electrical stimulation, the cell that initiated

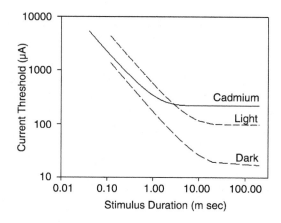

Figure 4.3 Strength–duration curves obtained from electrical stimulation of the retina under various conditions. The retina was stimulated by an electrode on the ganglion cell surface, and action potentials were recorded from retinal ganglion cells in response to stimulation. The dark condition represents the state in which the retina is easiest to excite with electrical stimulation. The light condition results in hyperpolarization of the photoreceptors thereby decreasing the excitability of that part of the retina. The addition of cadmium blocks synaptic transmission, limiting the response site to the ganglion cells.

the response was the cell for which depolarization resulted in action potential generation in the ganglion cells.

In a normal retina, the retinal ganglion cell response could be generated either by direct electrical stimulation of the ganglion cells or by electrical stimulation of photoreceptor or bipolar cells and subsequent activation of retinal ganglion cells through synaptic transmission. The presence of cadmium strongly inhibits synaptic transmission, thus precluding a response initiated by any cell except a retinal ganglion cell. The strength–duration curve generated with cadmium suggests that ganglion cells are more responsive at short pulse widths but quickly reach rheobase current. Light saturation of the retina hyperpolarizes the photoreceptor, making the bipolar cell more likely to respond than the photoreceptor. However, because the strength–duration curve for the light-bleached retina is different than that for the cadmium condition, the cell that initiates the response is likely to be different. This argues that, in the absence of photoreceptor response, the bipolar cells will initiate a response according to the strength–duration curve shown. Comparing the cadmium strength–duration curve to the light-bleached strength–duration curve, we conclude that ganglion cells respond better to short-duration pulses and bipolar cells respond better to longer duration pulses. Put another way, bipolar cells have a lower rheobase, but ganglion cells have a shorter chronaxie.

Other studies have examined the response of different cells to electrical stimulation. Weiland et al.[37] showed that, in a human model of photoreceptor degeneration, electrical stimulation of a normal retina resulted in a different psychophysical perception than electrical stimulation of a laser-damaged retina. McIntyre and Grill[12] have modeled the responses of spinal neurons that suggest a preference for soma stimulation with a cathodic first pulse and a preference for axonal stimulation with an anodic first pulse. Combined, these results suggest that, while proximity to the stimulating electrode is a major factor in determining which cell is stimulated, it is not the only factor and it is possible to selectively activate types of neurons through manipulation of the stimulus pulse.

It is also possible to control the site of the electrical stimulus by using multipolar stimulation — that is, by using other electrode sites as the current sink or current return, as opposed to a large distant electrode. Field theory shows that a dipole electric field is more concentrated than a monopole field. However, in practice, bipolar stimulation has not been shown in cochlear implants to improve the quality of the auditory sensation. In fact, some reports suggest that patients prefer monopolar stimulation.[38] Cuff electrodes, devices that encircle a nerve fiber bundle with multiple electrodes around the bundle, use current steering to target areas in the bundle. This is done by passing electrical current between electrodes at specific points around the nerve, thereby activating nerve fibers between the two electrodes. More sophisticated routines are possible that involve three to four electrodes to improve selectivity.

4.5   Practical examples and applications

In practice, commercial electrical stimulation systems implement some of the experimental and theoretical findings discussed above. Cochlear implants use charge-balanced, biphasic stimulation pulses to activate the auditory nerve. One important result from cochlear implant research was the effectiveness of pulses of short time duration. Early cochlear implants produced an electrical sinusoid at the same frequency as the sound wave it was trying to replicate. If a sound wave at 200 Hz was present, then the cochlear implant stimulated with a 200-Hz sine wave. It was shown, however, that short stimulus pulses of 0.1 msec presented at 200 pulses/sec were just as effective, in most cases, in replicating the sensation of a 200-Hz sound.[39–41] The advantage of using a 0.1-msec pulse vs. a 200-Hz (or 50-msec) sinusoid is the savings in electrical charge used. Because it was also noted that in some cases the sinusoid was more effective, some cochlear implants have the ability to stimulate with both pulse and sinusoid stimuli.

Another example of successful clinical application of neural stimulation is the deep brain stimulator (DBS). This device is currently being evaluated for the treatment of Parkinsonian tremor.[42] Seizures in the thalamic region of the brain are believed to cause some types of tremor that can render the afflicted individual unable to control movements. By electrically stimulating the thalamus, it is hypothesized that the seizures are moderated or depressed, allowing other areas of the brain to function properly to allow voluntary motor control. While the exact mechanism of

DBS is not clear, the results can be dramatic. In a matter of minutes, Parkinson's patient can go from being unable to stand due to tremor to being able to easily walk across a room.

Both the cochlear implant and deep brain stimulator demonstrate what should be a guiding principle for neural prosthetic devices. The purpose of the device is not to closely replicate the tissue it is replacing; rather, the purpose of the device is to replicate the lost function, regardless of how this is accomplished. We do not know how to build a cochlea, a thalamus, or a retina, but we do have some knowledge of how to electrically activate nerves. Using the established boundaries of safe stimulation, it is possible to achieve dramatic results in patients who otherwise would have no hope of a treatment.

References

1. Djourno, A. and Eyries, C., Prothese auditive par excitation electrique a distance du nerf sensorial a l'aide d'un bobinage inclus a demeure, *Presse Med.*, 35, 14–17, 1957.
2. Brindley, G. and Rushton, D., Implanted stimulators of the visual cortex as visual prosthetic devices, *Trans. Am. Acad. Ophthalmol. Otolaryngol.*, 78, OP741–OP745, 1974.
3. House, W.F., Cochlear implants, *Ann. Otol. Rhinol. Laryngol.*, 85(suppl. 27, pt. 2), 1–93, 1976.
4. Dobelle, W.H., Artificial vision for the blind by connecting a television camera to the visual cortex, *Am. Soc. Artificial Internal Organs J.*, 46, 3–9, 2000.
5. Dobelle, W.H., Quest, D.O., Antunes, J.L., Roberts, T.S., and Girvin, J.P., Artificial vision for the blind by electrical stimulation of the visual cortex, *Neurosurgery*, 5(4), 521–527, 1979.
6. Plonsey, R. and Barr, R.C., *Bioelectricity, A Quantitative Approach*, Plenum Press, New York, 1991.
7. Kandel, E.R., Schwartz, J.H., and Jessell, T.M., *Principles of Neural Science*, 3rd ed., Elsevier Science, New York, 1991.
8. Hodgkin, A. and Huxley, A., A quantitative description of membrane current and its application to conduction and excitation in nerve, *J. Physiol.*, 117, 500–544, 1952.
9. van Den, H.C. and Mortimer, J.T., Generation of unidirectionally propagated action potentials in a peripheral nerve by brief stimuli, *Science*, 206(4424), 1311–1312, 1979.
10. Reilly, J.P., Sensory responses to electrical stimulation, in Reilly, J.P., Ed., *Applied Bioelectricity: From Electrical Stimulation to Electropathology*, Springer-Verlag, New York, 1998, pp. 240–298.
11. McNeal, D.R., Analysis of a model for excitation of myelinated nerve, *IEEE Trans. Biomed. Eng.*, BME-23(4), 329–336, 1976.
12. McIntyre, C.C. and Grill, W.M., Selective microstimulation of central nervous system neurons, *Ann. Biomed. Eng.*, 28(3), 219–233, 2000.
13. Neher, E., Sakmann, B., and Steinbach, J.H., The extracellular patch clamp, a method for resolving currents through individual open channels in biological membranes, *Pflügers Arch.*, 375(2), 219–228, 1978.

14. Neher, E. and Sakmann, B., Single-channel currents recorded from membrane of denervated frog muscle fibres, *Nature*, 260(5554), 799–802, 1976.

15. Marshall, S. and Skitek, G., Electromagnetic concepts and applications, 2nd ed., Prentice-Hall, Englewood Cliffs, NJ, 1987.

16. Burton, M.J., Miller, J.M., and Kileny, P.R., Middle-latency responses. I. Electrical and acoustic excitation, *Arch. Otolaryngol. Head Neck Surg.*, 115(1), 59–62, 1989.

17. Neuman, M.R., Biopotential amplifiers, in *Medical Instrumentation: Application and Design*, Webster, J.G., Ed., Houghton-Mifflin, Boston, 1992, pp. 288–353.

18. Parsa, V., Parker, P.A., and Scott, R.N., Adaptive stimulus artifact reduction in noncortical somatosensory evoked potential studies, *IEEE Trans. Biomed. Eng.*, 45(2), 165–179, 1998.

19. Takashima, I., Kajiwara, R., and Iijima, T., Voltage-sensitive dye versus intrinsic signal optical imaging: comparison of optically determined functional maps from rat barrel cortex, *NeuroReport*, 12(13), 2889–2894, 2001.

20. Tominaga, T., Tominaga, Y., and Ichikawa, M., Simultaneous multi-site recordings of neural activity with an inline multi-electrode array and optical measurement in rat hippocampal slices, *Pflügers Arch.*, 443(2), 317–322, 2001.

21. Kohn, A., Metz, C., Quibrera, M., Tommerdahl, M.A., and Whitsel, B.L., Functional neocortical microcircuitry demonstrated with intrinsic signal optical imaging *in vitro*, *Neuroscience*, 95(1), 51–62, 2000.

22. Dragunow, M. and Faull, R., The use of c-*fos* as a metabolic marker in neuronal pathway tracing, *J. Neurosci. Meth.*, 29(3), 261–265, 1989.

23. Yoshida, K., Imaki, J., Fujisawa, H., Harada, T., Ohki, K., Matsuda, H. et al., Differential distribution of CaM kinases and induction of c-*fos* expression by flashing and sustained light in rat retinal cells, *Invest. Ophthalmol. Vis. Sci.*, 37(1), 174–179, 1996.

24. Saito, H., Miller, J.M., Pfingst, B.E., and Altschuler, R.A., Fos-like immunoreactivity in the auditory brainstem evoked by bipolar intracochlear electrical stimulation: effects of current level and pulse duration, *Neuroscience*, 91(1), 139–161, 1999.

25. Saito, H., Miller, J.M., and Altschuler, R.A., Cochleotopic *fos* immunoreactivity in cochlea and cochlear nuclei evoked by bipolar cochlear electrical stimulation, *Hearing Res.*, 145(1–2), 37–51, 2000.

26. West, D.C. and Wolstencroft, J.H., Strength-duration characteristics of myelinated and non-myelinated bulbospinal axons in the cat spinal cord, *J. Physiol. (London)*, 337, 37–50, 1983.

27. Rubinstein, J.T., Spelman, F.A., Soma, M., and Suesserman, M.F., Current density profiles of surface mounted and recessed electrodes for neural prostheses, *IEEE Trans. Biomed. Eng.*, BME-34(11), 864–875, 1987.

28. Lilly, J.C., Injury and excitation by electric currents: the balanced pulse-pair waveform, in *Electrical Stimulation of the Brain*, Sheer, D.E., Ed., Hogg Foundation for Mental Health, 1961.

29. McCreery, D.B., Agnew, W.F., Yuen, T.G.H., and Bullara, L., Charge density and charge per phase as cofactors in neural injury induced by electrical stimulation, *IEEE Trans. Biomed. Eng.*, 37(10), 996–1001, 1990.

30. McCreery, D.B., Agnew, W.F., Yuen, T.G., and Bullara, L.A., Comparison of neural damage induced by electrical stimulation with faradaic and capacitor electrodes, *Ann. Biomed. Eng.*, 16(5), 463–481, 1988.

31. McCreery, D.B., Yuen, T.G., Agnew, W.F., and Bullara, L.A., A characterization of the effects on neuronal excitability due to prolonged microstimulation with chronically implanted microelectrodes, *IEEE Trans. Biomed. Eng.*, 44(10), 931–939, 1997.

32. Brummer, S.B., Robblee, L.S., and Hambrecht, F.T., Criteria for selecting electrodes for electrical stimulation, theoretical and practical considerations, *Ann. N.Y. Acad. Sci.*, 405, 159–171, 1983.

33. Robblee, L.S., Lefko, J.L., and Brummer, S.B., Activated Ir: an electrode suitable for reversible charge injection in saline solution, *J. Electrochem. Soc.*, 130(3), 731–732, 1983.

34. Brummer, S.B. and Turner, M.J., Electrochemical considerations for safe electrical stimulation of the nervous system with platinum electrodes, *IEEE Trans. Biomed. Eng.*, 24(1), 59–63, 1977.

35. Greenberg, R.J., Analysis of electrical stimulation of the vertebrate retina: work towards a retinal prosthesis, The Johns Hopkins University, Baltimore, MD, 1998.

36. Greenberg, R.J., Velte, T.J., Humayun, M.S., Scarlatis, G.N., de JE, Jr., A computational model of electrical stimulation of the retinal ganglion cell, *IEEE Trans. Biomed. Eng.*, 46(5), 505–514, 1999.

37. Weiland, J.D., Humayun, M.S., Dagnelie, G., de JE, Jr., Greenberg, R.J., Iliff, N.T., Understanding the origin of visual percepts elicited by electrical stimulation of the human retina, *Graefes Arch. Clin. Exp. Ophthalmol.*, 237(12), 1007–1013, 1999.

38. Pfingst, B.E., Franck, K.H., Xu, L., Bauer, E.M., and Zwolan, T.A., Effects of electrode configuration and place of stimulation on speech perception with cochlear prostheses, *J. Assoc. Res. Otolaryngol.*, 2(2), 87–103, 2001.

39. Wilson, B.S., Finley, C.C., Lawson, D.T., and Zerbi, M., Temporal representations with cochlear implants, *Am. J. Otol.*, 1997. 18(6 Suppl), S30-S34.

40. Wilson, B.S., Finley, C.C., Lawson, D.T., Wolford, R.D., and Zerbi, M., Design and evaluation of a continuous interleaved sampling (CIS) processing strategy for multichannel cochlear implants, *J. Rehabil. Res. Dev.*, 30(1), 110–116, 1993.

41. Wilson, B.S., Finley, C.C., Farmer, J.C., Jr., Lawson, D.T., Weber, B.A., Wolford, R.D. et al., Comparative studies of speech processing strategies for cochlear implants, *Laryngoscope*, 98(10), 1069–1077, 1988.

42. Koller, W., Pahwa, R., Busenbark, K., Hubble, J., Wilkinson, S., Lang, A. et al., High-frequency unilateral thalamic stimulation in the treatment of essential and parkinsonian tremor, *Ann. Neurol.*, 42(3), 292–299, 1997.

chapter five

Extracellular electrical stimulation of central neurons: quantitative studies

Dongchul C. Lee, Cameron C. McIntyre, and Warren M. Grill

Contents

0-8493-1100-4/03/$0.00+$1.50
© 2003 by CRC Press LLC

5.1   Introduction

Electrical activation of the nervous system is a method to restore function to persons with neurological disorders due to disease or injury and a technique to study the form and function of the nervous system. Application of electrical stimulation, in the form of neural prostheses, is used to restore both motor and sensory functions. Although restoration of motor function has primarily been accomplished by activation of peripheral motor nerve fibers (i.e., last-order neurons), to restore complex motor functions it may be advantageous to access the nervous system at a higher level and use the intact neural circuitry to control the individual elements of the motor system,[22] which may be accomplished by intraspinal microstimulation.[17] Similarly, restoration of the sense of vision may be accomplished by intracortical stimulation in the visual cortex,[3,55] and restoration of the sense of hearing by stimulation of the ventral cochlear nucleus.[30]

In applications of central nervous system (CNS) stimulation, electrodes are positioned in geometrically and electrically complex volume conductors (the brain and spinal cord) containing cell bodies, dendrites, and axons in close proximity. When a stimulus is applied within the CNS, cells and fibers over an unknown volume of tissue are activated, resulting in direct excitation as well as trans-synaptic excitation/inhibition from stimulation of presynaptic axons and cell bodies. Fundamental knowledge of the interactions between the applied currents and the neurons of the CNS has been limited, and this void will impair the development of safe and effective future devices. In this chapter we review the issues that arise during stimulation of the CNS and use quantitative computational modeling to establish a foundation upon which to build future applications of electrical stimulation in the brain. We will focus on two fundamental questions that arise during CNS stimulation:

1.   What elements are stimulated by extracellular electrodes in the CNS?
2.   What methods will enable selective stimulation of different neuronal elements in the CNS?

5.1.1   What elements are stimulated in the CNS?

The question of what elements are stimulated by extracellular electrodes in the CNS was addressed in the seminal review of Ranck,[45] and we revisit and expand upon the findings set forth in that review. The neuronal elements of interest (Figure 5.1) include local cells around the electrode, including those

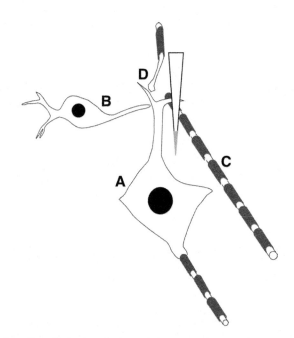

Figure 5.1 During electrical stimulation of the central nervous system, electrodes are positioned within heterogeneous populations of neuronal elements. Neuronal elements that can be activated by stimulation include local cells around the electrode projecting away from the region of stimulation (A), local cells around the electrode projecting locally (B), axons passing by the electrode (C), and presynpatic terminals projecting onto neurons in the region of the electrode (D).

projecting locally around the electrode as well as those projecting away from the region of stimulation, axons passing by the electrode, and presynaptic terminals projecting onto neurons in the region of the electrode. Effects of stimulation can be mediated by activation of any or all of these elements and include direct effects of stimulation of postsynaptic elements, as well as indirect effects mediated by electrical stimulation of presynaptic terminals that mediate the effects of stimulation via synaptic transmission.

Previous experimental evidence demonstrates that different neuronal elements have similar thresholds for extracellular stimulation and illustrates the need for the design of methods that would enable selective stimulation. *In vitro* measurements in cortical brain slices indicate that cells and fibers have similar thresholds for activation.[36,37] Similarly, *in vivo* measurements using microstimulation indicate that fibers and cells have similar thresholds for cathodic rectangular stimuli.[23,45,51] Recent computational studies of the excitation of CNS neurons indicate that, with conventional rectangular stimuli, axons of passage and local cells respond at similar thresholds.[32,33] Also, the thresholds for generating direct and synaptic excitation of neurons in the spinal cord,[23] red nucleus,[4] and cortex[26] were quite similar. Extracellular activation of type-identified spinal motoneurons indicated that the current amplitude necessary to induce repetitive firing with a monopolar electrode was not significantly different for type-S and type-F motoneurons,[58] and modeling results demonstrated a very small difference between the extracellular thresholds of different sized neurons.[33] Thus, with conventional stimulation techniques, the thresholds for activation of different neuronal elements are quite similar, and it is difficult to isolate stimulation of particular neurons.

5.1.2 Selective stimulation of CNS elements

A neural prosthesis using microstimulation of the CNS will require selective and controlled activation of specific neural populations. The "complexity of the spinal circuitry implies that if a supraspinal trigger could generate movement it would need to be a highly focused drive onto a select populations of interneurons."[7] Similarly, interpretation of physiological investigations employing microstimulation requires knowledge of the effects of stimulation on different neuronal elements. Application of extracellular currents may activate or inactivate (block) neurons and or axons dependent on their morphology, distance from the electrode, orientation with respect to the electrode, and discharge rate, as well as the stimulus parameters.[45] Effects on cells may differ from the effects on fibers of passage, and fiber activation will result in both antidromic and orthodromic propagation. While the technology for fabrication of high-density arrays of microelectrodes for insertion in the CNS has advanced greatly,[2,9] our knowledge of neuronal activation patterns has not. Advancing our understanding of neuronal excitation and determining what is required to target excitation we will provide useful design parameters for microelectrode arrays.

5.1.3 Computational modeling as a tool for understanding and design

Computational modeling provides a powerful tool to study extracellular excitation of CNS neurons. The volume of tissue stimulated, both for fibers and cells, and how this volume changes with electrode geometry, stimulus parameters, and the geometry of the neuronal elements are quite challenging to determine experimentally. Using a computer model enables examination of these parameters under controlled conditions. Neural modeling provides a powerful tool to address the effects of stimulation on all the different neural elements around the electrode simultaneously. Further, well-designed modeling studies enable generation of experimentally testable hypotheses regarding the effects of stimulation conditions on the pattern and selectivity of neuronal stimulation within the CNS. However, the strengths of modeling are tempered by the necessary simplifications made in any reasonable model. In turn, modeling should be coupled as closely as possible to experimental work enabling a synergistic analysis of results.

The utility of such an approach has been demonstrated by previous field-neuron models of cochlear stimulation which were able to replicate experimental activation patterns and document the effects of changes in electrode geometry and stimulus parameters.[14-16] Similarly, integrated field-neuron models of epidural spinal cord stimulation have been used to explain clinical patterns of paresthesias[10,11,57,59] and to design novel electrode geometries for selective stimulation of targeted neural elements.[25,60] These examples demonstrate the utility of computer-based modeling in understanding and controlling neural elements activated by electrical stimulation.

5.2   Quantitative models of CNS neurons

5.2.1   Physical basis of models: from cells to circuits

Neural cells in the CNS are highly varied in their form and function but all share the properties of excitability and polarization. Excitability is based on selective ion channels and polarization is based on the concentration differences of ions, generated by ionic pumps, across the cell membrane. According to Robertson's proposition,[52] all membranes or major portions of membranes have a common basic structure. This structure includes a lipid bilayer covered by non-lipid monolayers on both sides. Within the lipid bilayer reside proteins, and transmembrane proteins form the pores that enable ionic current across the membrane. The electrical properties of the membrane are similar to those of a parallel RC circuit, the so-called *passive membrane*, where the capacitor represents the lipid bilayer and the resistor represents the transmembrane ion channels.

The excitability of neurons comes from the ion channels. Ion channels contribute to changing the conductance across the membrane by passing specific ions. The conductance of some ion channels is dependent on the voltage across the membrane and thus can be considered as nonlinear resistors.[24] The nonlinear properties of ion channels, such as the sodium channel, can produce regenerative coupling to the transmembrane potential by the greatly increased conductance to sodium ions. This nonlinear property in conductance of specific ions is referred to as *active membrane*.

The passive and active properties are the basic components of the membrane throughout the neuron including the cell body, dendrites, and axon. The dendrite and axon are long cylindrical tubes filled with cytoplasm, which has a higher electrical conductivity than the extracellular fluid. Also, for axons and dendrites the electric current flowing through the membrane is much less than the current flowing parallel to the cylinder axis because the resistance of the membrane is much higher than the cytoplasmic resistance. Therefore, the nerve cell with passive membrane properties (i.e., not considering nonlinear ion channels) can be considered a good conductor insulated by a membrane that has high resistivity and a certain capacity. This analog bridges the investigation of electrophysiological properties of neurons to cable theory (Figure 5.2).

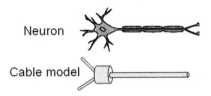

Figure 5.2 Typical neuron and cable model. The passive electrical properties of the dendrites, soma, and axon are similar to the core cable with insulation.

5.2.2   Geometric properties of a range of CNS neurons

The types of neurons in the CNS are exceedingly diverse and are classified into three large groups by shape as unipolar, bipolar, and multipolar neurons. Each group has common features of dendrite and axon structure with respect to the cell body. The morphology of neurons has an impact on their response to extracellular stimulation because the entire neuron is exposed to the electric field. Therefore, the response of the cell is modulated by influences from every branch exposed to the electric field. Table 5.1 shows examples of mammalian CNS neurons to indicate the range of cell structure and size.

5.2.3   Cable models of CNS neurons

Cable theory was first presented in 1855 by William Thomson, who provided the mathematical derivation and applications for submarine telegraphic cables. This theory included both steady-state and transient solutions for particular boundary conditions and initial conditions.[27,41] The solution was for a single spatial dimension that facilitated the theoretical treatment of transient as well as steady-state solutions. The theory was applied for nerve electronus by Matteucci in 1863 and further developed by Hermann in the 1870s.

5.2.4   Cable equation: continuous form

The continuous form of the cable equation is derived from a compartment model that consists of a series of compartments, each with a resistance and a capacitance.[41] Each compartment represents a segment of cable with the length of Δx, and all properties such as resistance and capacitance in the segment are lumped into one element for each (Figure 5.3). The cable equation is derived by application of Kirchoff's current law (KCL) — conservation of currents — which states that the difference in the currents (between i_{i1} and i_{i2}) in the axial direction is the current flowing through the membrane (Figure 5.4). Mathematically this is expressed as:

$$i_{i1} - i_{i2} = -\Delta i_i = i_m \Delta x$$

The membrane current ($i_m \Delta x$) is the sum of two components: the resistive current $V_m(\Delta x / r_m)$ and the capacitive current $c_m \Delta x ((\partial V_m)/(\partial t))$. Therefore, the membrane current is given by:

Table 5.1 Range of Geometrical Properties of CNS Neurons

Region	Neuron	Ref.	Mean Soma Diameter (µm)	Number of Dendrite Stems on Soma	Dendritic Terminal Length (µm)
Cerebellum	Purkinje cell	Roth and Hauser[54]	50 ~ 80	2 ~ 10	50 ~ 200
Thalamus	TCP neurons	Ohara and Havton[38]	11 ~ 20	4 ~ 8	28 ~ 3000
Hippocampus	Pyramidal cell	Bannister and Larkman;[5] Bilkey and Schwartzkroin[6]	15 ~ 30	2 ~ 8	300 ~ 1000
Retina	Ganglion cell	Sheasby and Fohlmeister[56]	20 ~ 30	3 ~ 7	50 ~ 1000
Spinal cord	Motor neuron	Rose et al.[53]	50 ± 10	10 ± 2	1,150 ± 304

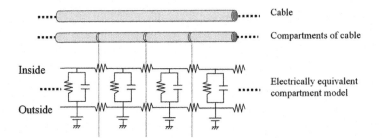

Figure 5.3 Compartment model of cable. Insulation (membrane) around core is equivalent to resistance in parallel with capacitance.

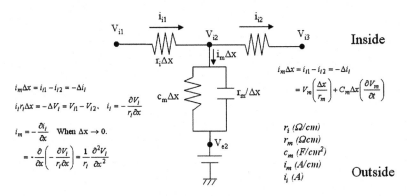

Figure 5.4 Application of Kirchoff's current law at a compartment of a cable model to derive the cable equation. The membrane current includes capacitive and resistive currents, and their sum is equal to the change in the axial current.

$$i_m \Delta x = V_m \left(\frac{\Delta x}{r_m} \right) + c_m \Delta x \left(\frac{\partial V_m}{\partial t} \right)$$

The equation is simplified by dividing Δx on both sides and the result is:

$$i_m = \frac{V_m}{r_m} + c_m \left(\frac{\partial V_m}{\partial t} \right)$$

The axial current, derived from Ohm's law as that current flowing between two consecutive nodes (V_{i1} and V_{i2} or V_{i2} and V_{i3}), is defined by potential difference divided by resistance and expressed as:

$$V_{i1} - V_{i2} = -\Delta V_i = i_i r_i \Delta x$$

$$i_i r_i = -\frac{\Delta V_i}{\Delta x}$$

By taking the limit $\Delta x \to 0$, the axial current is expressed by partial derivative:

$$i_i = -\frac{1}{r_i}\frac{\partial V_i}{\partial x}$$

Going back to KCL, defining conservation of the current, the membrane current with partial derivative form yields:

$$i_m = -\frac{\Delta i_i}{\Delta x}$$

$$= -\frac{\partial i_i}{\partial x}, \quad \text{when } \Delta x \to 0$$

Therefore, the membrane current is the partial derivative of the axial current with respect to x. Again, the axial current is also a partial derivative of the axial potential distribution by spatial variable x. Applying the axial current equation in partial derivative form $(-(\partial i_i/\partial x))$, the membrane current can be expressed as:

$$i_m = -\frac{\partial i_i}{\partial x} = -\frac{\partial}{\partial x}\left(-\frac{\partial V_i}{r_i \partial x}\right) = \frac{\partial^2 V_i}{r_i \partial x^2}$$

The transmembrane potential (V_m) is defined as $V_i - V_e$, so the intracellular potential V_i can be replaced by $V_m + V_e$. Finally, the current i_m is expressed as:

$$i_m = \frac{\partial^2 V_m}{r_i \partial x^2} + \frac{\partial^2 V_e}{r_i \partial x^2}$$

Combining with the earlier equation for transmembrane current as the sum of the resistive and capacitive current yields:

$$\frac{1}{r_i}\frac{\partial^2 V_m}{\partial x^2} + \frac{1}{r_i}\frac{\partial^2 V_e}{\partial x^2} = \frac{V_m}{r_m} + c_m\left(\frac{\partial V_m}{\partial t}\right)$$

$$\frac{1}{r_i}\frac{\partial^2 V_m}{\partial x^2} - \frac{V_m}{r_m} - c_m\left(\frac{\partial V_m}{\partial t}\right) = -\frac{1}{r_i}\frac{\partial^2 V_e}{\partial x^2}$$

The general single-dimension cable equation is a partial differential equation expressed as:

$$\lambda^2 \frac{\partial^2 V}{\partial x^2} - V - \tau\frac{\partial V}{\partial t} = F$$

where V is the transmembrane voltage as a function of x (spatial variable) and t (time), and $\lambda = \sqrt{r_m/r_i}$ and $\tau = r_m c_m$ are the space and time constants defined by the electrical properties of the neuron. The F represents a forcing function or input caused by synaptic conductance change, applied electric fields, and/or active membrane properties. This nonhomogeneous partial differential equation can be transformed to a homogeneous equation ($F = 0$) with additional initial conditions by the principle of superposition. Further simplification can be made for steady-state conditions by $(\partial V/\partial t) = 0$, which will yield a homogeneous linear second-order ordinary differential equation as $\lambda^2(\partial^2 V/\partial x^2) - V = 0$, the solution of which is a sum of two exponentials (cable of infinite length) or hyperbolic functions (cable of finite length). The solution for a finite cable is:

$$V = B_1 \cosh(\frac{x}{\lambda}) + B_2 \sinh(\frac{x}{\lambda})$$

where B_1 and B_2 are defined by boundary condition at $x = 0$ and $x = l$ (cable length). The equivalent solution for the infinite cable is:

$$V = A_1 \exp(\frac{x}{\lambda}) + A_2 \exp(-\frac{x}{\lambda})$$

where A_1 and A_2 are determined by conditions at $x = 0$ and $x = \pm\infty$. Transient solutions of the cable equation can be obtained by using separation of variables. For a finite-length cable with the boundary condition of sealed ends,[40,43,50] the solution can be expressed as an infinite series:

$$V = \sum_{n=0}^{\infty} B_n \cos(n\pi \frac{x}{\lambda}) \exp(-\frac{t}{\tau} - (\frac{n\pi\lambda}{l})^2 \frac{t}{\tau})$$

$$B_0 = (\frac{1}{l}) \int_0^l V(x, t = 0) dx$$

$$Bn = (\frac{2}{l}) \int_0^l V(x, t = 0) \cos(\frac{n\pi x}{l}) dx, n > 0$$

where l and τ are the length of cable and time constant of cable, respectively.

5.2.5 Cable equation: discrete form

The continuous form of the cable equation is limited to homogeneous structures with restrictive assumptions. The passive properties of the dendritic tree and unmyelinated axons with constant diameter can be modeled using the continuous cable equation. However, a typical dendritic tree has various diameters from stem to dendritic terminal, and myelinated axons have different membrane properties at the nodes of Ranvier and in the internodal segments. The nonhomogenous cable properties can be modeled using a discrete form of the cable equation that is a mathematical representation of the compartmental circuit model of a neuron.

The general discrete cable equation is directly derived from the continuous case. The partial derivative of the spatial variable is replaced by Δ, yielding:

$$\frac{1}{r_i}\frac{\Delta^2 V_m}{\Delta x^2} - \frac{V_m}{r_m} - C_m\left(\frac{dV_m}{dt}\right) = -\frac{1}{r_i}\frac{\Delta^2 V_e}{\Delta x^2}$$

This equation is only useful for compartmental models of homogeneous cables, but small modifications of the equation give rise to great flexibility for practical applications such as nonhomogeneous dendrites, cell bodies, and myelinated axons. In the discrete compartmental cable model, the resistance (axial and membrane) and capacitance are not required to be functions of Δx. Therefore, each variable — $1/(r_i\Delta x)$, $c_m\Delta x$, and $\Delta x/r_m$ — can be replaced by $G_a(k\Delta x)$ ($\Omega\angle^{1}$), $C_m(k\Delta x)$ (F), and $R_m(k\Delta x)$ (Ω). This yields the discrete cable equation applicable to a cable with any geometry. The steady-state solution of the discrete cable equation can be obtained from linear algebra; however, the transient solution must be obtained numerically.

As an example of application of the discrete cable equation, a myelinated axon is considered where the myelin has a very high resistivity (assumed to be an insulator), thus simplifying the model to include only the nodes of Ranvier and the axial resistance (Figure 5.5). Under this circumstance, the cable equation in differential form will be discrete in space and continuous in time. The membrane (nodal) current is defined similarly to the continuous case with a resistive current ($G_m V_m$) and a capacitive current ($C_m(dV_m/dt)$). The total current flowing through the node of Ranvier is expressed as:

$$I_{node} = G_m V_m + C_m\left(\frac{dV_m}{dt}\right)$$

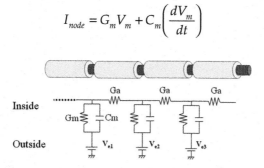

Figure 5.5 Compartmental model of a myelinated axon. The internodal segment is modeled by an axial resistor, because of the highly resistive myelin sheath. The node of Ranvier is lumped into a single compartment, which is a relatively small portion compared to the internodal segment (1.5 μm in length vs. 1 mm in length).

The axial current between successive nodes of Ranvier depends on the internodal conductance (G_a) and potential difference ($V_{i1} - V_{i2}$ or $V_{i2} - V_{i3}$) according to Ohm's law. Just as the membrane current was obtained from the difference of axial currents in the continuous case, the nodal current equation in the discrete model is:

$$I_{node} = I_i - I_2 = Ga \cdot (V_{i1} - V_{i2}) - Ga \cdot (V_{i2} - V_{i3})$$

$$= Ga \cdot (V_{i1} - 2V_{i2} + V_{i3})$$

$$= Ga \cdot (V_{m1} - 2V_{m2} + V_{m3}) + Ga \cdot (V_{e1} - 2V_{e2} + V_{e3})$$

where $V_m = V_i - V_e$. Combining this equation with the equation for the components of the membrane current yields:

$$G_m V_m + C_m \left(\frac{dV_m}{dt} \right) = Ga \cdot (V_{m1} - 2V_{m2} + V_{m3}) + Ga \cdot (V_{e1} - 2V_{e2} + V_{e3})$$

Rearranging terms and defining the second difference of the potential as:

$$Ga \cdot (V_{m1} - 2V_{m2} + V_{m3}) = Ga \cdot \Delta^2 V_m$$

yields the discrete cable equation for myelinated axon:

$$G_m V_m + C_m \left(\frac{dV_m}{dt} \right) - Ga \cdot (\Delta^2 V_m) = Ga \cdot \Delta^2 V_e$$

The final equation is similar to the general form of the cable equation with a forcing term of $G_a \cdot \Delta^2 V_e$. For a step change in the extracellular potential from steady state, as would occur by application of a rectangular current pulse, the forcing term will determine initial polarization pattern of the axon, because the $C_m(dV_m/dt)$ term will follow the sign of the forcing term.

5.2.6 Assumptions

The cable model and cable equation are good approximations to understand the properties of neurons, but they must be used with the following assumptions:

1. Any angular or radial dependence of V within the core or outside the membrane is neglected. This assumption is reasonable for steady-state solutions of small-diameter cables with high-conductivity cores corresponding to high cytoplasm conductivities. Low conductivity of the cable core will generate a voltage gradient within the cable in

the radial direction and violate this assumption. Practically, the high conductivity of the cytoplasm eliminates this effect. This assumption is also invalid under extracellular stimulation with large diameter cables such as used to represent the soma. The extracellular potentials are dependent on the angular position on the soma,[28,29] and the angular dependence of V is a function of the electrode to cell distance and of the electrode type. Therefore, it is valid only for relatively small diameter cables such as dendrites and axons or cables far from the stimulation electrode.

2. The core conductivity is uniform everywhere (homogeneous medium). This assumption implies that the ratio between the resistivity of cytoplasm and the cross sectional area is constant. For homogeneous cytoplasm, it implies constant diameter of the cable. Practically, the diameters of dendrites and axons may change along their paths. Thus, the morphology of neurons limits direct application of analytical solution of the continuous cable equation and requires discretization to account for geometric inhomogeneity.

5.2.7 Passive electrical properties of CNS neurons

In the nervous system, information is transported long distances by action potentials. The action potential is generated by depolarization of the membrane where active ion channels are present. The depolarization can be generated by any input, including synaptic current or current generated by an extracellular electric field. When the degree of polarization from a certain input (voltage or current) is under threshold the neuron will not fire an action potential and can be considered using the so-called *passive electrical properties*. The passive electrical properties are described by the input impedance, the time constant, and the resting potential. The input impedance of the neuron is defined by the relationship between the current applied by an intracellular electrode and the transmembrane voltage response. The input impedance determines how much the transmembrane voltage of neuron will change in response to a steady current:

$$\Delta V = I \times R_{in}$$

As the passive membrane is modeled as a resistance in parallel with a capacitance, the response of the membrane (voltage) takes time to reach steady state. The change in transmembrane potential of the passive membrane can be described by following equation:

$$\Delta V(t) = I_{injected} R_{in} (1 - e^{-t/\tau})$$

where τ is the membrane time constant given by the product of input resistance and input capacitance. The resting potential is determined by the ionic

Table 5.2 Range of Passive Electrical Properties of CNS Neurons

Region	Neuron	Ref.	Input Resistance (MΩ)	Time Constant (msec)	Resting Potential (mV)
Cerebellum	Purkinje cell	Raman and Bean;[44] Roth and Hausser[54]	59.3 ± 38.4	64.6 ± 17.2	-62 ± 3 with 300 nM TTX
Thalamus	Thalamocortical neuron	Turner et al.[61]	30 ~ 240	5 ~ 35	-60 ~ -73
Hippocampus	Pyramidal cell	Bilkey and Schwartzkroin[6]	23 ± 2	12.5 ± 0.97	-55.7 ± 1.28
Retina	Ganglion cell	Sheasby and Fohlmeister[56]	800 ~ 1600	47 ~ 85	-64 ~ -66
Spinal cord	Motor neuron	McDonagh et al.[31]	2.5 ~ 32	2.5 ~ 52	-60 ~ -83

concentration difference across the membrane and the states of the ion channels at rest. The balance of ion fluxes gives rise to the resting potential, which is quantified by the Goldman equation. It usually ranges from −60 to −70 mV. The resting potential is a reference point for measurements of changes in transmembrane potential. The range of passive electrical properties of example neurons presented in Table 5.2.

5.2.8 Intracellular vs. extracellular stimulation

Properties of neuronal excitation have been studied primarily by intracellular electrical stimulation by injecting current through a glass pipette electrode. The injected current will flow in the axoplasm and pass through membrane, and the neuron can again be modeled using a discrete compartmental cable model (Figure 5.6). The transmembrane voltage at each compartment is determined by current flowing through the membrane and axoplasm, and an analytical solution provides the profile of transmembrane voltage along the cable for different boundary conditions (Figure 5.7).

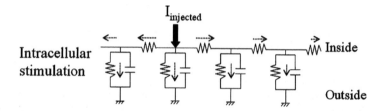

Figure 5.6 Electrically equivalent model to intracellular stimulation. Intracellular stimulation of the nerve cell or fiber is modeled as a cable with a current or voltage source connected to specific locations of the cable. The stimulation corresponds to the forcing term (F) in the cable equation.

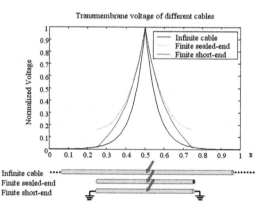

Figure 5.7 Analytical solution of the cable equation with three different boundary conditions to determine the profile of transmembrane potential generated by intracellular current injection.

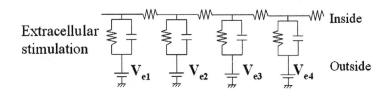

Figure 5.8 Electrically equivalent cable model under extracellular stimulation. The extracellular electrode will determine the extracellular potentials (V_e) along the cable which act as the sources to generate axial and transmembrane currents.

Application of current in the extracellular space (extracellular stimulation) creates a specific pattern of electric field and thus electric potentials along the cable (Figure 5.8). The extracellular potentials (V_e) along the cable are determined by the geometry of the cable, the source geometry and location, and the electrical properties of the extracellular space. As derived in the cable equation,[1,41,48] the second derivative of the extracellular potentials along the cable $(-(1/r_i)\,(\partial^2 V_e/\partial x^2))$ will produce current across the membrane and intracellular axial current and thus changes in transmembrane potential.

Mathematically this forcing term (*activating function*: $(1/r_i)\,(\partial^2 V_e/\partial x^2)$ is equivalent to current injection in each compartment in an equivalent model with intracellular current injections.[48,62] As we derived the continuous and discrete cable equations using Kirchoff's current law, the conservation of current at each node must be satisfied. Under the condition that the second derivative of V_e equals zero, the cable equation is homogeneous and consists of three terms:

$$\frac{1}{r_i}\frac{\partial^2 V_m}{\partial x^2} - \frac{V_m}{r_m} - C_m\left(\frac{\partial V_m}{\partial t}\right) = 0$$

The first term represents the axial current difference, the second represents the membrane resistive current, and the third represents the membrane capacitive current with unit of A/cm^2 Therefore, the additional term, $-(1/r_i)$ $(\partial^2 V_e/\partial x^2)$ arising from the extracellular potentials corresponds to a current (A/cm^2) source connected to each node, as shown in Figure 5.9. If a series of equivalent currents are injected along the cable, the effect of extracellular stimulation is equivalent to the effect of intracellular stimulation with a series of distributed intracellular current sources. The equivalent current at each node can be of anodic or cathodic polarity from a single extracellular electrode, because the forcing term ($G_a \cdot \Delta^2 V_e$) is determined by axon geometry and electrode type and location.[62]

The potentials generated by an extracellular electrode produce both inward and outward transmembrane current in the cable, as shown in Figure 5.10. The direction of transmembrane current depends on the location along the cable and creates regions of both depolarization and hyperpolarization

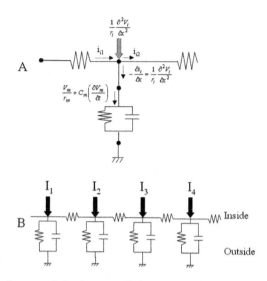

Figure 5.9 Equivalent model of extracellular stimulation with current injection in each compartment. (A) The equivalent current for each compartment is derived from the cable equation with the non-zero V_e using Kirchoff's current law. (B) The extracellular stimulation can be converted to an equivalent model with distributed intracellular current sources.

along the cable, in contrast to the unidirectional polarization resulting from intracellular stimulation (Figure 5.7).

5.2.9 *Source driving polarization*

Solution of the infinite cable equation provides the steady-state voltage distribution along the cable expressed as an exponential function. The change in voltage along the cable is determined by the space constant, which is a function of the ratio between the membrane and cytoplasmic resistivities (Figure 5.7). For realistic dendrites (finite cable with nonhomogeneous geo-

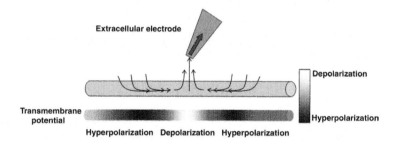

Figure 5.10 Current flow induced by an extracellular current source and polarization pattern (changes in transmembrane potential) along the cable. A cathodic extracellular point-source electrode will depolarize the cable (light) near the electrode and hyperpolarize both sides of the depolarized region (dark).

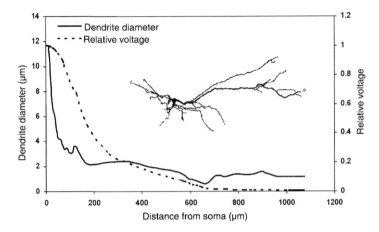

Figure 5.11 Simulation result from the real geometry of dendrite. The voltage profile along the real dendrite is determined by geometric information. This limits the application of theoretical cable equation to realistic dendrites or axon where nonlinear ion channels are located and activated by this voltage change.

metric structure), the voltage distribution depends on the cable geometry. Figure 5.11 illustrates the voltage profile along a long dendritic tree (dark line from cell body to the dendritic terminal) of a preganglionic parasympathetic neuron with a realistic morphology reconstructed from cat spinal cord by Morgan and Ohara.[35] The dendrite diameter changes along the tree and the transmembrane voltage depends on the cable diameter.

Under extracellular stimulation, the transmembrane voltage response depends on the orientation and geometry of the neuron, because the extracellular potential outside each compartment is determined by the type of source and the distance from it. As seen from the cable equation, the source driving polarization is the second derivative of the extracellular potential along the cable and must be non-zero to create changes in membrane potential in the neuron.

As an example the extracellular potential and transmembrane potentials of three myelinated axons under a point current (I) source were calculated using the discrete cable equation (Figure 5.12). The extracellular potential is inversely proportional to the electrode-to-cable distance (R) and is given by:

$$V_e = \frac{I}{4\pi\sigma_e R}$$

where σ_e is the conductivity of the extracellular space with unit of S/cm. The transmembrane potential generated by the point source is triphasic, which is approximately predicted by activating function (see Section 5.2.8). Anodic extracellular stimulation produces maximum hyperpolarization on the nearest point of the cable and two depolarization peaks next to it, while

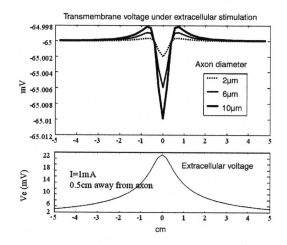

Figure 5.12 Transmembrane voltage of axons induced by extracellular current source and extracellular voltage along the axon. (A) The magnitude of transmembrane voltages (top traces) is proportional to axon diameter, because the larger diameter axon has a larger internodal distance, which will increase activating function by raising the second derivative $\Delta^2 V_e$ and conductance term G_a. (B) Extracellular voltage along the cable is inverse proportional to the electrode to cable distance.

cathodic extracellular stimulation produces the opposite pattern of polarization. Action potential initiation occurs at the point of maximum depolarization and will thus differ under anodic and cathodic stimulation.

5.2.10   Modeling of synaptic inputs

Excitation of neurons can be initiated by an extracellular electric field as described, but the behavior of the nerve cell is also modulated by synaptic inputs from presynaptic terminals that are exposed to the extracellular electric field. Because of the large number of connections, input from presynaptic terminals excited by extracellular stimulation can have strong postsynaptic effects (see Section 5.3.2).

Most rapid signaling between nerve cells in the CNS involves ionotropic transmission that occurs through synaptic connections. When an action potential arrives at a presynaptic terminal, depolarization results in release of neurotransmitters into the synaptic cleft. The released neurotransmitters diffuse to the postsynaptic cell membrane where they bind to the receptors. Receptors are activated by specific neurotransmitters and generate excitatory postsynaptic potential (EPSP) or inhibitory postsynaptic potential (IPSP).

As with other ion channels, activation of postsynaptic receptors changes channel conductance, which generates a postsynaptic current described by:

$$I_{syn} = f_{syn} \, \overline{g}_{syn} (V_{post} - E_{syn})$$

where f_{syn} is fractional transmitter release, $\bar{g}_{syn}$ is maximal postsynaptic conductance, and E_{syn} is the reversal potential.[8,13] A simple form of f_{syn} is modeled using an alpha function:[42]

$$f_{syn}(t) = \frac{t}{t_p} \exp(1 - \frac{t}{t_p})$$

This function has rapid rise to a peak at t_p and slow decay with time constant t_p and is a good empirical estimate of the postsynaptic conductance change. As synaptic inputs are modeled as current sources with nonlinear conductances, distribution of synaptic inputs on a modeled postsynaptic neuron corresponds to adding a forcing term in each compartment. These "indirect" synaptic forcing terms may alter the patterns of neural activation due to the "direct" forcing terms arising from extracellular stimulation and cannot be overlooked during CNS stimulation.

5.3   Properties of CNS stimulation

To make accurate inferences about anatomical structures or physiological mechanisms involved in electrical stimulation, one must know which neural elements the stimulus pulse activates. When stimulating within the CNS the electrode is placed within a complex volume conductor where there are three general classes of neurons that can be affected: local cells, axon terminals, and fibers of passage (Figure 5.1). Local cells represent neurons that have their cell body in close proximity to the electrode. Axon terminals represent neurons that project to regions near the electrode and make synaptic connections with local cells. Fibers of passage represent neurons where both the cell body and axon terminals are far from the electrode, but the axonal process of the neuron traces a path that comes in close proximity to the electrode. Each of these classes of neurons can be activated by stimulation with extracellular sources. However, activation of each different class can result in different physiologic and/or behavioral outputs. Experimental measurements indicate that local cells, axon terminals, and fibers of passage have similar thresholds for activation when stimulating with extracellular sources (see Section 5.1.1). Therefore, it is often difficult to determine the clear effects of extracellular stimulation.

5.3.1   Direct activation

The seminal review of Ranck[45] laid the groundwork for understanding the effects of electrical stimulation within the CNS; however, because of limitations in experimental techniques and the complex response of neurons to extracellular stimulation, our understanding of electrical stimulation of the CNS has advanced at a relatively slow pace. The use of multicompartment cable models of CNS neurons coupled to extracellular electric fields has given us the opportunity to address many important issues related to extracellular

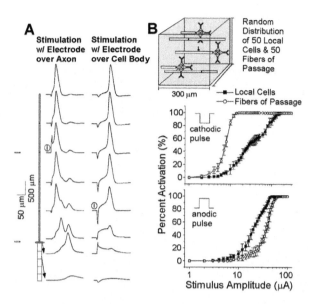

Figure 5.13 Direct excitation of neurons with extracellular stimulation. (A) Action potential initiation (API) and propagation for two different electrode locations (electrode-to-neuron distance of 100 μm) and cathodic stimulus pulses with durations of 0.1 msec. The left and right columns correspond to the responses from electrodes located over the axon or cell body, respectively. Each row shows the transmembrane voltage as a function of time at the segment of the neuron shown to the left. The site of API is noted by the circled *i*. (Modified from McIntyre and Grill.[32]) (B) Input–output relations for populations of neurons stimulated by a monopolar electrode. Excitation was studied using randomly distributed populations of 50 local cells and 50 fibers of passage. Percentages of activated neurons (mean ± one standard deviation from three different random distributions) are displayed as a function of stimulus amplitude for a monophasic cathodic stimulus (pulse duration [pd] = 0.20 msec) and a monophasic anodic stimulus (pd = 0.20 msec). (Modified from McIntyre and Grill.[33])

stimulation, including the site of action potential initiation and the effects of changes in stimulus parameters on activation patterns.[18,19,32–34,46,47] The analysis of modeling and experimental findings results in four general conclusions regarding the effects of stimulation within the CNS (Figure 5.13):

1. When stimulating local cells with extracellular sources, action potential initiation (API) takes place in a node of Ranvier of the axon relatively far away from the electrode (Figure 5.13A).
2. When stimulating axon terminals and fibers of passage with extracellular sources, API takes place in a node of Ranvier relatively close to the electrode (Figure 5.13A).
3. Anodic stimuli are more effective in activating local cells than cathodic stimuli (Figure 5.13B).
4. Cathodic stimuli are more effective in activating fibers of passage and axon terminals than anodic stimuli (Figure 5.13B).

It should always be noted, however, that activation of any neuron with extracellular electric fields is dependent on four main factors:

1. Electrode geometry and the electrical conductivity of the tissue medium — The response of the neuron to stimulation is dependent on the electric field generated by the electrode, which is dependent on the size and shape of the electrode. In addition, the inhomogeneous and anisotropic electrical properties of the CNS tissue medium affect the shape of electric field.[21] Therefore, both the type of electrode used and the region of the CNS where it is inserted will affect the neural response to stimulation.

2. Stimulation parameters — Changes in stimulation parameters can affect the types of neurons activated by the stimulus and the volume of tissue over which activation will occur. The four primary stimulation parameters are the polarity, duration, and amplitude of the stimulus pulse and the stimulus frequency. In general, alterations in the stimulus pulse duration and amplitude will affect the volume of tissue activated by the stimulus, and alterations in the stimulus polarity and frequency will affect the types of neurons activated by the stimulus.

3. Geometry of the neuron and its position with respect to the electrode — In general, the closer the neuron is to the electrode the lower the stimulation current necessary for activation. However, complex neural geometries such as dendritic trees and branching axons result in a large degree of variability in current–distance relationships (threshold current as a function of electrode-to-neuron distance) of the same types of neurons. Therefore, the orientation of the neural structures with respect to the electrode is of similar importance to the geometric distance between them, especially for small electrode-to-neuron distances.

4. Ion channel distribution on the neuron — Axonal elements of a neuron consist of a relatively high density of action-potential-producing sodium channels compared to cell bodies and dendrites. As a result, the axonal elements of a neuron are the most excitable and regulate the neural output that results from application of extracellular electric fields. However, while the cell body and dendrites may not be directly responsible for action potential spiking that results from the stimulus, they do contain several types of calcium and potassium channels that can affect neuronal excitability on long time scales when trains of stimuli are used.

5.3.2 *Indirect effects*

Previous experimental and modeling results have shown that the threshold for indirect, or trans-synaptically evoked, excitation or inhibition of local cells stimulated with extracellular sources is similar to (in some cases, dependent on electrode location less than) the threshold for direct excitation of local cells (see Section 5.1.1). Indirect excitation or inhibition of local cells is the result of stimulation-induced release of neurotransmitters that results from the activation of axon terminals activated by the stimulus. Axon terminals are activated at low stimulus amplitudes relative to local cells, especially when cathodic stimuli are used. Therefore, when considering the effect of the stimulus on local cells near the electrode it is probable that a large number of axon terminals are activated, resulting in high levels of synaptic activity on the dendritic trees of local cells.

Stimulation-induced trans-synaptic activity can be predominantly excitatory, predominantly inhibitory, or any relative mix of excitation and inhibition, depending on the types and numbers of synaptic receptors activated. Therefore, the interpretation of the effects the stimulation on the neuronal output of local cells is made up of two components: (1) the direct effect of the extracellular electric field on the local cell, and (2) the indirect effect of the stimulation-induced trans-synaptic excitation and/or inhibition. As a result, an action potential can be generated either by the stimulus pulse itself or by indirect synaptic activation. However, activation of axon terminals by extracellular stimuli is nonselective to excitatory or inhibitory neurotransmitter release.

In general, the indirect effects of extracellular stimulation of local cells result in a biphasic response of a short period of depolarization followed by a longer period of hyperpolarization. This biphasic response is the result of the interplay between the time courses of the traditionally fast excitatory synaptic action and the traditionally slow inhibitory synaptic action. The role of indirect effects on the output of local cells can be enhanced with high-frequency stimulation (Figure 5.14). If the interstimulus interval is shorter than the time course of the synaptic conductance, the indirect effects will summate. Because inhibitory synaptic action traditionally has a longer time course than excitatory synaptic action, the effect of this summation is hyperpolarization of the cell body and dendritic arbor of the local cell when high-frequency stimulus trains are used. This hyperpolarization can limit the neuronal output that results from direct effects from the stimulus and can functionally block the ability of the neuron to integrate non-stimulation-induced synaptic activity during the interstimulus interval. However, because action potential initiation takes place in the axon of local cells in response to the direct effects of the stimulation, local cells will still fire action potentials in response to each stimulus pulse given that the stimulus amplitude is strong enough (Figure 5.14C).

5.4 Selective stimulation

Microstimulation in the CNS can activate neurons with greater specificity than is possible with larger electrodes on the surface of the spinal cord or brain. The potential thus arises for electrical activation of intact neuronal circuitry, and, in turn, generation of distributed and controlled physiological outputs

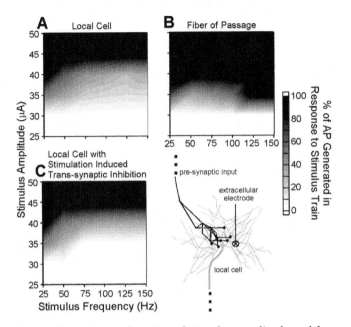

Figure 5.14 Neuronal output as a function of stimulus amplitude and frequency. Neuronal output (percentage of stimuli in a 500-msec stimulus train that generate propagating action potentials in the neuron models) was quantified for direct excitation of a local cell (A) and a fiber of passage (B) from a train of charge-balanced, cathodic-phase first symmetrical biphasic stimuli (100 μsec per phase). Both neurons had a threshold for activation from a single pulse of 34 μA. (C) Influence of stimulation-induced trans-synaptic inhibition on the neuronal output of the local cell. Each terminal of the presynaptic input was activated by a stimulus, and in turn synaptic conductances representative of GABAergic inhibition were applied to the dendrites of the local cell following each stimulus in the train. (Modified from McIntyre and Grill.[34])

for the study of the neural control of function or for application in neural prostheses. However, in many regions of the CNS, local cells, axon terminals, and fibers of passage are intermingled in close proximity to the electrode. In general, only one class of neurons is the target population to achieve the desired output from the stimulus. Yet, the stimulation used to activate the target neurons (e.g., local cells) can also result in activation of the other classes of neurons around the electrode (e.g., axon terminals, fibers of passage). Therefore, techniques that can enable selective activation of target populations of neurons with little to no activation of non-target populations can improve our understanding of experimental results using electrical stimulation of the CNS and provide important tools for application in neuroprosthetic devices.

Previous modeling and experimental work have shown that local cells have lower thresholds for activation with anodic stimuli, while axonal elements (axon terminals and/or fibers of passage) have lower thresholds with cathodic stimuli (Figure 5.13B);[32,33,45] however, chronic application of electrical stimulation within the nervous system requires the use of biphasic stimuli because of

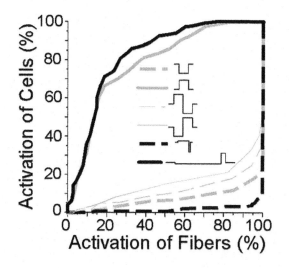

Figure 5.15 Selective activation of targeted neuronal populations via alterations in the stimulus waveform. Using randomly distributed populations of neurons (Figure 5.1B), we examined the effect of changes in the stimulus waveform on the relative activation of local cells compared to fibers of passage. Plotted is the percentage activation of local cells as a function of the percentage activation of fibers of passage for six different stimulus waveforms. The stimulus waveforms (from top to bottom in Figure 5.5; see caption) consisted of a monophasic cathodic pulse (pulse duration [pd] = 0.2 msec), a monophasic anodic pulse (pd = 0.2 msec), a symmetrical anode first biphasic pulse (pd = 0.2 msec for each phase), a symmetrical cathode first biphasic pulse (pd = 0.2 msec for each phase), an asymmetrical anode first biphasic pulse (pd = 0.2 msec for anodic phase; 0.02 msec for cathodic phase), and an asymmetrical cathode first biphasic pulse (pd = 1 msec for cathodic phase; 0.1 msec for anodic phase). (Modified from McIntyre and Grill.[33])

issues related to tissue damage and electrode corrosion.[39] In general, when biphasic stimuli are used, local cells or axonal elements will be activated during the anodic or cathodic phases of the stimulus, respectively, resulting in low selectivity for activation of a target population (Figure 5.15).[33] Alterations in the stimulus frequency and/or stimulus waveform represent techniques that can enable enhanced selectivity of either local cells or axonal elements.

5.4.1   Effect of stimulus frequency

Mammalian neurons exhibit both depolarizing and hyperpolarizing after-potentials that follow an action potential spike. The time course of these afterpotentials is different in the cell body and axon, and these afterpotentials affect the threshold for generation of subsequent impulses.[34] Myelinated axons exhibit a relatively long-duration (~15 msec), high-amplitude (~5 mV) depolarizing afterpotential (DAP) followed by a long-duration (~80 msec), low-amplitude (~1 mV) afterhyperpolarization (AHP). Neuronal cell bodies traditionally exhibit a shorter duration, lower amplitude

DAP that is followed by a pronounced AHP on the order of ~5 to 10 mV that reaches its maximum ~10 to 20 msec after the action potential spike. During the DAP of the myelinated axon, the threshold to generate another action potential is decreased, and during the AHP of the neuronal cell body the threshold to generate another action potential is increased. The overlap in time course of these afterpotentials for the exploitation of biophysical differences in local cells and fibers of passage to enhance selectivity.

Figure 5.14 shows the responses of a neuron with its cell body near the electrode compared to the response of a fiber of passage to symmetrical biphasic stimulus trains.[34] Maps of the percent of stimuli that generated propagating action potentials during the stimulus train as a function of stimulus amplitude and frequency were generated. Both the local cell (Figure 5.14A) and fiber of passage (Figure 5.14B) can fire in response to the first stimulus in the train when the stimulus amplitude is greater than or equal to 34 µA. However, the near-threshold stimulus amplitudes, the ability of either the local cell or the fiber of passage to follow the stimulus train in a one-to-one ratio is affected by the different time courses and amplitudes of the afterpotentials in the two neurons.

These results demonstrate that modulation of the frequency of the stimulus train can enhance selectivity between activation of cells and fibers of passage within the CNS. However, it should be noted that while the selectivity of fibers of passage can be increased with high-frequency stimulation, local cells can still respond to the stimulus, albeit at a lower average rate. The limited output of the local cells from high-frequency stimulation could still be great enough to generate a functional activation of their efferent target, and, conversely, driving fibers of passage over 100 Hz may exceed the physiological limits for those neurons and result in unexpected or unwanted affects. In addition, the output of local cells at high stimulus frequencies is dependent not only on its direct excitation characteristics but also on the role of indirect trans-synaptic influences (Figure 5.14C). Therefore, while modulation of stimulation frequency can enhance selectivity of fibers of passage over cells, this technique is not especially effective without the augmentation of alterations in the stimulus waveform.

5.4.2 *Effect of stimulus waveform*

The stimulating influence of extracellular electric fields is related to the second difference of the extracellular potential distribution on the surface of the individual neurons, and this stimulating influence will cause regions of both depolarization and hyperpolarization in the same cell.[32,46,47] In general, when stimulating local cells API occurs in the axon of the neuron, relatively far from the electrode, while when stimulating axonal elements (axon terminals and/or fibers of passage) API occurs in a region of the fiber relatively close to the electrode (Figure 5.13A). Stimulus waveforms can be used that exploit the nonlinear conductance properties of the neural elements of local cells and axonal elements. Due to geometrical and biophysical factors, API in local cells takes place in neural elements that are hyperpolarized by cathodic stimuli and depolarized by anodic stimuli.[32] As a result, monophasic anodic stimuli are more effective in activating local cells than monophasic cathodic stimuli (Figure 5.13B); however, fibers of passage are stimulated more effectively with cathodic stimuli than anodic stimuli (Figure 5.13B). Yet, chronic stimulation requires the use of biphasic stimulus waveforms.[39] Therefore, asymmetrical charge-balanced biphasic stimulus waveforms have been developed to selectively activate either local cells or axonal elements.[33]

Stimulus waveforms capable of selectively activating either local cells or axonal elements use a long-duration, low-amplitude, pre-pulse phase followed by a short-duration, high-amplitude, stimulation phase. The long-duration, pre-pulse phase of the stimulus is designed to create a subthreshold depolarizing pre-pulse in the neural element, where excitation will take place in the non-target neurons, and a hyperpolarizing pre-pulse in the neural element, where excitation will take place in the target neurons.[33] The effect of this subthreshold polarization is to decrease the excitability of the non-target population and increase the excitability of the target population via alterations in the degree of sodium channel inactivation.[20] Therefore, when the stimulation phase of the waveform is applied (opposite polarity of the pre-pulse), the target neuronal population will be activated with enhanced selectively compared with monophasic stimuli (Figure 5.15). Further, charge balancing is achieved as required to reduce the probability of tissue damage and electrode corrosion.

Figure 5.15 shows the effects of changing the stimulus waveform on the activation of populations of local cells and fibers of passage randomly distributed around the stimulating electrode.[33] As seen in Figure 5.13B, monophasic anodic or cathodic stimuli result in selective activation of local cells or fibers of passage, respectively. However, when symmetrical charge-balanced biphasic stimuli are used, selectivity is diminished. Asymmetrical, charge-balanced, biphasic, cathodic-phase first stimulus waveforms result in selective activation of local cells, and asymmetrical, charge-balanced, biphasic, anodic-phase first stimulus waveforms result in selective activation of fibers of passage. However, even with the appropriate stimulus waveform it should always be noted that selective activation of either local cells or axonal elements is affected by the stimulation-induced, trans-synaptic influences on the local cells.

5.5   Conclusion

Electrical stimulation of the CNS is a powerful tool to study neuronal connectivity and physiology, as well as a promising technique to restore function to persons with neurological disorders. However, the complexity of central volume conductors (brain and spinal cord) and the neuronal elements (cells, axons, dendrites) therein has limited our understanding of CNS stimulation. To understand the results of studies employing CNS stimulation requires knowledge of which neuronal elements are affected by stimulation, and optimizing neural prosthetic interventions requires techniques that enable selective stimulation of targeted neuronal populations. Computational modeling is a powerful tool to address both the question of what neuronal elements are activated by CNS stimulation and to design optimal interventions.

Acknowledgments

Preparation of this chapter was supported by NIH Grant R01-NS-40894.

References

1. Altman, K.W. and Plonsey, R., Development of a model for point source electrical fibre bundle stimulation, *Med. Biol. Eng. Comput.*, 26, 466–475, 1988.
2. Anderson, D.J., Najafi, K., Tanghe, S.T., Evans, D.A., Levy, K.L., Hetke, J.F., Xue, X., Zappia, J.J., and Wise, K.D., Batch fabricated thin-film electrodes for stimulation of the central auditory system, *IEEE Trans. Biomed. Eng.*, 36, 693–698, 1989.
3. Bak, M., Girvin, J.P., Hambrecht, F.T., Kufta, C.V., Loeb, G.E., and Schmidt, E.M., Visual sensations produced by intracortical microstimulation of the human occipital cortex, *Med. Biol. Eng. Comput.*, 28(3), 257–259, 1990.
4. Baldissera, F., Lundberg, A., and Udo, M., Stimulation of pre- and postsynaptic elements in the red nucleus, *Exp. Brain Res.*, 15, 151–167, 1972.
5. Bannister, N.J. and Larkman, A.U., Dendritic morphology of CA1 pyramidal neurones from the rat hippocampus. II. Spine distributions, *J. Comp. Neurol.*, 360, 161–171, 1995.
6. Bilkey, D.K. and Schwartzkroin, P.A., Variation in electrophysiology and morphology of hippocampal CA3 pyramidal cells, *Brain Res.*, 514, 77–83, 1990.
7. Burke, D., Movement programs in the spinal cord [commentary], *Behav. Brain Sci.*, 15, 722, 1992.
8. Calabrese, R., Hill, A., and VanHooser, S., Realistic modeling of small neuronal circuits, in *Computational Neuroscience*, DeSchutter, E., Ed., CRC Press, Boca Raton, FL, 2001, pp. 259–288.
9. Campbell, P.K., Jones, K.E., Huber, R.J., Horch, K.W., and Norman, R.A., A silicone-based three-dimensional neural interface: manufacturing processes for an intracortical electrode array, *IEEE Trans. Biomed. Eng.*, 38, 758–767, 1991.
10. Coburn, B., Electrical stimulation of the spinal cord, two-dimensional finite element analysis with particular reference to epidural electrodes, *Med. Biol. Eng. Comput.*, 18, 573–584, 1980.
11. Coburn, B., A theoretical study of epidural electrical stimulation of the spinal cord. II. Effects on long myelinated fibers, *IEEE Trans. Biomed. Eng.*, 32, 978–986, 1985.
12. Coburn, B. and Sin, W.K., A theoretical study of epidural electrical stimulation of the spinal cord. I. Finite element analysis of stimulus fields, *IEEE Trans. Biomed. Eng.*, 32, 971–977, 1985.
13. Destexhe, A., Mainen, Z., and Sejnowski, T., An efficient method for computing synaptic conductances based on a kinetic model of receptor binding, *Neural Comput.*, 6, 14–18, 1994.

14. Finley, C.C., Wilson, B.S., and White, M.W. Models of neural responsiveness to electrical stimulation, in *Cochlear Implants: Models of the Electrically Stimulated Ear*, Miller, J.M. and Spelman, F.A., Eds., Springer-Verlag, New York, 1990, pp. 55–96.

15. Frijns, J.H.M., de Snoo, S.L., ten Kate, J.H., Spatial selectivity in a rotationally symmetric model of the electrically stimulated cochlea, *Hearing Res.*, 95, 33–48, 1996.

16. Frijns, J.H.M., de Snoo, S.L., and Schoonhoven, R., Potential distributions and neural excitation patterns in a rotationally symmetric model of the electrically stimulated cochlea, *Hearing Res.*, 87, 170–186, 1995.

17. Giszter, S.F., Grill, W.M., Lemay, M.A., Mushahwar, V., and Prochazka, A., Intraspinal microstimulation, techniques, perspectives and prospects for FES, in *Neural Prostheses for Restoration of Sensory and Motor Function*, Moxon, K.A. and Chapin, J.K., Eds., CRC Press, Boca Raton, FL, 2001, pp. 101–138.

18. Greenberg, R.J., Velte, T.J., Humayun, M.S., Scarlatis, G.N., and de Juan E., A computational model of electrical stimulation of the retinal ganglion cell, *IEEE Trans. Biomed. Eng.*, 46, 505–514, 1999.

19. Grill, W.M. and McIntyre, C.C., Extracellular excitation of central neurons, implications for the mechanisms of deep brain stimulation, *Thalamus Related Syst.*, 1, 269–277, 2001.

20. Grill, W.M. and Mortimer, J.T., Stimulus waveforms for selective neural stimulation, *IEEE Eng. Med. Biol.*, 14, 375–385, 1995.

21. Grill, W.M., Modeling the effects of electric fields on nerve fibers influence of tissue electrical properties, *IEEE Trans. Biomed. Eng.*, 46, 918–928, 1999.

22. Grill, W.M., Electrical activation of spinal neural circuits, application to motor-system neural prostheses, *Neuromodulation*, 3, 89–98, 2000.

23. Gustafsson, B. and Jankowska, E., Direct and indirect activation of nerve cells by electrical pulses applied extracellularly, *J. Physiol.*, 258, 33–61, 1976.

24. Hodgkin, A.L. and Katz, B., The effect of sodium ions on the electrical activity of the giant axon of the squid, *J. Physiol. (London)*, 108, 37–77, 1949.

25. Holsheimer, J., Nuttin, B., King, G.W., Wesselink, W.A., Gybels, J.M., and de Sutter, P., Clinical evaluation of paresthesia steering with a new system for spinal cord stimulation, *Neurosurgery*, 42, 541–547, 1998.

26. Jankowska, E., Padel, Y., and Tanaka, R., The mode of activation of pyramidal tract cells by intracortical stimuli, *J. Physiol.*, 249, 617–636, 1975.

27. Kelvin, W.T., On the theory of the electric telegraph, *Proc. Roy. Soc. London*, 7, 382–399, 1855.

28. Lee, D. and Grill, W., Polarization of a spherical cell in a non-uniform electric field: transient response and comparison with polarization in a uniform field, in press.

29. Lee, D. and Grill, W., Polarization of a spherical cell in a nonuniform electric field: steady state analysis, *Ann. Biomed. Eng.*, 29(suppl.), S-128, 2001.

30. McCreery, D.B., Shannon, R.V., Moore, J.K., and Chatterjee, M., Accessing the tonotopic organization of the ventral cochlear nucleus by intranuclear microstimulation, *IEEE Trans. Rehabil. Eng.*, 6, 391–9, 1998.

31. McDonagh, J.C., Gorman, R.B., Gilliam, E.E., Hornby, T.G., Reinking, R.M., and Stuart, D.G., Electrophysiological and morphological properties of neurons in the ventral horn of the turtle spinal cord, *J. Physiol. Paris*, 93, 3–16, 1999.

32. McIntyre, C.C. and Grill, W.M., Excitation of central nervous system neurons by nonuniform electric fields, *Biophys. J.*, 76, 878–888, 1999.

33. McIntyre, C.C. and Grill, W.M., Selective microstimulation of central nervous system neurons, *Ann. Biomed. Eng.*, 28, 219–233, 2000.

34. McIntyre, C.C. and Grill, W.M., Extracellular stimulation of central neurons: influence of stimulus waveform and frequency on neuronal output, *J. Neurophysiol.*, 88, 1592–1604, 2002.

35. Morgan, C.W. and Ohara, P.T., Quantitative analysis of the dendrites of sacral preganglionic neurons in the cat, *J. Comp. Neurol.*, 437, 56–69, 2001.

36. Nowak, L.G. and Bullier, J., Axons, but not cell bodies, are activated by electrical stimulation in cortical gray matter. I. Evidence from chronaxie measurements, *Exp. Brain Res.*, 118, 477–488, 1998.

37. Nowak, L.G. and Bullier, J., Axons, but not cell bodies, are activated by electrical stimulation in cortical gray matter. II. Evidence from selective inactivation of cell bodies and axon initial segments, *Exp. Brain Res.*, 118, 489–500, 1998.

38. Ohara, P.T. and Havton, L.A., Dendritic architecture of rat somatosensory thalamocortical projection neurons, *J. Comp. Neurol.*, 341, 159–171, 1994.

39. Pudenz, R.H., Bullara, L.A., Jacques, S., and Hambrecht, F.T., Electrical stimulation of the brain. III. The neural damage model, *Surg. Neurol.*, 4, 389–400, 1975.

40. Rall, W. and Rinzel, J., Branch input resistance and steady attenuation for input to one branch of a dendritic neuron model, *Biophys. J.*, 13, 648–687, 1973.

41. Rall, W., Cable theory for neurons, in *Handbook of Physiology*, Kandel, E.R., Ed., American Physiological Society, Bethesda, MD, 1977, pp. 39–97.

42. Rall, W., Distinguishing theoretical synaptic potentials computed for different soma-dendritic distributions of synaptic input, *J. Neurophysiol.*, 30, 1138–1168, 1967.

43. Rall, W., Time constants and electrotonic length of membrane cylinders and neurons, *Biophys. J.*, 9, 1483–1508, 1969.

44. Raman, I.M. and Bean, B.P., Ionic currents underlying spontaneous action potentials in isolated cerebellar Purkinje neurons, *J. Neurosci.*, 19, 1663–1674, 1999.

45. Ranck, J.B., Which elements are excited in electrical stimulation of mammalian central nervous system: a review, *Brain Res.*, 98, 417–440, 1975.

46. Rattay, F., Analysis of the electrical excitation of CNS neurons, *IEEE Trans. Biomed. Eng.*, 45, 766–772, 1998.

47. Rattay, F., The basic mechanism for the electrical stimulation of the nervous system, *Neuroscience*, 89, 335–346, 1999.

48. Rattay, F., Analysis of models for extracellular fiber stimulation, *IEEE Trans. Biomed. Eng.*, 36, 676–682, 1989.

50. Rinzel, J. and Rall, W., Transient response in a dendritic neuron model for current injected at one branch, *Biophys. J.*, 14, 759–790, 1974.

51. Roberts, W.J. and Smith, D.O., Analysis of threshold currents during microstimulation of fibers in the spinal cord, *Acta Physiol. Scand.*, 89, 384–394, 1973.

52. Robertson, J.D., Unit membranes, in *Cellular Membranes in Development*, Locke, M., Ed., Academic Press, New York, 1964.

53. Rose, P.K., Keirstead, S.A., and Vanner, S.J., A quantitative analysis of the geometry of cat motoneurons innervating neck and shoulder muscles, *J. Comp. Neurol.*, 239, 89–107, 1985.

54. Roth, A. and Hausser, M., Compartmental models of rat cerebellar Purkinje cells based on simultaneous somatic and dendritic patch-clamp recordings, *J. Physiol.*, 535, 445–472, 2001.

55. Schmidt, E.M., Bak, M.J., Hambrecht, F.T., Kufta, C.V., O'Rourke, D.K., and Vallabhanath, P., Feasibility of a visual prosthesis for the blind based on intracortical microstimulation of the visual cortex, *Brain*, 119(pt. 2), 507–522, 1996.

56. Sheasby, B.W. and Fohlmeister, J.F., Impulse encoding across the dendritic morphologies of retinal ganglion cells, *J. Neurophysiol.*, 81, 1685–1698, 1999.

57. Sin, W.K. and Coburn, B., Electrical stimulation of the spinal cord: a further analysis relating to anatomical factors and tissue properties, *Med. Biol. Eng. Comput.*, 21, 264–269, 1983.

58. Spielmann, J.M., Laouris, Y., Nordstrom, M.A., Robinson, G.A., Reinking, R.M., and Stuart, D.G., Adaptation of cat motoneurons to sustained and intermittent extracellular activation, *J. Physiol.*, 464, 75–120, 1993.

59. Struijk, J.J., Holsheimer, J., van Veen, B.K., and Boom, H.B.K., Epidural spinal cord stimulation: calculation of field potentials with special reference to dorsal column nerve fibers, *IEEE Trans. Biomed. Eng.*, 38, 104–110, 1991.

60. Struijk, J.J., Holsheimer, J., Spincemaille, G.H.G.H., Gielen, F.L., and Hoekema, R., Theoretical performance and clinical evaluation of transverse tripolar spinal cord stimulation, *IEEE Trans. Rehab. Eng.*, 6, 277–285, 1998.

61. Turner, J.P., Anderson, C.M., Williams, S.R., and Crunelli, V., Morphology and membrane properties of neurones in the cat ventrobasal thalamus *in vitro*, *J. Physiol.*, 505(pt. 3), 707–726, 1997.

62. Warman, E.N., Grill, W.M., and Durand, D., Modeling the effects of electric fields on nerve fibers: determination of excitation thresholds, *IEEE Trans. Biomed. Eng.*, 39, 1244–1254, 1992.

chapter six

Semiconductor-based implantable microsystems[1]

Wentai Liu, Praveen Singh, Chris DeMarco, Rizwan Bashirullah, Mark S. Humayun, and James D. Weiland

Contents

[1] This research is supported by grants from NSF, NIH, and Department of Energy.

6.1   Introduction

The field of integrating a dysfunctional biological subsystem with microelectronic systems is an emerging and fast growing one. The main functionality of the implantable microelectronics are (1) biochemical sensor for sensing of vital life signs, oxygen, pH, glucose, temperature, and toxins by miniaturized chemical and bio-sensors; (2) telemetry, to provide a local or remote link for data/audio/visual signals; and (3) actuation, to provide life saving, to restore lost function, and augment normal function. Implantable and/or wearable microelectronics can be used to replace lost function or to monitor physiological conditions if the interface with living tissue can be properly achieved. Some success has been obtained, including the cochlear prosthesis, implantable stimulators for the central nervous system to alleviate pain and reduce the unwanted tremor associated with diseases such as Parkinson's, the vagus nerve implant for reducing the seizure activity of epilepsy, a functional neuromuscular implant to restore motor mobility for paraplegia, and the visual prosthesis for restoring eyesight. Furthermore implantable or wearable microelectronics could potentially augment the existing human senses of hearing, sight, touch, smell, and taste beyond the current abilities of the human body. This ability would be useful for individuals in hostile environments, such as soldiers and firefighters, or in dedicated situations, such as doctors performing microsurgery. In particular, we believe very strongly that just as novel drug and gene therapies have future roles to play in curing human disease so does the field of implantable or wearable microelectronics.

Today, the scale of micron technology is compatible with the cell dimension, while the scale of deep-submicron and nanotechnology is compatible with molecular dimensions. Clearly the advances of micro- and nano-fabrication would greatly benefit the development of the core implantable technology. Thus, with the development of ultra-low-power CMOS (complementary metal-oxide semiconductor) microelectronics, advances in micro-fabrication and microelectromechanical systems (MEMS) technology, these implantable technologies will eventually enable the realization of an integrated system for stimulation and data collection of electrical and/or biochemical activity over long time scales in freely moving subjects. The keys toward efficient and successful implantable microelectronics are miniaturization, extremely low power dissipation, biocompatibility, and implant durability. Implantable devices with smartness and mobility require an integration of information technology and wireless technology. Thus, implantable microelectronics provide a challenge in material, device fabrication, and design technique. In order to capture this apparent wealth of opportunity, a multidisciplinary team is necessary.

The core implantable technologies include biocompatible materials, interface of device and living tissue, miniaturized passive and active devices, hermetic sealing/packaging, energizing mechanism, heat and EM propagation modeling in living systems, wireless link of distributed sensors and actuators, and signal processing. In this chapter, the issues and perspectives of the implantable technology have been exemplified by the presentation of the intraocular prosthesis project. The intraocular prosthesis project was conceived a decade ago and requires a multidisciplinary research effort toward an implantable prosthetic device. The rehabilitative device is designed to replace the functionality of defective photoreceptors in patients suffering from retinitis pigmentosa (RP) and age-related macular degeneration (AMD). In the next section, the system building blocks for prostheses are presented.

6.2 Building blocks of an implant microsystem

Implant microsystems usually share a common framework: an external signal processing unit for biological sensory information (sound, image, etc), a bidirectional telemetry unit, an internal signal processing unit, a stimulus generator/driver, and an electrode array for interfacing to tissue or nerves. Figure 6.1 shows a typical implanted microsystem. Various applications

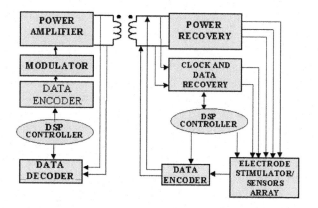

Figure 6.1 Building blocks of implantable microsystems.

require different implementation substrates, such as ceramics, polyimide, thin-film, and silicon; however, the future implanted microsystems would be fully integrated, motivated by the factors of power dissipation, size, and packaging. Modern silicon technology does offer the advantage of capability of integration, thus a silicon implant has the potential of becoming the dominant one in future implant applications. Because the implant is the mechanism for interfacing with living tissues, care must be exercised to prevent toxicological damage and infection. Thus, biocompatibility is of great concern and accordingly a hermetic package is a necessity in the assembly of an implanted microsystem.

6.2.1 Bidirectional telemetry unit

Developing implantable prostheses are usually partitioned to include some implanted components and some exterior components, at least until such time as the prostheses have well matured. It seems ideal that a prosthesis consisting of neuro-stimulators or neuro-recorders would have all components fully implanted such that communication with any exterior devices is minimized or even eliminated. Accordingly, for biocompatibility, a more stringent power dissipation budget would be necessary to support a fully implantable prosthesis. However, a number of advantages can be derived from partitioning a prosthesis into parts internal and external to the body, beyond merely simplifying the implantable unit. Primary benefits of exterior electronics, where possible, include the decreased risk of adverse reaction to implanted materials, lower internal heat dissipation, and ease of refinement and upgrade of signal-processing algorithms and functionality in the exterior components. This is the case for the NCSU/JHU/USC epiretinal prosthesis.[1]

Previously, invasive tethering to implanted electronics using percutaneous connectors has been used to facilitate the studies of long-term biocompatibility and the investigation of the necessary signal preprocessing needed to enhance the performance of implants;[2] however, in all cases of chronic prostheses, but especially with the ocular prostheses (owing to the eye's fragility), percutaneous connectors are unsuitable. Therefore, this section summarizes the common types of noninvasive connections to bio-implants and the current progress of biotelemetry as it relates to the power and communication needs of prostheses.

The use of transcutaneous, or wireless, connections is justified due to the many complications that are typical for percutaneous schemes. Percutaneous connectors (physically wired links between implanted and exterior prosthesis components) raise the risk of infections due to a perpetual breach of the body, through which wires must pass. Depending on mechanical anchoring, such as to bone, percutaneous connectors may restrict movement of a prosthesis in the tissues in which it is implanted and with which it interfaces electrically. These tissues may be free to move with respect to a percutaneous connector with obvious complications. In the case of ocular prostheses, rapid eye movement can break any penetrating wires passing through the scleral wall, for example, that might otherwise be used to connect an intraocular prosthesis to external electronics. Furthermore, dislodging of the implant due to external tethering on the wires could occur. In the case of most all prostheses, implantable batteries are undesirable because of the associated replacement surgery (except where charging can be initiated from outside of the body, as from coils, for example). Even renewable cells have limited recharge cycles. Furthermore, as a consequence of the biological environment, implanted batteries must be enclosed in the implant encapsulant, which complicates replacement.

A wireless link is a viable approach to support the required data bandwidth and simultaneously provides adequate power to an implanted prosthesis. For example, devices functioning as stimulators also require configuration and stimulus data to function. Furthermore, in the case that the implant also performs neuro-recording and/or monitors biofunctions or device status, or performs self-diagnostics, then there is also the need to transmit data to exterior devices. This functionality is termed *back-telemetry*. Therefore, in the general case, the success of prostheses for the long term requires a telemetry link, which can provide adequate power to all implanted electronics and support a means to communicate information bidirectionally. Three types of telemetry are considered to have the potential of meeting the goals of simultaneous power and data delivery to implants. These are optical, magnetic (inductive), or electromagnetic (radiofrequency, RF). Each of these is expounded upon in the following sections.

6.2.2   Optical telemetry

Optical telemetry appears primarily in research groups that are addressing ocular prostheses, as the cornea and crystalline lens are usually transparent. This form of telemetry uses a source of high-energy light, such as a laser, to excite the photoelectronics that form part of the implant. The MIT/Harvard epi-retinal prosthesis group[3,4] and the epi-retinal prosthesis group at Fraunhafer, Germany,[5] have been researching optical telemetry.[6]

The architecture of the first-generation MIT/Harvard prosthesis employed a photodiode array and stimulator attached to either side of a thin-film polyimide electrode array and anchored near the vitreous base. A supply level of 300 µA at 7 V was derivable with exposure to a 30-mW, 820-nm laser. A second-generation architecture has been proposed with improved electronics mounted in the crystalline lens position on a polymer annulus. Stimulation data were transmitted via intensity modulation of the external laser. No back-telemetry facilities appear to have been provided.

6.2.3   Inductive/magnetic telemetry

Inductive telemetry has been the standard means of wireless connection to implanted devices for years.[7,8] It is based on the mutual magnetic coupling of two proximal, coaxial, and coplanar coils. The secondary coil in this arrangement is implanted along with the stimulator/neuro-recorder, which it services, while the primary coil remains exterior. A low-frequency "carrier" (i.e., usually 1 to 10 MHz) is driven onto the primary coil, which can then be coupled onto the secondary coil for the transfer of power. Furthermore, this power carrier can be modulated to transmit a datastream to the implant. This arrangement forms the basis for inductive coupling, while the precise details of how the primary coil is excited and modulated and how DC power and data are recovered on the secondary (implanted) side are varied and continue as topics of research.

Two major disadvantages, however, are noted regarding magnetic coupling and are simply characteristics of the coils. These impact the development of the primary-coil excitation and modulation circuits on the exterior (non-implanted) side:

- *Radiation compliance* — It is obviously desirable to radiate energy only toward the secondary coil. In reality, the primary coil radiates energy widely in many directions, particularly with no ferrous core to concentrate the magnetic flux, as in a transformer. This omnidirectional radiation pattern imposes unnecessary drain of the system batteries. It also potentially imparts EMI to other nearby electrical devices, and care must be taken to meet the regulatory guidelines. Inevitably, various strengths of electric and magnetic fields are emitted from the primary coil. Thus, deposition of electromagnetic and thermal energy in the proximity of the implant is a serious safety issue in these kinds of telemetry systems. The design of the telemetry link must comply with the IEEE safety standard with respect to the human exposure to the radiofrequency electromagnetic fields.[9]
- *Coupling problem* — Due to the implantation, the coils must be disjoint at a distance and subsequently cannot be coupled, or linked, by a common ferrous material but are rather "air-cored." Therefore, the mutual inductance, or coupling coefficient, could be low. High magnetic field strength in the primary coil is required in order to induce sufficient energy in the secondary coil to power the implant. Therefore, three criteria are used to optimize the inductive link performance for use in the context of bio-implants: *power-transfer efficiency, driving efficiency,* and *carrier-modulation bandwidth.* Each of these issues is covered separately with attention given to how they have been addressed in the research literature.

6.2.3.1 Power-transfer efficiency

Several studies have been conducted to measure the mutual inductance between two air-coupled coils. Coupling strength is dependent on coil loading, excitation frequency, number of turns, coaxial alignment, coil separation, coil geometry, and angular alignment. Most of these factors are subject to variation in prostheses. Typical values for coil coupling are between 0.01 and 0.1. Power transfer efficiency can be improved with an increase in the coil coupling coefficient.[10] Power efficiency is achieved most easily when the primary and secondary coils are series or parallel resonated with a capacitance of appropriate value, $C = 1/[(2\pi f)^2 L]$, where L is the self-inductance of the coil.

6.2.3.2 Driving efficiency

Modulation techniques for data delivery are usually performed using amplifier circuits chosen for their high driving efficiency. This efficiency reflects the additional power consumption associated with the driving amplifier for energizing the primary coil. In general, a low coupling coefficient can be compensated by driving the primary coil at high magnetic field strength. In this case, it is essential for practical battery life that the driving amplifier of the coil be as efficient as possible.

Switch-mode DC–DC converters such as the buck, boost, Cuk, and Sepic achieve a high efficiency by retaining power which is not used by the load during each cycle. Rather than dissipating this unused energy as heat, as in linear or shunt regulators, these switch-mode devices trade the energy back and forth between electric and magnetic fields by resonating the coils. For inductive links, this also achieves a high magnetic field strength in the primary coil, while retaining the efficiency.

The circuit topology of choice for driving the resonated primary coil is the class-E type of amplifier, first characterized by Sokal.[11,12] In these original contexts, the amplifier is shown to drive a resistance load as illustrated in Figure 6.2a. This amplifier differs from class-A through class-D types of amplifiers in that the active device operates as a switch rather than as a current source.

The class-F amplifier shown in Figure 6.2b also boasts high-efficiency operation using the concept of harmonic termination to shunt frequency components, which would cause high-power dissipation in the switch.[14] However, it is generally more difficult to design than the class-E amplifier and requires a correctly sized transmission line, as in Figure 6.2b, or else a theoretically infinite number of resonated inductive elements to achieve the high-efficiency condition.

Because of the low coupling coefficient, it is expected that minor shifts in the load impedance on the secondary side will be negligibly reflected onto the primary and thus will not affect class-E performance significantly. However, it has been pointed out that the close proximity of the primary coil to metallic objects can alter self-inductance and equivalent resistance or deformations in coil geometry may heavily disrupt proper class-E operation.[15] Therefore, autocompensation of the class-E frequency is proposed to affect changes in component values (particularly shifts in L_2) using a feedback controller to maintain high operating performance. The technique for this involves tapping the magnetic field of L_2 with a sensing coil in order to monitor the zero-crossing of the inductor current.

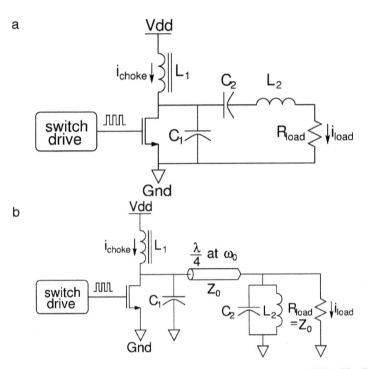

Figure 6.2 Amplifier topology: (a) class E, (b) class F. (From Lee, T.H., *The Design of CMOS Radio Frequency Integrated Circuits*, University Press, Cambridge, U.K., 1998. With permission.)

6.2.3.3 Data-modulation bandwidth

Many micro-stimulator designs documented recently in the literature have stimulus instructions encoded into packets for transmission using digital communication protocols.[16–20] It is advantageous for advanced signal-processing hardware to remain outside of the body with only the stimulus unit and associated hardware implanted. Therefore, once a link design has been established that can support the power demands of the implant, capabilities for transmitting the datastream must be added. For a low data rate, this is commonly accomplished by modulating the power carrier using a conventional scheme, such as amplitude, frequency, or phase-shift keying. For power carrier modulation, amplitude shift keying (ASK) appears as the dominant modulation scheme for inductive links, as discussed below.

Ziaie et al.[21] noted that the voltage developed across inductor L_2 used as the primary inductor is linearly related to the class-E amplifier supply voltage, V_{dd}. This motivates one to perform ASK power-carrier modulation by switching the supply voltage between two levels using the circuit configuration of Figure 6.3. Capacitor C_r is used to filter switching noise from the amplifier supply. A data bandwidth of 80 kb/sec using this technique is reported.

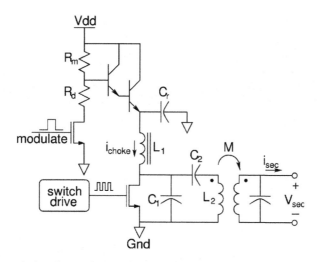

Figure 6.3 Supply level switching in a class-E amplifier.

Troyk and Schwan[15] propose a method of amplitude modulating the power carrier radiated from inductor L_2 in the class-E amplifier. By introducing a slight duty-cycle shift into the feedback controller, which produces the gate drive for the switching field effect transistor (FET), the class-E amplifier is moved slightly off of the optimum operating frequency. As reported, 0.1% changes in oscillation frequency can produce upwards of 10% decrease in primary coil current. A disadvantage to power carrier modulation in resonated, high-Q circuits such as the class-E amplifier described by Troyk and Schwan[15] is that the forced deviations in operating frequency do not track quickly. It is evident from their research that 5 to 10 cycles, at best (for 760-kHz operation), are required for the carrier to settle into a new steady-state amplitude. This imposes constraints on achievable data bandwidth. Another limitation of the ASK modulation described by Troyk and Schwan[15] is that in pushing the amplifier away from the class-E frequency, the amplifier is forced away from optimum power, or driving, efficiency.

This approach is further improved using a technique called *suspended carrier modulation,* which provides a means of performing on/off shift keying (OOSK) with greater achievable bandwidth than with the ASK modulation technique described previously. One notable disadvantage of this scheme is that the implant is not externally powered during carrier suspension and must draw power from its internal supply capacitance because the primary coil carries no current during these times. Mueller and Gyurcsik[22] have reported another scheme for performing enhanced bandwidth ASK modulation of power carriers in high-Q amplifiers. The scheme is called *half-cycle amplitude shift keying* (HCASK) and is claimed to have a potential for the transmission of 2 b/cycle of the power carrier.

6.2.3.4 *Secondary side power recovery*

On the receiver side, inside of the body, DC power is usually recovered in conventional ways using a rectifying diode. Secondary coils are either series or parallel resonated to maximize power-transfer efficiency from primary to secondary. Power recovery schemes are designed for single or dual supply lines. Charge-balanced stimulation is used to prevent chronic damage. Output drivers of the H-bridge type, as illustrated in Figure 6.4a, only work well

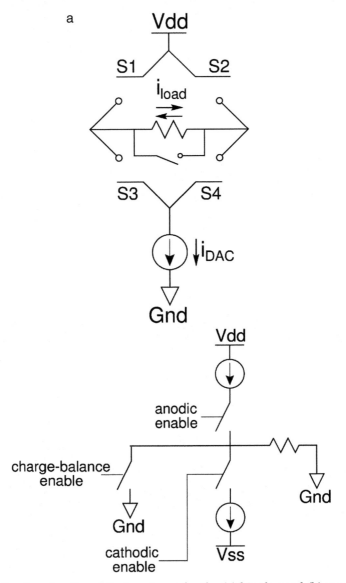

Figure 6.4 Stimulus circuit architecture types for the (a) bipolar and (b) monopolar electrode configurations.

when using bipolar electrodes. Although this can be accomplished with a single supply rail, V_{dd} relative to *Gnd*, two electrode contacts (output and independent return) are required per stimulus output. Monopolar electrodes (shared return) on the other hand, are better suited for use with a positive V_{dd} rail and negative V_{ss} rail relative to *Gnd* to actively source and sink stimulus currents to the load, as shown in Figure 6.4b.

A concern with inductive link designs is that changes in coil coupling and changing load conditions may induce variable and sometime excessive voltage at the secondary side. There are three potential solutions to this problem. The first and less often attempted approach is to monitor the supply levels and to detune the resonance of the secondary coil to lower the coupling coefficient and power-transfer efficiency. A second approach is to instruct the exterior electronics on the primary side via back-telemetry to lower the power transfer. A third solution is an on-chip supply regulator.

For example, in Ziaie et al.,[21] a parallel resonated secondary coil drives a half-wave rectifier. This is used with a reversed Zener diode regulator to recover V_{dd} and *Gnd*. This may be an undesirable arrangement for some tissue types as the Zener is expected to generate heat as it shunts unwanted power. Gudnason et al.,[24] Von Arx and Najafi,[25] and Nardin and Najafi[26] use bridge rectifiers with Zener diode regulators. Jones and Normann,[17] Tanghe and Wise,[18] and Kim and Wise[19] have used dual-voltage supplies with push/pull stimulus-circuits for monopolar electrodes. Ziaie[21] has improved upon the use of Zener diodes as regulators, using them to bias a NPN BJT in series with the supply rail.

Conventional P–N diodes used as rectifiers can account for appreciable losses and heating because of the inherent 0.7-V drop during conduction. Schottky diodes can be used to reduce the power loss as they offer a lower voltage drop during conduction (typically 0.5 V). The disadvantage with Schottky diodes is that they are not as easily integrated with stimulus circuits. A further rectification strategy used with implantable devices is the synchronous rectifier constructed from a metal-oxide semiconductor field effect transistor (MOSFET).[27] With this approach, FETs emulate the rectifying function of diodes by switching on and off accordingly with gate control automatically derived from the received carrier. With voltage drops as low as 0.1 V, the synchronous rectifiers can offer heat reduction by as much as 50%.

6.2.3.5 *Primary side data encoding and secondary side data recovery*

With a realizable carrier modulation scheme in place that can coexist with the power transfer to the implant, the issue remains of encoding data for transmission. Usually implanted micro-stimulator integrated circuits (ICs) do not possess an onboard timing reference; therefore, the asynchronous transmission of datastreams is not feasible as in established schemes such as RS232, RS488, USB, etc. The datastreams are typically synchronized to an external clock, which must also be supplied to the implant.

Quadrature modulation schemes afford the possibility of modulating two separate carriers with the clock and data signals, respectively. However, this is unattractive because the power carrier may be the only carrier available. Therefore, an alternative approach is to encode the clock and data into a new digital signal, which is subsequently used to modulate the carrier. A well-known scheme for performing this is the Manchester encoding algorithm, which is trivial to encode (on the primary side) but is more difficult to decode on the secondary side, where the silicon area for circuit implementation is more limited.

Another possible approach is to encode the clock and data into a common signal through the pulse-width modulation of a digital square signal according to the data. This technique has been described by Najafi,[28] where the carrier is amplitude modulated with short and long pulses to encode logical "0" and "1" data. The pulse-width modulation approach is also used in the Retina-3 micro-stimulator of the NCSU/JHU/USC epiretinal prosthesis.[1,16] Positive clocking edges are directly reproduced in the encoded signal. Following each positive edge, subsequent pulse widths are defined according to the datastream content. Logic "0" data are represented with a duty cycle of 50% and logic "1" data are represented with duty cycles of 25 and 75%. This encoding ensures a zero DC level in the signal. The NCSU approach to recovering the clock and data waveforms on the secondary side is a two-stage circuit, consisting first of an ASK demodulator followed by a delay-locked loop (DLL) in the second stage (refer to Section 6.3.6.5).

6.2.3.6 Radiofrequency telemetry

Radiofrequency telemetry is similar to inductive coupling except that a traveling electromagnetic wave is employed with antennas rather that a stationary magnetic field as in inductive coupling. Work is apparently limited due to (1) antenna size with respect to carrier frequency, and (2) dielectric loss of human tissues with respect to carrier frequency. Because the human body is primarily of water content, the permitivities and conductivities of the tissues will be similar. Table 6.1 provides an indication of how the dielectric

Table 6.1 Dielectric Properties of Selected Eye Tissue: Aqueous/Vitreous Humor

Frequency (Hz)	Conductivity (S/m)	Relative Permittivity	Penetration Depth (m)
1E6	1.501	84	0.4115
1E7	1.502	70.01	0.1316
1E8	1.504	69.08	0.04656
1E9	1.667	68.88	0.02702
1E10	15.13	57.87	0.002739
1E11	77.39	7.001	0.0002305

Note: The properties of the ocular tissues listed here were taken from the *Dielectric Properties of Human Tissues* at http://niremf.iroe.fi.cnr.it/tissprop/, where permittivity and conductivity were calculated online for various frequencies of relevance to biotelemetry.

properties of selected eye tissues vary with frequency. Conductivity is shown to increase with frequency, which means that as the carrier wavelength decreases it will penetrate to increasingly shallower depths.

6.2.3.7 Back telemetry

Back telemetry refers to the capability of a prosthesis to transmit information from implanted electronics to the exterior of the body. This is commonly implemented in one of three ways. In active telemetry, coil-based inductive links can again be used. If sufficient transparency allows, as in ocular prostheses, optical telemetry can be used. As an alternative to active methods, passive telemetry can be used. Our approach has been to develop back telemetry using the power link itself. This work is summarized below.

6.2.3.7.1 Smart bidirectional telemetry system. Current technologies for the telemetry between units external and internal to the human body are generally based on low-frequency inductive links, because at low frequencies the magnetic field can well penetrate the human body. While suitable for low data rate applications, this approach has the inherent limitation associated with the Q factor that causes a narrow data bandwidth not capable of supporting the stimulation of thousands of electrodes at the image rate of 60 frames/sec or more required in the future retinal prosthesis. The image rate of 60 frames/sec is necessary to obtain a nonflickering image when stimulating remaining retinal neurons of a blind retina.

To accommodate the data required by the increasing number of electrodes while still wirelessly providing system power, the next generation retinal prosthetic telemetry unit currently being developed at NCSU is designed to operate as a multifrequency telemetry link under the ISM-band constraint. The proposed scheme is a hybrid system that separates the data transmission from the power delivery by allocating different frequencies for the power and data carriers. The solution is unique compared to conventional RF transceivers because of severe constraints in area, power dissipation, and limited intraocular space for off-chip components such as a local crystal oscillator. Powering of the system is still through coil coupling, leading therefore to a novel mixed low-/high-frequency system.

The use of a separate data carrier permits power-efficient solutions that actively use back-telemetry as a method to optimize the overall quality of the communication link. This means that, through detectable changes of the internal unit, the primary unit will obtain valuable information on the instantaneous power level necessary to the internal unit and possible malfunctions of stimulating electrodes sensed by impedance variations. In particular, it is important to keep the received power fairly constant despite coupling variations. Coil separation variations and misalignments due to the displacement of the extraocular unit and/or eye movement exert a negative effect on the inductive coupling coefficient. If the power transfer of the inductive link is highly dependent on the relative position of the coils, appreciable losses in power conversion efficiencies result due to voltage regulation within the

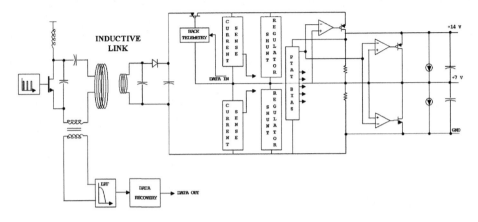

Figure 6.5 Block diagram of the inductive link, which consists of a power transmitter with back-telemetry detection circuitry, as well as secondary unit rectification, telemetry, and regulation analog front-end circuitry.

implant. The loss of energy in the implanted device, dissipated as heat, has a detrimental effect on the retina due to long-term increase of temperature. On the other hand, the power transfer efficiency should not fall below the minimum required to produce a stable voltage supply to ensure proper device operation.

Figure 6.5 shows the inductive link with back-telemetry and voltage-regulation analog front end. The power transmitter for the inductive link uses a high-efficiency class-E driver to drive the transmitter coil. Rectification is achieved via a single-phase rectifier and an off-chip charge storage capacitor. The voltage regulators generate a dual rail supply of ±7 V for the electrode stimulators and a 3-V supply for the digital logic. An initial shunt-type regulator provides current regulation and coarse voltage regulation. A series-type regulator provides the necessary fine voltage regulation required for subsequent stages. The analog front-end also includes the passive back-telemetry unit for reverse data communication. The back-telemeter uses load modulation to transmit binary ASK data on the power carrier. Data detection on the primary side is accomplished by sensing the current in the inductor and low-pass-filtering the analog waveform.

6.2.3.7.2 Back telemetry operation. Communication from the implanted device back to the extraocular unit is accomplished by using the backscatter load modulation technique, also known as *reflectance modulation*. Two techniques are used for reflectance modulation: ohmic modulation and capacitive modulation. Essentially, both modulation techniques are based on the concept of changing the loaded quality factor, Q, of the resonant circuit to produce amplitude modulation on the received carrier. The operation of the back-telemetry unit is shown in Figure 6.6. In order to produce a detectable change in amplitude on the primary side, the back-telemeter modulates

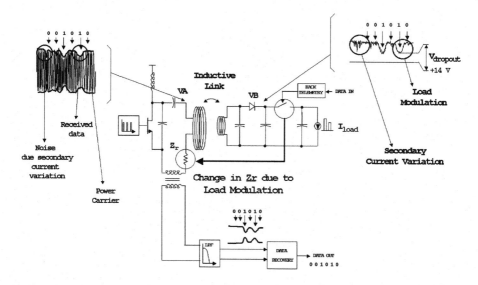

Figure 6.6 Operation of the back-telemetry unit.

the switch to change the load impedance seen by the secondary resonant circuit, thereby changing its quality factor. This change in Q of the secondary resonant tank is sensed on the primary side via the change in the reflectance impedance, Z_R. The change in impedance, in turn, modulates the amplitude of the RF carrier on the primary side, which can be demodulated for data recovery by sensing the current in its resonant circuit.

The use of the reflectance impedance technique for back-telemetry brings about some effects linked with voltage regulation that requires careful attention. First, in general, the change in Q of the implanted device via ohmic or capacitive modulation will disturb the voltage in the storage capacitor used to supply charge to the regulator and the rest of the chip. The voltage variations must be kept lower than the drop-out voltage of the voltage regulator (series type) to ensure proper functionality of the chip, as shown in Figure 6.6. On the other hand, larger modulation indexes are required to facilitate the data detection at the external receiver, especially if the coupling coefficients are relatively low. Larger charge storage capacitors can minimize the voltage variations at the expense of a lower overall quality factor, reducing the overall efficiency of the inductive link. Second, there is an optimal power transfer that minimizes the power dissipated as heat in the voltage regulators. If too much power is delivered to the implanted device, the voltage regulators will "burn" an excessive amount of power. Similarly, if the power delivered is too low, the voltage regulators will operate in the "drop-out" region.

To resolve these problems, we are developing a novel closed-loop system between the power transmitter and the implanted unit via back-telemetry that will be used to guarantee optimal power transfer and system functioning. By sensing the current through the shunt regulator in the implanted unit, it is possible to infer power transfer variations due to the relative displacement of the units. By counteracting the power fluctuations via an increase or decrease of the transmitted power, it is possible to maintain the power delivered to the implanted unit at an optimal level. Thus, an optimal power supply regulation will be established despite relative displacements and misalignments of the coils as well as the circuit malfunction in the intraocular unit. Unlike traditional inductive links, the proposed system will aim to minimize the radiated power while ensuring proper device operation. In addition to regulating power transfer variations due to relative coil displacements, the proposed active telemetry link will monitor the on-chip power consumption. For instance, if the implanted device is operating on standby mode where the power consumption by the internal circuitry is very low, the overall Q of the system will be high. A high Q implies large induced voltages, which result in excessive amount of dissipated power as heat in the voltage regulators. Ensuring minimum radiated power and heat dissipation will be accomplished via the proposed active feedback system. Further improvements will be achieved by efficient design of voltage regulators, power rectification, analog amplifiers, and on-chip band-gap reference cells.

Preliminary simulation results of the back-telemeter and voltage regulation circuitry are shown in Figure 6.7. The reverse telemetry data are used to modulate the Q of the secondary resonant coil of the implanted device, which gives rise to "bursts" in the 135-kHz power carrier that is detectable on the transmitting coil outside the eye. Unlike typical implementations of forward/reverse telemetry links, the reverse telemetry link can be continuously operational because the forward data telemetry is operated at a much higher frequency. The energy bursts on the power carrier are isolated from the 7/14-V power rails via continuous voltage regulation. Notice that the voltage regulator provides stable power supply rails of 7/14 V despite the

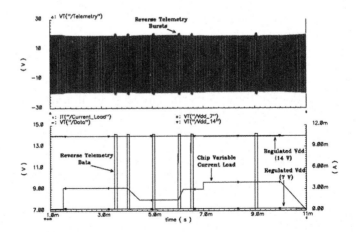

Figure 6.7 Operation of reverse-telemetry and voltage-regulation units.

internal current consumption variations and sharp on/off loading transitions (i.e., standby mode to fully functional). In particular, the telemetry/regulation analog front end is designed to provide high regulation efficiency under worst-case current loading variations and reverse telemetry. Simulation results show that the non-dropout-mode voltage regulation results in a worst-case peak power supply variation of 219 mV, for a regulation efficiency of approximately 1.5%.

Because the back-telemeter periodically updates the external unit of induced voltage levels to obtain improved power regulation efficiency, the on-chip controller internal to the eye uses a synchronization protocol for resource sharing of the back-telemeter. Following the data synchronization of the telemetry link, a 12-bit data sequence is continuously sent to update the external unit; 2 bits are allocated for the power level diagnostics, and the remaining 10 bits are used to update the electrode impedance levels and chip temperature. The back-telemeter controller is synchronized to the on-chip clock used for the digital logic.

6.2.4   Electro–bio interface

The electro–bio interface is a critical issue and may vary significantly from implant to implant. Sensing and stimulation of tissues are two types of interfaces, as described in next two sections.

6.2.4.1   Sensing interface

Sensing the biological signal may require invasive or noninvasive methods. An example of noninvasive sensing is the electroencephalogram (EEG), in which brain activity is recorded using electrodes placed on the head. In invasive interfaces biosensors are implanted through surgery. Currently research also is directed to make interfaces using cultured neuron probes. Neurons are cultivated over the electronic devices and their activities can be recorded. When implanted, the probe neuron will make synaptic connections to the host neuron and will provide very effective signal transfer.

6.2.4.2   Stimulating interface

Electrodes are usually the interface between stimulus circuit and the biological cells. The requirement can vary for implants. For instance, in the retinal prosthesis project, the number of electrodes vary from 10 to 1000. The diameter of each electrode is 50 to 400 µm. Because of limited mechanical support inside the eye, the array structure has to be light weight and it has to be shaped to fit the curvature of the retina and should exert little or no pressure on the retina.

The material for the electrode should be of high conductivity. It should be able to deliver current at high charge densities without corroding or dissolving in the saline water environment. Charge injection capacity, electrochemical stability, and mechanical strength of the electrode are considered. Noble metals such as gold (Au), platinum (Pt), or iridium (Ir) or their compounds (e.g., IrO_2) are generally used because of their chemically inert properties. Safe maximum charge densities are around 100 to 1000 $\mu C/cm^2$. Other factors to be considered are shape of electrode, polarizability, and frequency response. Usually small-area electrodes are preferred for better resolution and uniform charge density.

6.2.5 Analog/digital circuitry and digital signal processing

Sensors are used to convert mechanical, optical, and chemical signals to an electrical signal. The input analog signal is converted to digital data for processing and/or transmission. Based upon the requirements, analog/digital (A/D) circuitry of varying dynamic range, resolution, and speed is used. Data processing is done at a rate from a few kilobytes per second to a few megabytes per second (Table 6.2).[28a] For example, it is required in the frequency analysis of signals in the cochlear implants and the mapping of the input image to have a set of electrodes in the retinal prosthesis. Complex data processing may be required for some systems. For example, in motion prostheses for smooth movements of parts, command signals from muscles as well as from numerous other sensors have to be processed and sent to numerous motion controllers.

A typical system may consist of many analog circuit components such as filters, amplifiers, D/A circuits, A/D circuits, and current mirrors. Implant systems are either open loop or closed loop. Feedback may be used overall or locally and may or may not have biological components. The stimulus generated from the circuit is usually a voltage or current waveform. The required output may vary in shape, amplitude, and frequency. Some of the circuit design constraints may be a minimum input and output impedance, a specific gain, maximum or minimum power and area, low distortion, and limited variation on outputs. In many implants, low power circuit is a prime requirement for implanted chips. This is because of limited energy availability or maximum power dissipation capability in the human body. Failure analysis is a must to analyze the chances of a failure. An example of failure may be shorting of the output to the power supply. Safeguards are designed to prevent any damage to the cells in case of a failure.

Table 6.2 Data Channels for Different Implants

Implantable Device	RF–Channel Utilization and Data Bit Rate
Heart pacemakers	Only during configuration and medical checks; 10 kb/sec acceptable.
Phrenic nerve pacemaker	Continuous but low frequency pulses; tens of kb/sec acceptable.
Current cochlear implants	Continuous; <300 kb/sec by the moment.
Bladder control stimulators	3–4 times/day or continuous; <300 kb/sec.
Hand prosthesis	Continuous, as high as possible; 1.25 Mb/sec acceptable.
Visual prosthesis	Continuous, as high as possible; >2Mb/sec acceptable.

6.2.6   Hermetic packaging

Prosthetic implants need to be protected from the hostile ionic environment of extracellular fluids for the lifetime of an implant recipient, which at times may be expected to be 100 years. It should be noted that many of the insulating materials that are commonly used may not be employed in the saline water environment. Flow or presence of electric charge because of electrical excitation can also facilitate some chemical reactions. Most of the electronics are made of materials that are not compatible with the body. For any electronic implant to work reliably for long time, the implants must be hermetically sealed.

6.2.6.1   Failures and mechanisms

An implant can fail in a number of ways.[29] Lead wire insulation failure, encapsulation failure, failure at the interface between materials, substrate corrosion, surface dielectric corrosion, MOS gate contamination, P–N junction contamination, and failure of metal and polysilicon interconnects can be reasons for failure. The failure mechanism can be chemical attack, dissolution of coating into the body fluid, condensation of water, mobile ions and other biochemicals along interfacial planes, electrochemical reactions to insulators leading to dissolution into body fluids, and electrochemical corrosion of the silicon substrate. Careful accelerated testing of the packing should be done. Many chemical reactions are possible for packaging degradation. The dominant chemical reaction and dominant mechanism for a particular reaction may change with elevated temperatures. The properties of the packing material may also change with temperature increase.

6.2.6.2   Hermetic packaging techniques

The hermetic packaging not only should encapsulate the electronics but also should provide connections. Polymers such as silicon and fluoropolymers have been successful in accelerated life testing. Silicon is used to achieve hermetic packaging in many implants. Stainless steel and titanium are used to encapsulate in some of the implants. Polyamide substrates have also been reported to provide hermetic packaged wires and inductors. Research in protein polymer coating, electronically conducting polymers (polyaniline), and ionically conducting polymers (polyethylene oxide) is being done. A particular technique with electrostatic bonding of a custom-made glass capsule and a supporting substrate has been reported.[30] This sealed feed-through technology allows the transfer of electrical signals through polysilicon conductor lines located at the silicon substrate.

6.3   Implant systems

Over the years, many implants have been developed and some of them have been very successful. With the advent of powerful and miniaturized technology, improvement in the old systems and development of more sophisticated implants are underway.

6.3.1   Cardiac pacemaker implant

While many people have an abnormal heartbeat, for most of people it is harmless. But, for some people, an irregular heartbeat can be a matter of life or death. The cardiac pacemaker[31] is an electronic device that operates during the abnormal period to stimulate the heart. The pacemaker has two parts: the generator and the leads. At the generator, a battery and information on the stimulation are stored. The wires from the generator, passing through a large vein, are anchored in the heart. Electrical impulses delivered through these wires stimulate the heart. The pacemaker can detect when the heart beat falls below a certain rate and start stimulation. Similarly it turns stimulation off at a certain set rate. The weight of a modern pacemaker is less than 30 g. Average lifetime of the battery is 7 to 8 years and it has to be replaced through surgery. The condition of the implant is regularly monitored by a healthcare professional, and reconfiguration of devices is done if required.

6.3.2   Implant for epilepsy

The vagus nerve stimulator (VNS), under the brand name NeuroCybernetic Prosthesis, works to inhibit seizures by stimulating the vagus nerve in the neck. The VNS, a small pacemaker-like device, is implanted in the seizure patient's chest muscle, just below the skin. It consists of a battery generator and lead wire. The lead wire runs under the skin to the neck where it attaches to the electrodes, which in turn stimulate the vagus nerve. The vagus nerve runs signals from the body to the brain. At regular intervals, the VNS sends a small electrical impulse to the brain interrupting seizure activity. The vagus nerve stimulation increases blood flow in both right and left thalami, which organize sensory messages to the higher brain. Thus, the seizure activity is subdued through stimulating the thalami and regulating thalamic processing.

6.3.3   Cochlear implants

Deafness can be the result of many causes. Deafness that is the result of defects in the outer and middle ear, but not further up in the sound sensing system, can be corrected by a hearing aid. Cochlear implants[32,33] are used for those people who do not have sensory hair cells and their hearing loss cannot be corrected by the use of a hearing aid. The cochlear implant electrodes stimulate the auditory nerve. For people with a defect in auditory nerve, auditory brainstem implants (ABIs) are required; however, ABIs are still in the developmental stages. Initially, cochlear implants were developed to imitate the lost function of the cochlea. Understanding of sound information processing by the cochlea and brain has contributed to the enhancement in the implant effectiveness. For example, it has been found that the sound-sensing cells are organized in the cochlea tonotopically. The apex of the cochlea, which senses low frequency of sound, is very thin and winding, hence not accessible to the implant. But, fortunately, the brain is able to fill in the frequency gap if enough overtones are present.

Several factors have been found to affect the effectiveness of the implant. For example, people with deafness acquired after learning speech and language have performed better with implants than those acquiring them before learning speech and language. The implant is more effective for younger people because their brains are more adaptive to the implants. Brains of young children have been found to make new neuron connections. With the help of patients, studies have been conducted to improve the effectiveness of the implants. Some of them are early adapting to music and frequency shifting. Work is being done to increase the number of stimulus sites, decrease the size of implant, reduce the power consumption, apply MEMS technology, and provide more signal processing.

6.3.4   Mobility implants

Artificial human body parts have been in use for a long time. They have progressed from wood to metal to plastic to composite, lightweight material. Electrical and mechanical systems have been assembled in these artificial parts to duplicate the motion of the human body.[34] Lately, attempts have been made to interface these prosthetic devices with better human control. While sophisticated robotic parts have been manufactured, communication between the parts and humans has been the weakest link.

This communication interface requires the expression of control by the person in some part of the body, sensing that expression, processing the expression, and subsequently controlling the artificial parts. While sometimes the expression can be easily detected, surgery may be used to implant the sensors in an alternate region. The complexity of processing the information and controlling the motion of these robotic parts poses a big challenge. Smooth movement, as in the human body, requires sensing and controlling at a large number of points. While full success has not been achieved, steady progress is being made.

Controlling robotic parts directly through the brain is showing great promise. The brain commands are sensed noninvasively with an electroencephalogram or an implanted electrode. Being noninvasive, the EEG is attractive but for the current technology it provides a very slow rate of information (25 bits per minute) and hence is unsuitable for complex movement control. Implanted chips can gather more information but are too bulky to provide a sufficient number of electrodes.

6.3.5   Visual prostheses

Blindness robs millions of individuals of their keen sense of vision. Several groups around the world are evaluating the feasibility of creating visual perception by electrical stimulation of the remaining retinal neurons in patients blinded by photoreceptor loss.[35–39] Several neuro-stimulator devices have been designed and fabricated for electrical stimulation of tissues. The cortical prosthesis group at the University of Utah has developed a stimulator that accompanies a penetrating electrode array for eliciting visual sensations through cortical stimulation. The neuro-prosthesis group at the University of Michigan has also produced a number of stimulation devices.[18] Additional work is in progress among various other researcher groups toward the development of retinal prostheses.[5,39–42] Although it is good to compare these approaches, for the sake of brevity we are discussing here a typical approach represented by our project. The Retinal-Prosthesis Group at North Carolina State University, Johns Hopkins University, and University of Southern California has developed several generations of stimulation ICs. These ICs are designed to deliver the currents for retinal stimulation, as were determined by clinical studies conducted on the visually impaired with RP and AMD.[39,43]

6.3.6   A prosthetic microsystem for an intraocular prosthesis

The human eye with its nearly 100 million photoreceptors is a complex system that allows us to enjoy high-resolution imagery full of colors. The system functions when light is focused by the cornea and crystalline lens within the eye on to the retina. The retina is highly structured, composed of multiple cell layers, each with a specific function. The photoreceptor cell layer captures light energy and converts it into electrochemical signals. These signals then excite bipolar cells located in the second layer of neurons in the retina. The bipolar cells subsequently transmit the signal to retinal ganglion cells whose axons collectively form the optic nerve that connects the retina to the visual centers of the brain. Other cells such as the horizontal and amacrine cells form connections between neurons within a layer, adding to the computational power of the retinal network of neurons. This biological neural network is a sophisticated image processor, capable of compressing information from 100 million photoreceptors into 1 million retinal ganglion cells.

While various causes for blindness exist, AMD is the leading cause of blindness in individuals 60 years or older, with 200,000 pairs of eyes left legally blind each year. RP has an incidence of 1 in 4000 and in the United States alone afflicts 100,000 people. Currently, no treatment exists for either RP or AMD. We have shown that in these diseases, despite near total loss of the photoreceptors, the remaining retinal neurons remain intact and patients could regain vision. It is proposed that artificial vision be provided by the prosthetic system with *multiple-unit artificial retina chipset* (MARC) implanted within the eye. The system would have an external camera (mounted in a glasses frame) to acquire an image and convert it into an electrical signal. This signal would be wirelessly transmitted to an implanted chip, which would electrically stimulate the cells of the retina that have not degenerated. The pattern of electrical stimulation would be controlled to produce the perception of an image; hence, the sense of vision could be partially restored with such an implantable electronic device.

6.3.6.1 Electrical Perceptions Experiments

In 15 human volunteers who had RP and AMD, intraocular prostheses were tested. It was shown that the remaining neurons in a non-seeing eye can be activated by an electrical signal, resulting in the perception of light. The human tests were performed in an operating room, where the volunteers were under local anesthesia. At first tests were done with a single electrode, and then patterns of electrical signals generated by an external computer were applied to the retina via the electrode arrays. Using a 25-electrode array, at first rows and columns of electrodes and then patterns of electrodes outlining a large letter or simple geometric shapes such as a rectangle were stimulated. In all experiments it took several minutes before the patient could confidently identify the artificially created spot of light. Once they recognized the first dot of light they made quick progress. As expected, the 25-electrode array provided poor resolution; however, it was sufficient for all the patients tested to date to see forms such as a large letter and simple geometric shape. It was also established that flicker-free vision could be achieved with sufficiently high applied pulse rate. Our results from tests in blind human volunteers demonstrate that form vision is possible using controlled pattern electrical stimulation of the retina.

6.3.6.2 Progress Summary

Concomitant with the human tests, we have been engineering a MARC, a completely implantable system capable of creating hundreds of individual spots of stimulus. Specifically, we had to develop a chip that would be small enough to be placed within the eye and yet could stimulate the retina through a large number of electrodes. The currently developed chip Retina 3.55 is 4.6 mm × 4.7 mm and can stimulate 60 electrodes with a safe and effective charge, and the design of a 1000-channel stimulator is progressing. A prototype image-acquisition system using a miniature CMOS camera and an image-processing system using a very-large-scale integration (VLSI) image-processing board have been built. Wireless communication electronics using a radiofrequency link have been developed that can supply the required data transmission rate and power coupling. Stimulating electrode materials are being tested to assess their ability to withstand corrosion within the eye.

Of equivalent importance to the engineering development of the device is the biological testing that will show that a retinal implant is well tolerated in the eye. Although intraocular retinal surgical procedures are commonly performed there is no precedence to attaching a device to the delicate, paper-thin retina. The device could tear the retina, cause intraocular bleeding, or dislodge because of eye movements. A metal alloy tack is used to secure the implant. In animal experiments, the implants have remained affixed to the retina through 11 months and caused no structural of functional damage. Experiments in progress will examine the effects of electrical stimulation on the retina.[35]

6.3.6.3 *Architecture of the Retina-3.55 micro-stimulator design*

Functional specifications for chip operation require excitation currents up to 600 µA amplitude delivered to retinal tissue with a characteristic impedance of approximately 10 kΩ.[44] Variable stimulation rates of 50 to 60 Hz and higher are desired to achieve flicker-free vision as well as definable pulse widths and inter-phase delays on the order of 1 to 5 msec. Stimulation pulses are produced in the standard biphasic fashion with anodic-leading or cathodic-leading charge-balanced pulses for compatibility within the biological environment. To achieve greater flexibility in programming the stimulus schedule, the 60 current driver circuits are independently programmable and connect directly with 60 output channels (or pads). This provides a single stimulation frame, and thus an intended visual experience in the form of an 8 × 8 pixelated image (minus the four corners).

6.3.6.3.1 *Communication protocol and subsystems.* The instructions that specify the operation of the stimulator are defined digitally and loaded into the IC serially through a *data input* and an accompanying *clock input*. A *configuration frame* packet format is defined for specifying the full-scale output current magnitude and stimulus timing, including pulse widths and interphase delay. Another packet format, designated as the *data frame*, is defined for specifying desired stimulation current amplitudes per phase, or for requesting charge cancellation, for all driver circuits in real time, thereby constituting a single image frame. Both of these frames are 1024 bits in length, which is convenient for delimiting frame boundaries as packets are loaded. A unique 16-bit synchronization word identifies the beginning of a configuration frame. If the cyclic redundancy check (CRC) and checksum signatures are verified, the configuration data are latched into internal registers. Otherwise, the data are ignored and subsequently overwritten as new packets are shifted into the IC. As with the configuration packet, a data packet is initiated with a unique 16-bit synchronization word, checked, and latched.

6.3.6.3.2 *Error-detection subsystem.* Data errors could have unintended consequences, such as generating biphasic current signals which are not properly charge balanced. Due to the risks associated with processing erroneous data, CRC and checksum error detection mechanisms are used. A 32-bit *cyclic redundancy check* (CRC) computation engine is used. To increase the robustness, a 16- bit checksum error detection is also used. The chip will not process any runtime data until a configuration packet is first received without error and latched, as the timing profile will not be defined until then. Retina-3.55 does not have back-telemetry at this time to support the communication of status or other data back to the extraocular host controller; therefore, the external system cannot know when error-free configuration has been achieved. The interim solution to this is retransmission of configuration-packets at a regular interval. If a valid configuration has been latched, then subsequent error-free runtime data packets will be latched for stimulus generation.

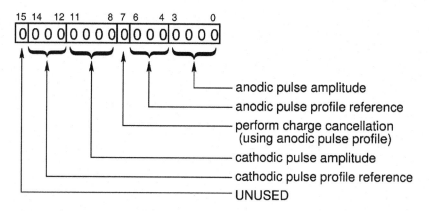

Figure 6.8 Retina-3.55 driver subpacket format.

6.3.6.3.3 Pulse profile generator design. The pulse profile generators produce eight global, independently programmable (via a configuration-packet) pulsed timing references. These exist in the form of 8-bit (start time, stop time) pairs. Each parameter represents a count of clock cycles, relative to the start of each of data frame. The generation of the pulse profiles is done by a counter and comparator circuit, which compares the counter output with the start time and stop time.

6.3.6.3.4 Stimulus current driver design. Each of the 60 current drivers is capable of producing a biphasic stimulus current using information provided in a 16-bit data subpacket as shown in Figure 6.8. A conceptual model of the biphasic current driver is as shown in Figure 6.4b. The anodic current is sourced to the load by an ideal current source referenced to a positive supply of V_{dd}. Similarly, the cathodic current is sunk from the load by an ideal current source referenced to a negative supply of V_{ss}. The electrode array[44] might be attached to the retina using retinal tacks. Due to implementation in the N-well process, to minimize body effects all digital logic circuits and low-voltage analog circuits operate between V_{ss} (logic "0") and *Gnd* (logic "1"). V_{dd} is used only in the output stage of the driver circuit. The driver employs two NMOS binary-weighted current-mode digital-to-analog converters (DACs) to produce the anodic and cathodic currents. The DAC design is as shown in Figure 6.9.

To date, the standard DAC arrangement for producing biphasic currents consists of a stacked PMOS-DAC/NMOS-DAC configuration driving a monopolar output.[18,19] This structure may not support the high voltage required in our design and may require alternative transistor structures that can support high drain-source voltages without modification of the process, such as high-voltage transistors.[45,46] Therefore, the current-driver structure is modified so that the DAC currents are mirrored and scaled into an output circuit of high-voltage compliance. The circuit for this is shown Figure 6.10. Each current driver contains two 8-to-1 multiplexers for programmable selec-

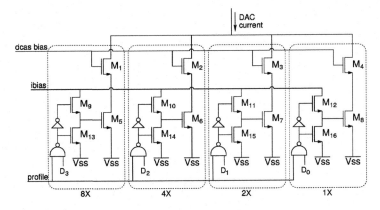

Figure 6.9 Current-mode DAC circuit detail.

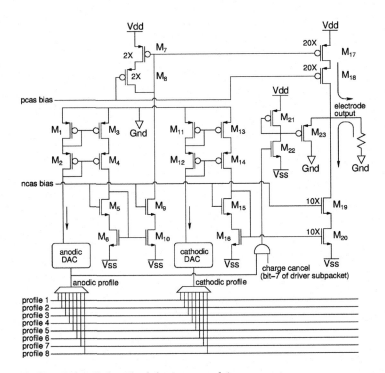

Figure 6.10 Circuit-level detail of the current driver.

tion of any of the eight pulse timing profiles. One component of the driver circuit is a charge cancellation mechanism, which is intended to limit any unintentional accumulation of charge on the electrode.

6.3.6.3.5 DAC biasing circuit. The wide-swing cascode current mirrors in the NMOS DACs and the biphasic amplifier stage receive their bias voltages from a central biasing circuit. The schematic of the bias circuit is provided in Figure 6.11. The biasing circuit can be considered in three sections, which are labeled in Figure 6.11 as the *bootstrap reference*, the *DAC biasing* section, and the *biphasic amplifier* section. The bootstrap reference is a circuit that settles into the intersecting operating point of a linear resistance and a nonlinear transconductance.[47] The devices are sized to achieve a 20 µA master reference current, which becomes the basis for other currents produced on-chip. The DAC biasing section taps the mirrored current of the bootstrap reference to establish two biases for the two NMOS DACs in the drivers. Using two bits from a prior latched configuration packet ($Iref_1$, $Iref_0$), the DAC biasing section can be tuned to mirror the bootstrap reference with gains of 1×, 2×, or 3×. This ultimately permits global full-scale current output in the driver's biphasic amplifiers to be programmed at 200, 400, or 600µA. The driver DACs employ wide-swing cascade mirrors using the associated biases of *dcas* (DAC cascode bias) and *ibias* (DAC current-source bias).

6.3.6.4 Experimental measurements

The Retina-3.5 stimulator IC, which consists of the design described herein without the CRC and checksum error detection logic, was implemented in the AMI 1.2-µm CMOS and occupies an area of 4.7 mm × 4.6 mm. As Retina-3.55 has not been fabricated at this time, experimental measurements are taken from the Retina-3.5 IC design, containing the same stimulus driver core. The

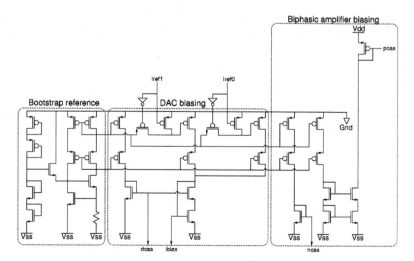

Figure 6.11 Central-bias circuit for DACs and biphasic amplifier.

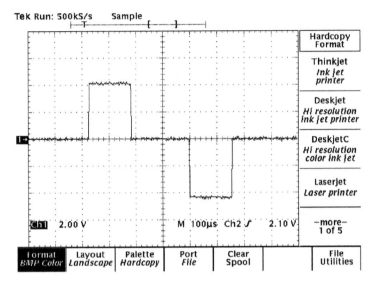

Figure 6.12 Charge-balanced biphasic pulse produced from Retina-3.5.

driver circuit is characterized in terms of linearity and pulse amplitude matching, current sensitivity to power supply variations, and effectiveness of the
charge cancellation mechanism and power consumption requirements.

    6.3.6.4.1 Output flexibility and example waveforms. A typical biphasic
stimulus current generated from Retina-3.5 is shown in Figure 6.12. For this
figure, the chip is operated from +5 V/−5 V rails and is driving a 10-kΩ load
referenced to ground. This is a transient plot of the voltage across the load
resistance. The stimulator IC is processing 100 frames per second in this
example. Figure 6.13a,b illustrates how the anodic/cathodic pulse amplitudes can be independently specified. It should be pointed out that in order
to generate a targeted current amplitude of 600 µA, the supply voltages must
be increased beyond +5 V/−5 V (assuming a 10-kΩ characteristic load impedance). Although pulses are achievable at +7 V/−7 V, the matching between
anodic and cathodic pulses degrades due to limitations in the output stage
design for such high supply voltages in the AMI 1.2 µm CMOS. These
supplies are beyond those rated for the process; however, the gate oxides
and drain/source junctions have been found to be tolerant of +7 V/−7 V
rails without permanent damage.

6.3.6.5 Telemetry circuitry: data link

    6.3.6.5.1 ASK demodulator. The prototype implantable device
receives both power and data via an inductive link. Amplitude modulation
instead of frequency modulation is chosen in order to reduce the circuit
complexity and power consumption and is able to sustain the data rate of
the requisite functions. The ASK modulation scheme governs the amplitude

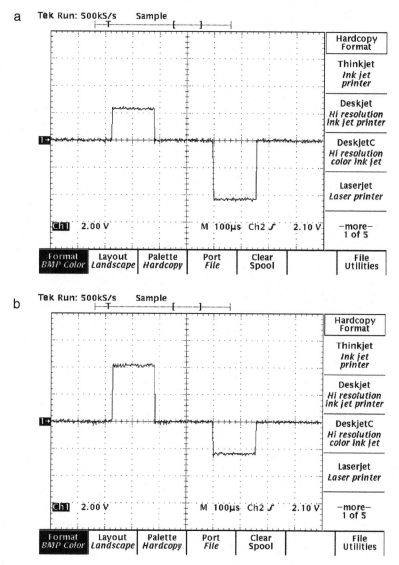

Figure 6.13 Independent amplitude variability of the (a) anodic phase and (b) cathodic phase.

of the carrier signal according to the desired digital data. Because the recovered power is also derived from this amplitude, the average transferred power could depend on the transmitted digital data pattern. To avoid this data dependency, we first encode the data to be transmitted using the alternate mark inversion pulse-width modulation (PWM) scheme, which subsequently modulates the power carrier as shown in Figure 6.14. The system is designed for PWM data ranging from 25 to 250 kb/sec with a carrier frequency ranging from 1 to 10 MHz.

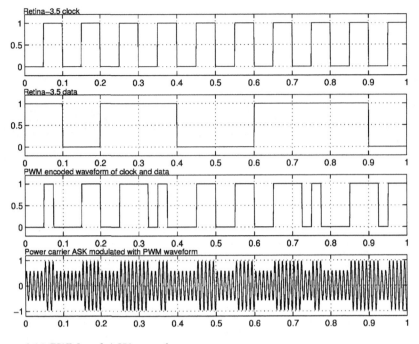

Figure 6.14 PWM and ASK waveforms.

On the receiver side, the power carrier is rectified and filtered to obtain the base-band carrier envelope containing the PWM data. The envelope is further filtered in order to provide a low-ripple DC voltage for the chip power supply. The unfiltered carrier envelope is passed to the ASK demodulator circuit of Figure 6.15 to extract the digital (rail-to-rail) PWM waveform.

The demodulator is a comparator with a predefined amount of hysteresis in which one input is derived from the envelope of the modulated carrier (A in Figure 6.15). The other input is derived from the average of that signal (B in Figure 6.15). Transistors M_0 and M_1 provide level shift of the input signal to the common mode range of the differential amplifier.

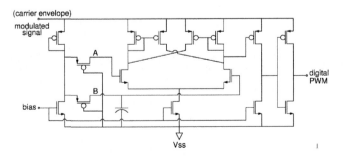

Figure 6.15 ASK demodulator circuit.

Transistor M_2 with the 10 pF capacitor C_0 serves as a low-pass filter and provides the average of an input signal. Both signals are applied to the differential amplifier with the cross-connected active load of M_3 to M_6. The hysteresis is achieved via positive feedback. The demodulator is designed to process the carrier envelope with a ripple between 6.5 to 7.5 V. It has a hysteresis of 500 mV, so as to ensure that the likelihood of an extra transition in the output waveform due to noise is minimized. The output waveform of the demodulator is the digital PWM with voltage swing between 0 and 7 V.

6.3.6.5.2 Clock and data-recovery circuit. The rising edges of the PWM signal from the demodulator are fixed in time periodically and serve as a reference in the derivation of an explicit clock signal. On the other hand, the data are encoded by the position of the falling transition at each pulse. In an alternate mark inversion encoding scheme, a zero is encoded as a 50% duty-cycle pulse, and ones are alternately encoded by 40 or 60% duty-cycle pulses. This eliminates the need for a local clock oscillator such that the clock and data recovery circuit is simplified. It also provides an average coupled power which is essentially independent of the data.

A diagram of the clock and data-recovery unit is shown in Figure 6.16. It consists of a delay-locked loop (DLL) and decoder logic. The DLL consists of a phase-frequency detector (PFD), a charge pump, a loop filter, and a voltage-controlled delay line. The 36-stage delay line is locked to one period of the PWM waveform. The waveform is decoded by an XNOR gate, the inputs of which are the tapped-out signals at the 15th and the 21st stages. The positions for the tapped signals correspond to the duty-cycle percentages used in the PWM waveform.

A challenge for the DLL design is to make the lockable frequency as low as 25 kHz without consuming a large chip area. A three-state PFD is used with the additional delay at the reset path to reduce the dead-zone effect. The voltage-controlled delay line is based on current-starved inverters made of low W/L transistors. The charge pump current is 3.5 μA provided by matched wide-swing current sources. A unity gain amplifier is included to prevent the ripple distortion of the control voltage due to charge sharing. The 200 pF loop filter capacitor is integrated on the

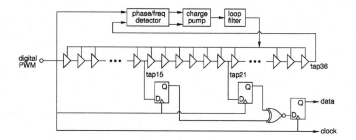

Figure 6.16 Clock and data recovery.

chip. The initial condition of the DLL must be set up correctly to guarantee a correct locking condition. The very first input to the PFD after
reset must be the rising edge of the PWM waveform fed back from the
delay line. This can be enforced by using a simple pulse swallow circuit
which would only ignore the very first pulse of the incoming PWM
waveform prior to entering the PFD. The loop filter capacitor is initialized
to full charge when power is applied. Consequently, the delay line starts
with the smallest delay. The loop then initiates the discharge of the
capacitor, resulting in a decrease of the control voltage and an increase
of the delay. This process continues until the delay is exactly equal to
one period of the incoming PWM waveform. In this way, locking to a
subharmonic of the PWM waveform is prevented. The DLL is always
stable because it is a first-order system.

6.3.6.5.3 Communication circuit. The chip is designed only to receive
externally initiated communication; no back-telemetry is implemented.
Although intended for a data rate of 25 to 250 kb/sec, our measurements
show that the ASK and PWM demodulator circuits can operate in excess of
1 Mb/sec. However, at this high data rate, the modulation index needs to
be increased up to 30%. Furthermore, if the ratio of carrier frequency to the
data rate is more than 40, they can be independent of one another. Figure 6.17
shows the measured communication waveforms; Ch1 is the carrier envelope
(top waveform), Ch2 is the PWM output from the demodulator, and Ch3 is
the recovered non-return to zero (NRZ) data. Note that there are two clock
periods of latency between the NRZ output and the PWM input, owing to
the flipflops in Figure 6.16.

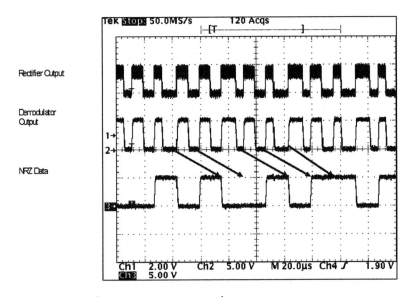

Figure 6.17 Measured communication waveforms.

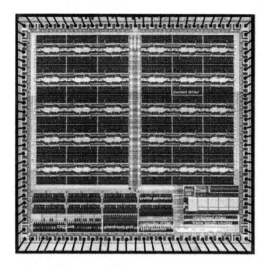

Figure 6.18 Fabricated chip (Retina 3.55).

6.3.6.6 Chip implementation

A die photograph of the Retina-3.55 IC is shown in Figure 6.18. The chip was implemented in the AMI 1.2-μm, two-metal, two-poly CMOS process through the MOSIS facility, with die physical dimensions of 4.6 mm × 4.7 mm. A peripheral ring of 104 100-μm × 100-μm pads encloses the stimulator core. Power and control signal pads are allocated along the bottom edge of die to facilitate simple connection, with a separate adjacent IC (not discussed here) providing power carrier rectification and filtering and supply regulation. (Continued development of the stimulator design in context of the inductive powering scheme and wireless telemetry would likely see the carrier rectifier and filtering and supply regulation co-integrated with the stimulator circuits.) The 60 stimulus-current output pads are distributed along the remaining three sides of the die.

6.4 Conclusion

The field of semiconductor-based implants is becoming more and more exciting. Today, advanced semiconductor technologies offer devices at the micron level virtually at the same or smaller size than the biological cell size, the complexity of system on a chip is in the millions of devices, and the speed of devices can be in the range of several hundred megahertz. Moreover, further reduction in device size and advances in the field of microelectromechanical systems (MEMS) will help to further miniaturize the implants. Improvements in low power circuit design and higher digital signal processing capability are desired. It is also to be noted that projects such as retinal prostheses require expertise in many fields, including chemistry, electronics, materials, and biomedicine. Having the capability to produce sophisticated, low-risk medical procedures for implants and people's willingness to adopt high-tech implants are the driving forces behind new implant development. While cochlear implants, pacemakers, artificial limbs, and portable dialysis have become stable technologies, work on projects such as retinal prostheses, brain implants, and neural recording is in progress.

References

1. Liu, W., McGucken, E., Vichienchom, K., Clements, M., de Juan, E., Jr., and Humayun, M.S., Dual unit visual intraocular prosthesis, *Proc. of the 19th Int. Conf. of the IEEE Engineering in Medicine and Biology Society*, 1997, pp. 2303–2306.

2. Rucker, L. and Lossinsky, A., Percutaneous connectors, 30th Neural Prosthesis Workshop, NINDS, NINCD, NIH, October 12–14, 1999.

3. Wyatt, J.L. and Rizzo, J.F., Ocular implants for the blind, *IEEE Spectrum*, 33, 47–53, 1996.

4. Rizzo, J.F. and Wyatt., J.L., Prospects for a visual prosthesis, *Neuroscientist*, 3, 251–262, 1997.

5. Eckmiller, R., Learning retina implants with epiretinal contact, *Ophthalmic Res.*, 29, 281–289, 1997.

6. Gross, M., Buss, R., Kohler, K., Schaub, J., and Jager, D., Optical signal and energy transmission for a retina implant, *Proc. First Joint BMES/EMBS Conf.*, 1, 476, 1999.

7. Zierhofer, C.M. and Hochmair, E.S., High-efficiency coupling-insensitive transcutaneous power and data transmission via an inductive link, *IEEE Trans. Biomed. Eng.*, 37(7), 716–722, 1990.

8. Zierhofer., C.M., A class-E tuned power oscillator for inductive transmission of digital data and power, *Proc. 6th Mediterranean Electrotechnical Conf.*, 1, 789–792, 1991.

9. *IEEE Standard for Safety Levels with Respect to Human Exposure to Radio Frequency Electromagnetic Fields, 3KHz–3GHz*, IEEE Standard C 95.1., 1999 ed.

10. Zierhofer, C.M. and Hochmair, E.S., Coil design for improved power transfer efficiency in inductive links, *Int.Conf. Eng. Med. Biol.*, 4, 1538–1539,1997.

11. Sokal, N.O. and Sokal, A.D., Class-E: a new class of high-efficiency tuned single-ended switching power amplifier, *IEEE J. Solid-State Circuits*, SC-10(3), 1975.

12. Sokal, N.O., Class-E: high-efficiency switching-mode tuned power amplifier with only one inductor and one capacitor in load network — approximate analysis, *IEEE J. Solid-State Circuits*, SC-16(4), 1981.

13. Lee, T.H., *The Design of CMOS Radio Frequency Integrated Circuits*, University of Cambridge Press, Cambridge, U.K., 1998.

14. Raab., F.H., Class-F power amplifiers with maximally flat waveforms, *IEEE Trans. Microwave Theory Techniques*, 45(11), 2007–2012, 1997.

15. Troyk, P.R. and Schwan, M.A.K., Closed-loop class E transcutaneous power and data link for MicroImplants, *IEEE Trans. Biomed. Eng.*, 39(6), 589–599, 1992.

16. Clements, M., Vichienchom, K., Liu, W., Hughes, C., McGucken, E., DeMarco, C., Mueller, J., Humayun, M., de Juan, E., Jr., Weiland, J., and Greenberg, R., An implantable neuro-stimulator device for a retinal prosthesis, *Int. Solid-State Circuits Conf.*, February 1999, pp. 216–217.

17. Jones, K.E. and Normann, R.A., An advanced demultiplexing system for physiological stimulation, *IEEE Trans. Biomed. Eng.*, 44(12), 1210–1220, 1997.
18. Tanghe, S.J. and Wise, K.D., A 16-channel CMOS neural stimulating array, *IEEE J. Solid-State Circuits*, 27(12), 1819–1825, 1992.
19. Kim, C. and Wise, K.D., A 64-site multishank CMOS low-profile neural stimulating probe, *IEEE J. Solid-State Circuits*, 31(9), 1230–1238, 1996.
20. Liu, W., Vichienchom, K., Clements, M., DeMarco, S.C., Hughes, C., McGucken, E., Humayun, M.S., de Juan, E., Jr., Weiland, J.D., and Greenberg, R., A neuro stimulus chip with telemetry unit for retinal prosthetic device, *IEEE J. Solid-State Circuits*, 35(10), 1487–1497, 2000.
21. Ziaie, B., Nardin, M.D., Coghlan, A.R., and Najafi, K., A single-channel implantable microstimulator for functional neuromuscular stimulation, *IEEE Trans. Biomed. Eng.* 44(10), 909–920, 1997.
22. Mueller, J.S. and Gyurcsik, R.S., Two novel techniques for enhancing powering and control of multiple inductively-powered biomedical implants, *Proc. Int. Symp. Circuits Systems*, 1, 289–292, 1997.
23. Jun, Z.X. and Hong, Z., A full-custom-designed receiver-stimulator chip for amultiple-channel hearing prosthesis, *Proc. Int. Symp. on VLSI Technology, Systems, and Applications*, 1993, pp. 288–291.
24. Gudnason, G., Bruun, E., and Haugland., M., An implantable mixed analog/digital neural stimulator circuit, *Proc. Int. Symp. Circuits Systems*, 5, 375–378. 1999.
25. Von Arx, J.A. and Najafi, K., A wireless single-chip telemetry-powered neural stimulation system, *IEEE Int. Solid-State Circuits Conf.*, 1999, pp. 214–215.
26. Nardin, M. and Najafi., N., A multichannel neuromuscular microstimulator with bi-directional telemetry, *8th Int. Conf. Solid-State Sensors and Actuators, Eurosensors IX, Transducers Ö95*, 1, 59–62, 1995.
27. Matsuki, H., Yamakata, Y., Chubachi, N., Nitta, S.I., and Hashimoto, H., Transcutaneous DC–DC converter for totally implantable artificial heart using synchronous rectifier, *IEEE Trans. Magnetics*, 32(5), 5118–5120, 1996.
28. Von Arx, J.A. and Najafi, K., A wireless single-chip telemetry-powered neural stimulation system, *IEEE Int. Solid-State Circuits Conf.*, 1999, pp. 214–215.
28a. Marin, D., Troosters, M., Martinez, I., Valderrama, E., and Aguilo, J., New development for high performance implantable stimulators: First 3 Mbps up to 4.4 Mbps demodulator chip through a wireless transcutaneous link, *MicroNeuro '99*, pp. 120–126.
29. Edell, D.J., NIH supported work. 1996 #Final report.
30. Cheng, L. and Najafi, K. A hermetic glass-silicon package formed using localized aluminum/silicon–glass bonding, *J. Micromech. Syst.*, 2001.
31. Machines in our hearts: the cardiac pacemaker, the implantable defibrillator, and American Healthcare.
32. *Cochlear Implants in Adult and Children*, NIH consensus statement, May 15–17, 1995, pp. 1–30, http://text.nlm.nih.gov/nih/cdc/www/100txt.html.
33. Spelman, F.A., The past, present and future of cochlear prostheses, *IEEE Eng. Med. Biol.*, 27–33, 1999.
34. The bionic man: restoring mobility, *Science*, 295, 2002.
35. Humayun, M.S., de Juan, E., Jr., Weiland, J. et al., Pattern electrical stimulation of the human retina, *Vision Res.*, 39, 2569–2576, 1999.

36. Liu, W., Humayun, M., de Juan, E., Jr. et al., Retinal prosthesis to benefit the visual impaired, in *Intelligent System and Techniques in Rehabilitation Engineering*, Teodorescu, N., Ed., CRC Press, Boca Raton, FL, 2000, chpt. 2.

37. Wyatt, J. and Rizzo, J., Ocular implants for the blind, *IEEE Spectrum*, 47–53, 1996.

38. Chow, A. et al., The subretinal microphotodiode array retinal prosthesis, *Ophthalmic Res.*, 195–198, 1998.

39. Zrenner, E., Stett, A. et al., Can subretinal microphotodiodes successfully replace degenerated photoreceptors?, *Vision Res.*, 2555–2567, 1999.

40. Rizzo, J.F., Loewenstein, J., and Wyatt, J., Retinal degenerative diseases and experimental theory, in *Development of an Epiretinal Electronic Visual Prosthesis: The Harvard-Medical Massachusetts Institute of Technology Research Program*, Kluwer Academic, Norwell, MA, 1999, pp. 463–470.

41. Yagi, T., Ito, Y., Kanda, H., Tanaka, S., Watanabe, M., and Uchikawa, Y., Hybrid retinal implant: fusion of engineering and neuroscience, in *Proc. 1999 IEEE Int. Conf. Systems, Man, Cybernetics*, 4, 382–385, 1999.

42. Chow, A.Y. and Chow V.Y., Subretinal electrical stimulation of the rabbit retina, *Neurosci. Lett.*, 225, 13–16, 1997.

43. Humayun, M.S., de Juan, E., Jr., Weiland, J.D., Dagnelie, G., Katona, S., Greenberg, R., and Suzuki, S., Pattern electrical stimulation of the human retina, *Vision Res.*, 39, 2569–2576, 1999.

44. Majji, A.B., Humayun, M.S., Weiland, J.D., Suzuki, S., D'Anna, S.A., and de Juan, E., Jr., Long-term histological and electrophysiological results of an inactive epi-retinal electrode array implantation in dogs, *Invest. Ophthalmol. Visual Sci.*, 40(9), 2073–2081, 1999.

45. Decemberlercq, M., Clement, F., Schubert, M., Harb, A., and Dutoit, M., Design and optimization of high-voltage CMOS devices compatible with a standard 5V CMOS technology, in *Proc. of the IEEE 1993 Custom Integrated Circuits Conference*, 1993, pp. 24.6.1–24.6.4.

46. Ballan, H. and Declercq, M., *High Voltage Devices and Circuits in Standard CMOS Technologies*, Kluwer Academic, Dordrecht/Norwell, MA, 1999.

47. Allen, P.E. and Holberg, D.R., *CMOS Analog Circuit Design*, Holt, Rinehart, & Winston, New York, 1987.

chapter seven

Silicon microelectrodes for extracellular recording

Jamille F. Hetke and David J. Anderson

Contents

0-8493-1100-4/03/$0.00+$1.50
© 2003 by CRC Press LLC

7.1   Introduction

When envisioning a neural prosthesis, the image that emerges is of electrical stimulation through one or several implanted electrodes to restore lost body function. While stimulation is the basis for restoration of sensory and motor function, recording is also essential in the development and employment of a prosthesis. First, extensive physiological recording research must be performed to investigate and perhaps map an area into which a stimulation device will be implanted. The foundation on which a cochlear implant performs, for example, is the tonotopic organization of the cochlea. This map, studied and verified in part using extracellular recording, became the basis for the design of the stimulation prosthesis. Similarly, maps have been identified in primary visual cortex which encode features such as orientation, direction, and color.[1–3] Once an implantation target has been chosen, extracellular recordings are often used to validate efficacy and optimize the design of implanted stimulation electrodes and protocols. In addition to its use as a basic research tool, the recording device is also an essential component of closed-loop prostheses currently being developed for spinal cord injuries. In these devices, implanted recording electrodes record biological control signals to drive implanted stimulating electrodes. An example of a closed-loop prosthesis currently under development is a motor prosthesis in which a recording array, implanted in the motor cortex, sends control signals to a stimulation device that controls a robotic arm.[4–7] Indeed, extracellular recording electrodes play a significant role in the world of neural prostheses. This chapter will begin with a history of recording microelectrodes fabricated using thin-film techniques and will describe several of the current research projects aimed at developing these electrodes. It will then focus on silicon microprobes being developed at the University of Michigan and their design, fabrication and use.

7.2   Extracellular recording

Extracellular recording is one of the most widely used techniques for studying the nervous system at the cellular level. When a neuron receives appropriate stimuli from other cells, its membrane depolarizes and causes ionic currents to flow in the surrounding cytoplasm. The voltage drop associated with this extracellular current, or *action potential*, can be measured if a suitable electrode is located near the active neuron. An extracellular action potential is typically about 50 to 500 microvolts (μV) in amplitude, with a frequency content from 100 Hz to about 10 kHz. In order to record these neural signals, the electrode must pass through the extracellular space and approach the active neuron without damaging it or other cells interacting with it. For this reason, it is critical that the recording electrode be as small and noninvasive as possible.

One type of recording electrode is the glass micropipette. These devices are formed by heating and pulling a 1- to 2-mm diameter glass capillary into two pieces. Commercial pipette pullers permit control of the temperature and force with which the capillary is pulled and therefore control the taper of the resulting tip down to 0.1 µm. Beveling of the tip can also be performed to even more precisely define the tip diameter and impedance. The pulled pipette is filled with an electrolyte solution such as KCl to form a conductive link to the tissue, and a large-area reversible electrode, inserted in the solution from the top of the pipette, is used to couple to the external world. The electrical properties of glass electrodes have been reviewed by Schanne.[8] In general, the equivalent circuit for such a device is dominated by the series resistance of the fluid-filled tip and the shunt capacitance of the pipette wall. This forms a low-pass filter which measures DC and low-frequency potentials well but rarely permits a frequency response above 1 kHz. Glass micropipettes are therefore typically used for intracellular recording.

The preferred method for detecting action potentials extracellularly is with a metal microelectrode. Traditionally, these electrodes have been formed by electrolytically sharpening[9,10] or mechanically beveling[11,12] a small-diameter metal wire (e.g., tungsten, stainless steel, platinum) to a fine tip (<1 µm) and then insulating it, leaving only the tip exposed. Alternatively, microwires can be formed from preinsulated fine wires that have been cut to expose the cross-sectional area at the end of the wire. The electrical properties of metal microelectrodes have been detailed by Robinson.[13] Basically, metal microelectrodes function by forming a capacitive interface to the aqueous tissue medium, allowing the detection of changes in the potential field created by the extracellular currents. The electrodes described in the remainder of the chapter are of the metal type.

7.3 Multichannel recording

While the recordings from a single electrode site can reveal characteristics of one or a few cells, they cannot give information about how networks of cells work together to process information. This requires the use of arrays of microelectrodes to study temporal and spatial relationships between groups of neurons. Although only single-channel separation methods existed in the 1960s, the use of such techniques was not disseminated.[14] The mathematical tools for analysis of multichannel records were being developed by Gerstein and his colleagues[15–17] well before the recording methods were available. These studies revealed the importance of observing the activity and interaction of many neurons simultaneously and led to numerous attempts to construct arrays of microelectrodes by mounting wire electrodes on a common structure. These types of electrodes provided a multiplication of simultaneous data collection channels and the potential for detecting relationships among cells. Many reviews exist for this technology and the philosophy driving it.[18] From a signal processing point of view, the separation of neural events recorded on one of the channels depended on various methods of waveform discrimination, a method that was well developed by the early 1980s.[19] An important advance in multichannel recording occurred when multielectrodes were fabricated with closely spaced sites that had overlapping recording volumes. This allowed neurons to be separated by spatial distribution as well as waveform. The stereotrode, introduced by McNaughton et al.,[20] was a close-spaced, two-site wire electrode that could accomplish discrimination based on spatial distribution of signals. Drake[21] showed the greater discrimination power of a larger number of sites placed close together on silicon substrates, the principle being that more sites remove spatial ambiguity and provide greater noise immunity by averaging multiple sources. Later, a four-wire device was introduced as the tetrode by McNaughton's group.[22,23] Detailed refinement of the method with quantitative evaluation came from Gray.[24] A more general approach to electrodes having correlated activity was developed by Gozani.[25]

Various methods of electrode fabrication, including gluing the shanks of metal wire electrodes together, welding glass pipettes together, and depositing thin films on silicon substrates, have been used to achieve multichannel recordings. Microwire electrode arrays are still used extensively today for both acute and chronic extracellular recording[26–28] and can be obtained commercially with a variety of different metals and insulators, varying array configurations, and even with independently positionable electrodes. Vendors who offer these arrays and associated instrumentation include FHC, Inc. (www.fh-co.com), NB Labs (www.nblabslarry.com), Thomas Recording (www.thomasrecording.de), and Alpha Omega Engineering (www.alpha-omega-eng.com).

7.4 Photoengraved microelectrodes

Although arrays of bundled metal microelectrodes have proven to be very useful for studying neural circuits, they do have several disadvantages. First, their added volume introduces more tissue damage than does a single conductor electrode. In addition, the exact geometrical configuration of such hand-built arrays is not reproducible. The ability to increase the number of electrode recording sites without increasing the volume of the array is an attractive one and spurred the onset of a number of attempts to create a microelectrode array using high-precision photolithographic techniques employed in the microelectronics industry. Using these methods, a single recording site can be made about as small as the tip of the finest wire electrode. However, the microelectrode shank, the portion that supports the recording sites and displaces the tissue, can carry multiple recording sites and can be at least as small as a single-wire electrode. In general, these techniques are the same as those used to create integrated circuits and therefore utilize similar substrate, conductor, and insulating materials. Fabrication typically starts on a wafer substrate and the electrode features are added using a number of photolithographically patterned thin-film layers that are defined by etching. These methods are attractive because they result in highly reproducible, batch-processed devices that have features defined to within less than ±1 μm. In addition, many of the techniques are compatible with the inclusion of on-chip circuitry. The latter is an important characteristic because, as the density of sites increases, so does the number of interconnect leads and packaging complexity. In fact, multiplexing becomes imperative as the lead count approaches 32 or more, especially if the electrode is to be chronically implanted. In addition, on-chip buffering and amplification can reduce crosstalk, noise coupling and signal attenuation, improving the overall signal-to-noise ratio.

7.4.1   Early attempts

The exploration of photoengraved microelectrodes began in the mid-1960s with efforts at Stanford University.[29,30] This silicon substrate structure consisted of an array of gold electrodes and conductors insulated by a thin layer of silicon dioxide (Figure 7.1). These probes were capable of recording single-

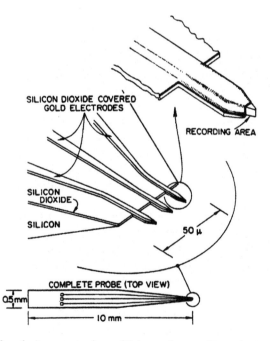

Figure 7.1 An early photoengraved, multichannel recording electrode. This device was based on a silicon substrate and had gold electrodes and conductors insulated with silicon dioxide. (From Wise, K.D. et al., *IEEE Trans. Biomed. Eng.*, 17, 238, 1970. With permission.)

unit activity from rat cortex[31] and paved the way for future developments. Since that time, a number of efforts have been directed toward the development of batch-fabricated, photolithographically defined microelectrode arrays. They have differed in their substrate, interconnect, and dielectric materials and in methods used for shaping the device. Silicon, tungsten, molybdenum, glass, and polyimide have been among the substrate materials that have been explored. Dielectrics that have been investigated include polyimide, silicon oxide, silicon nitride, and glass. A variety of materials including gold, platinum, tungsten, tantalum, and nickel have been used to form electrode sites. While many of these early attempts resulted in probes that were capable of recording extracellular activity, they were often low in yield and quality due to difficult fabrication sequences and imprecise methods for shaping of the substrate. Details on these early electrodes can be found in References 32 to 46.

7.4.2   Recent developments

A number of groups continue to develop and improve photoengraved microprobes using advanced techniques and materials. Recent and ongoing efforts include the Utah Electrode Array, arrays based on silicon-on-insulator (SOI) wafer processing, flexible polyimide arrays, and the Michigan probe. For a more inclusive list of recent developments, see References 47 to 65.

7.4.2.1   The Utah Electrode Array

Dr. Richard Normann and his colleagues at the University of Utah have developed a microelectrode array with a high density of penetrating shafts and referred to as the Utah Electrode Array, or UEA.[52-54] Each shaft is 1 to 1.5 mm long and projects down from a 0.2-mm-thick glass/silicon composite base (Figure 7.2). The device is formed from a monocrystalline block of silicon using a diamond dicing saw and chemical sharpening. The resulting silicon shafts are electrically isolated from one another with a glass frit and from the surrounding tissue with deposited polyimide or silicon nitride. The tip-most 50 to 100 μm of each shaft is coated with platinum to form the electrode site.

Arrays have been fabricated with up to 100 (10×10) penetrating shafts that are spaced on 400 μm centers. Interconnection to the electrode sites is accomplished by bonding either individual, insulated 25-μm wires or a multilead polyimide ribbon cable to bond pads on the top of the array. While the shafts are sharp, they are dense and can cause significant tissue dimpling during normal insertion. A high-velocity insertion technique using a pneumatic device was developed to alleviate this problem.[55] Although the UEA was originally designed and has been successfully used for acute and chronic recording in cat cortex,[56,57] a modified design has now permitted use in cat peripheral nerve.[58,59] A variety of acute and chronic versions of the device, along with instrumentation associated with their use, are available commercially from Bionic Technologies (www.bionictech.com).

Figure 7.2 The Utah Electrode Array (UEA). This 10×10, silicon-based array has platinum electrode sites on the tip of each shaft. The shafts are spaced on 400-μm centers. (From Nordhausen, C.T. et al., *Brain Res.*, 726, 129, 1996. With permission.)

7.4.2.2   SOI-based fabrication

Another relatively new set of research efforts is directed toward development of planar neural electrodes fabricated using SOI wafers.[60–62] SOI wafers are manufactured with an oxide layer buried a specified distance below the top silicon surface (6 to 25 μm for these neural devices). This oxide layer provides an etch-stop that accurately defines the final thickness of the electrodes. Metal conductors and electrode sites are defined photolithographically and are insulated by deposited layers of silicon nitride and silicon dioxide. Deep reactive ion etching (DRIE) is used to etch through the entire thickness of the device from the front, stopping on the buried oxide layer. The backside of the wafer is then patterned and etched through using DRIE to the other side of the buried oxide layer. The electrodes are finally released by etching the buried oxide in buffered HF.

The final thickness of the shanks of these devices is determined by the depth of the buried oxide layer. Shanks from 6 to 25 μm have been fabricated. Bond pad regions are typically left at full wafer thickness (>500 μm). At the time of this writing, single and multiple shank acute designs have been fabricated with up to 32 sites and have been used successfully for acute recordings. In anticipation of chronic use, one group has added a process step to include surface topology on the devices in an effort to improve mechanical anchoring in the tissue.[60] In addition to being compatible with the inclusion of on-chip circuitry, this type of fabrication offers the advantage that all of the process steps are compatible with current foundry techniques.

7.4.2.3   Polyimide electrodes

An alternative approach that is being investigated for chronic multichannel recording is flexible substrate electrodes. While a flexible substrate may be difficult to insert into the brain without an insertion aid, it has been hypothesized that once it is successfully implanted its long-term performance may be superior to silicon because its elastic modulus more closely mimics that of the brain and it may therefore reduce trauma due to micromotion. Advancements in the material polyimide, originally investigated in the early 1980s as an electrode substrate,[39,40,42–44] have caused it to be recently revisited by several groups.[63–65] Basically, these planar structures consist of metal traces sandwiched between upper and lower layers of polyimide. The process begins with a silicon wafer that acts as a carrier throughout the process. Depending on how the devices are to be released from the wafer, a sacrificial layer may first be deposited on the wafer. Polyimide is then spun on to the desired thickness. If the polyimide is photodefinable it is next exposed and developed to define the base of the final device. If the polyimide is not photodefinable, RIE can be used for etching. Metal is deposited, patterned, and etched to form the interconnects. The top layer of polyimide is spun on, exposed, and developed to define the openings to the underlying metal which will become the recording sites and bond pads. Finally, the devices are released from the wafer either by dissolving the sacrificial layer in a wet etch[65] or by peeling them off with tweezers.[63]

Although it has not yet been attempted, these devices may be compatible with the inclusion of on-chip circuitry. For the interim, the Fraunhofer Institute has developed a novel technique for high-density interconnection of hybrid integrated circuit (IC) chips to polyimide devices.[66] In addition, unlike the two silicon-based devices described above, an integrated flexible cable can be extended off the electrode for chronic connection to the external world via a percutaneous connector. The final devices have thicknesses less than 20 μm. The Fraunhofer devices are designed for chronic interfacing with peripheral nerves. Rousche and his colleagues[65] are working toward a chronic intracortical array and have made three-dimensional structures by folding the two-dimensional devices into the appropriate shape (Figure 7.3). Both types of arrays have successfully obtained chronic recordings. Devices fabricated by Rousche et al. have "wells" included in the substrate that are designed to be seeded with bioactive materials to improve biological acceptance of the device.

7.4.2.4   The Michigan probe

Our group at the University of Michigan has developed a planar microelectrode array useful for both recording and stimulation. The probes, for which the basic structure is shown in Figure 7.4, are based on a silicon substrate, the thickness and shape of which are precisely defined using boron etch-stop micromachining.[67,68] The substrate supports an array of conductors that are insulated by thin-film dielectrics. Openings through the upper dielectrics are inlaid with metal to form the electrode sites for contact with the tissue, and the bond pads for connection to the external world. The Michigan probe is the focus of the remainder of this chapter.

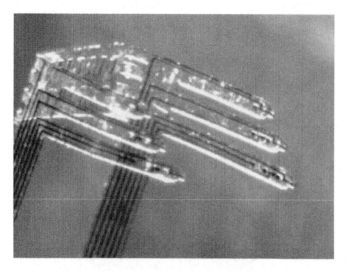

Figure 7.3 A flexible polyimide electrode array. This 12-channel device, developed at Arizona State University, has been folded to achieve a three-dimensional structure. The shanks are spaced on 100-µm centers. (From Rousche, P.J. et al., *IEEE Trans. Biomed. Eng.*, 48, 361, 2000. With permission.)

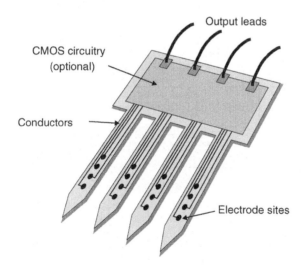

Figure 7.4 Basic structure of the Michigan silicon probe.

7.5   The Michigan probe

7.5.1   Fabrication

While the Michigan probe process is compatible with the inclusion of on-chip CMOS (complementary metal-oxide semiconductor) circuitry for signal conditioning and multiplexing and neural recordings have been made using

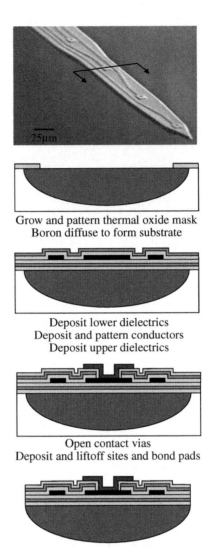

Grow and pattern thermal oxide mask
Boron diffuse to form substrate

Deposit lower dielectrics
Deposit and pattern conductors
Deposit upper dielectrics

Open contact vias
Deposit and liftoff sites and bond pads

Etch field dielectrics
Release probes in EDP

Figure 7.5 Fabrication sequence for the Michigan passive probe (not to scale). Fabrication requires only single-sided processing and uses eight masks.

these active arrays,[69] this chapter centers on the passive electrode array, which does not include circuitry. The Michigan passive probe process, outlined in Figure 7.5, starts with a silicon wafer of standard orientation <100> and thickness (500 μm). The wafer is thermally oxidized and this oxide is then selectively removed using photolithography to define the shape of the probe. Boron is diffused into the silicon in these open areas. Because the boron-doped silicon acts as an etch stop during the final release of the probes, the final thickness of the devices is defined by the depth of this diffusion. A standard deep diffusion results in a final device thickness of 15 μm. After diffusion, the masking oxide is stripped and a tri-layer stack of dielectrics (silicon dioxide, silicon nitride, and another layer of silicon dioxide) is deposited using low-pressure chemical vapor deposition (LPCVD). These dielectrics insulate the lower substrate from the thin-film interconnects that will be deposited in the next step. They are put down in this configuration and in thicknesses that result in a neutral mechanical stress condition (3000, 1500, 3000 Å) so that a flat device results in the end.

Next, the thin-film interconnect material that will lead from the electrode sites to the bond pads is deposited, patterned using photolithography, and etched using reactive ion etching (RIE) to form the multiple leads. The material typically used for recording interconnects is 8000 Å of phosphorus-doped polysilicon. The traces are then insulated from above using the same tri-layer stack of dielectrics used for lower insulation. Contact vias are next opened using a combination of wet (HF) and dry (RIE) etching to expose the interconnect traces for metallization which will form the electrode sites and bond pads. These sites and bond pads are formed using sputtering and then a lift-off process that removes the deposited metal from the wafer everywhere except in the electrode site and bonding pad areas. Typically, the electrode sites are made from iridium with a thin underlayer of titanium for adhesion, and the bond pads are made from gold with a chromium adhesion layer. Other site materials, including gold, titanium nitride, carbon, and platinum have been used depending on the intended application. It should be noted here that if iridium is used for the sites, the sites can be electrochemically activated and are suitable for stimulation as well as recording.[70–72] The next step in the process is to remove the dielectrics surrounding the probe shapes, which is accomplished using RIE.

The final process steps are performed to release the probes from the wafer. First the wafer is thinned from the backside using an unmasked etch in a hydrofluoric-nitric acid mixture. The thinned wafer is finally placed in an anisotropic silicon etch composed of ethylenediamine, pyrocatechol, and water (EDP). EDP etches away all of the undoped silicon and stops on the boron-doped areas. It does not attack any of the other exposed materials used in the probe process. Once they are removed from EDP, the probes are rinsed clean in solvents and deionized water and are ready for sorting and packaging.

One additional masking step can be added to the above process to allow the formation of a sharper probe tip, a thinner substrate, and/or an integrated ribbon cable for chronic use.[73] This extra step, a shallow boron diffusion, is performed just after the deep diffusion. The rest of the process (i.e., deposition and patterning of the leads, dielectrics, and sites) is carried out as described above. While the deep diffusion results in a final device thickness of about 15 μm, the shallow-diffused device thickness is only about 5 μm. The probe substrate is therefore thick enough to penetrate the pia of most animals; the shallow-diffused region, if used to form an integrated cable, is thin and flexible and minimizes tethering forces on an implanted probe. Probe tips formed using shallow diffusion have a radius of about 1 µm and enter the tissue with minimal dimpling. In addition, these sharp tips may be useful for penetrating tougher structures such as nerve and spinal cord.

The Michigan passive probe process with shallow boron diffusion uses eight masks. It is capable of yields in excess of 90%. Typically, 10 to 15 different designs are included on a given mask set and, depending on device sizes, a 4-inch silicon wafer can produce well over 1000 probes. Passive devices with lengths of 1.5 mm to 5 cm and widths as narrow as 5 µm have been fabricated with as many as 96 sites.

7.5.2   Design considerations

One of the attractive features of the Michigan and other planar photoengraved probes is the ability to customize design for specific experiments. The substrate can have any two-dimensional shape with single or multiple shanks, electrode sites can be of any surface area and can be placed anywhere along the shank(s) at any spacing, tips can be made very sharp or blunt, and features such as holes and barbs can be included for special applications such as sieve electrodes.[74–76] The NIH NCRR-sponsored University of Michigan Center for Neural Communication Technology (CNCT) has been providing probes to investigators since 1994. At the time of this writing, over 150 designs have been fabricated through the CNCT, a subset of which comprise a catalog of basic designs. A number of scientific papers and presentations, listed at www.engin.umich.edu/facility/cnct/papers.html, have resulted from use of these probes. The design freedom offered by this technology makes the devices attractive to researchers in a broad range of applications; however, there are some practical and process-enforced limitations on probe features that must be considered when selecting or designing a probe.

7.5.2.1   Shank length

When selecting or designing a probe, a first characteristic an investigator typically considers is shank length. This is the length of the device that will be inserted into tissue or the maximum depth of the intended target. While shank length can be considered a process-limited feature due to the maximum limit imposed by the size of the wafer, it is actually more practically limited as defined by stiffness and strength. Typical 15-µm-thick, 100-µm-wide probe shanks longer than about 6 mm may bend during insertion, causing the target to be missed. As described by Najafi and Hetke,[77] when choosing a shank length one must consider the buckling load, which defines the stiffness of the probe as it is pressed against the tissue, and the maximum stress at the bottom of the probe shank, which defines the strength of the probe. The buckling load is proportional to t^3W/L^2 and the maximum stress is proportional to t/L^2. Therefore, to design a probe that is stiff and strong with the given thickness $t = 15$ µm, the substrate width (W) should be

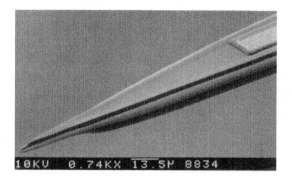

Figure 7.6 A sharp probe tip formed using a combination of shallow and deep boron diffusions. Probes with these sharp tips have been used to penetrate tough structures such as peripheral nerve and spinal cord.

increased, and the length (L) should be the minimum necessary to reach the intended target.

Insertion properties can also be improved by increasing the sharpness of the tip.[78] As described above, this can be achieved by using a shallow boron diffusion at the probe tip which results in a structure that is no more than 1 to 2 μm in diameter. In addition to the shallow diffusion, a very sharp taper angle (<10 degrees) can be used to improve penetration characteristics. Figure 7.6 shows a probe tip that is shallow diffused and has a taper angle of 18 degrees that has successfully penetrated sciatic and auditory nerves in cat, structures which are not penetrable with standard deep diffused tips.

7.5.2.2 Shank width

Taking into consideration the above discussion on shank length and the relationship of shank width to the strength and stiffness of long shanks, one should still try to minimize the width of the shank to make the device as noninvasive as possible. Shank width is a process enforced limitation that is controlled by the diffusion mask and lateral diffusion. As the width of the diffusion opening decreases, lateral diffusion of boron underneath the mask eventually limits the minimum achieveable shank width for a given substrate thickness.[79] Using an EDP etch alone, the minimum shank width achievable for a standard deep-diffused 15-μm-thick shank is about 15 μm because, as the mask width decreases, so does the thickness. Narrow "scaled" probes can be realized, however, using shallow-diffusion and/or an additional masking and etching step. To do this, deep RIE is used to etch into the silicon substrate and trim off the lateral diffusion, resulting in a narrower device than would be realizable using the EDP etch alone. Scaled probes have been fabricated with shanks as narrow as 5 μm. While 1.5-mm-long, 5-μm-thick scaled shanks have successfully penetrated the guinea pig pia mater and recorded from cortex, the mechanical practicality of these tiny devices has yet to be proven.

Another issue related to shank width is the interconnect lead width and spacing. Although submicron feature sizes are standard in industry, one must consider electrical crosstalk between very close leads. When more electrodes are to be accommodated on a narrower shank, the width of interconnect lines and the spacing between the lines must be reduced, resulting in an increased coupling capacitance. As described by Najafi et al.,[79] for probes with line widths and spacing as small as 1 μm, the crosstalk approaches 1%. This is acceptable for recording devices, as effects will be negligible compared to background noise. Even with feature sizes as small as 0.25 μm, the crosstalk is still less than 4%.

7.5.2.3 Substrate thickness

Substrate thickness is a feature that is of importance when it comes to insertion into tough tissue or into deep structures. The standard Michigan probe process is capable of producing probes 5 to 15 μm in thickness based on the time and temperature of the boron diffusion and hence the depth of the etch-stop. Probes 15 μm thick and 5 mm long and tapered from 15 to 120 μm in width are capable of penetrating the pia with minimal bending in most preparations. If the device is much longer than this, however, the buckling force is lower. If a very long shank is absolutely required by the application, a stronger, stiffer device can be achieved by forming a "box-beam" substrate. This type of device uses the same fabrication process as the Michigan chemical delivery probe[80] to form an open channel within the silicon substrate (Figure 7.7). This channel results in a box-beam structure that is inherently stiffer and stronger than a normal beam because the second moment of its cross-section is larger. This structure is being investigated for its efficacy in penetrating deeper and/or tougher structures, including the inferior colliculus, spinal cord, and cochlear nucleus in cats.

Figure 7.7 A probe with three channels, spaced 20 μm apart, buried within its substrate. While these probes are being developed for chemical delivery, they are also useful for applications where a stiffer device is necessary. (From Chen, J. et al., *IEEE Trans. Biomed. Eng.*, 44, 760, 1997. With permission.)

7.5.2.4 Site spacing

Another characteristic one considers when choosing a probe design is the center-to-center spacing between the recording sites. This feature is basically unlimited; the spacing between sites can be as small as a few microns to as large as several hundred microns or more. The choice will depend on the location of intended use as well as the type of data to be collected. For example, for purposes of spike sorting, signal improvement, and/or cell imaging, correlated signals are desired on more than one site. A general rule of thumb for these types of recordings is that the sites should be spaced at 50 μm or less.[21] Close-spaced sites can be in the form of a linear array, or in a tetrode configuration as shown in Figure 7.8. Sites that are spaced 100 μm apart will show some correlated as well as uncorrelated activity. When the sites are spaced at 200 μm or more, little or no coherency occurs between sites.[21] Arrays with this larger spacing, for example, can be used to investigate the layered organizations of cortical tissue.

7.5.2.5 Site area

The choice of site area for recording is an electrical consideration as it is related to thermal noise and the ability to record low-level signals and hence influences the achievable signal-to-noise ratio. There is a trade-off when choosing a site area. Although perhaps more selective in the signals it detects, a smaller site imposes a higher impedance, causing signal attenuation and added noise. In contrast, a larger site offers a lower impedance at the expense of being less selective and therefore picks up signals from a larger population of cells. While site areas on Michigan probes ranging from 70 to 4000 μm²

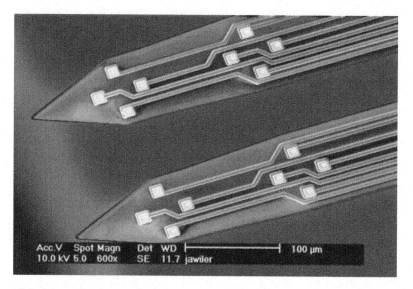

Figure 7.8 A two-shank tetrode array; sites are spaced at 25 μm on the diagonal, and shanks are spaced at 150 μm.

have successfully recorded neural spikes, the current typical recording site on a Michigan probe is just under 200 μm^2. The average 1-kHz impedance on a 200-μm^2 iridium site is around 2 MΩ and, with appropriate amplifiers, the baseline noise level is about 15 μV_{RMS}. This site area has evolved over many years as a "best" choice for maximizing signal-to-noise ratio and, in fact, is still evolving. It is likely that a single best choice for all brain regions does not exist due to variations in cell size and density. A systematic study is currently underway to investigate the recording quality of different site areas in a variety of brain structures.

If the electrode site is to be used for stimulation, a larger site size may be required to safely deliver the charge required for the application. In addition, the charge capacity of the site can be increased by an order of magnitude by forming iridium oxide through electrochemical activation.[70,72] This procedure involves cycling the electrode potential between positive and negative limits, which results in a porous, hydrous, multilayer oxide film that has a high charge capacity and is exceptionally resistant to dissolution and corrosion during stimulation.

7.5.3 Three-dimensional arrays

The Michigan probe process results in devices that are planar and two dimensional. Because neural systems are three dimensional, however, in order to fully instrument the target tissue volume it is important to be able to realize three-dimensional arrays. Such three-dimensional structures can be constructed from the two-dimensional components using microassembly techniques.[81,82] The procedure is based on inserting multiple two-dimensional probes into a silicon micromachined platform that is intended to sit on the cortical surface. Silicon spacers are inserted over the probes to hold them parallel to one another. Gold tab bond pads on the probes are bent at right angles to form a flat contact with mating pads on the platform. Ultrasonic bonding of these pad pairs is performed to establish electrical contact between the probes and the platform. Exposed connections are then potted in silicone rubber. An integrated cable on the platform carries the signals to a percutaneous connector.

Although these three-dimensional arrays are dense, they occupy less than 1% of the tissue volume into which they are inserted. Each shank of the array is only 15 μm thick. Recording arrays having up to 8×16 shanks on 200-μm centers have been fabricated, constructed, and used to record acute single units from guinea pig cortex. Following the methods of the Utah group, a dynamic insertion tool is used to insert the arrays into cortex with minimal traumatic injury. Unlike the UEA, these arrays can have multiple sites along each shank. This makes the need for on-chip electronics obvious; the signal leads must be multiplexed prior to lead transfer between the probe and the platform in order to make the transfer feasible. A device with on-chip CMOS signal processing circuitry has been created and used to record acute neural activity (Figure 7.9).

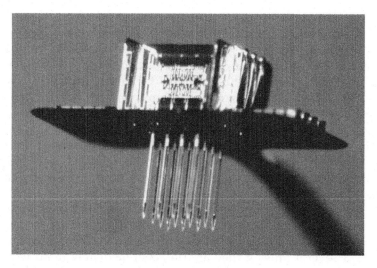

Figure 7.9 A three-dimensional silicon array that has been microassembled from planar components. This device has on-chip CMOS preamps and buffers. (From Bai, Q. et al., *IEEE Trans. Biomed. Eng.*, 47, 281, 2000. With permission.)

7.5.4 Acute probes

Acute recordings are made on anesthetized animals and continue anywhere from several hours to several days. These experiments are less complex than chronic experiments because the surgery is simpler, the animal is not awake during recordings, and the electrode assembly is not subjected to the corrosive physiological environment for an extended time. The acute Michigan probe assembly is shown in Figure 7.10. The probe is electrically connected to a printed circuit board using ultrasonic bonding, and the exposed connections are insulated with silicone rubber. This package is easy to handle

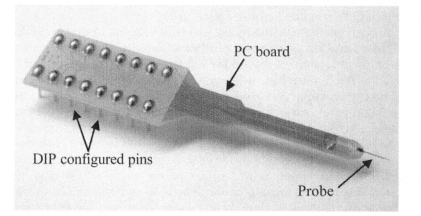

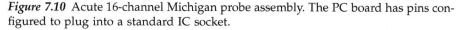

Figure 7.10 Acute 16-channel Michigan probe assembly. The PC board has pins configured to plug into a standard IC socket.

and use in most acute preparations. The pins on the circuit board are configured to plug into a standard IC socket which makes the package easily adaptable to most existing systems. In addition, several commercial vendors have developed headstage amplifiers that mate directly to Michigan probes. At the time of this writing, these vendors include Plexon, Inc. (www.plexoninc.com), Neuralynx, Inc. (www.neuralynx.com), and Tucker-Davis Technologies (www.tdt.com).

These acute headstages are intended to be mounted on a micromanipulator to permit the attached probe assembly to be lowered precisely into the brain. Prior to insertion, the dura must be pierced to permit probe penetration. Buckling of the shank can be minimized by inserting the probe shank(s) at an angle normal to the brain's surface. Once the device is lowered into the desired area, warm agar solution may be used over the exposed tissue to prevent dessication and minimize pulsation artifact. Because acute probes are not typically fixed in place with cement or acrylic, they can often be recycled several times if they are rinsed immediately upon withdrawal from the tissue with deionized water.

7.5.5 *Chronic recordings*

Chronic recordings are required in a variety of studies including those related to learning, adaptation, and plasticity. Stable chronic recordings for periods extending over years are critical if an electrode is to be used for studying or implementing a neural prosthesis. These recordings are more difficult to obtain than acute recordings not only because the animals are typically awake and sometimes behaving, but also because the electrode/tissue interface must remain stable over time. The natural response of the tissue to the foreign electrode body is to encapsulate it in a cellular sheath, which isolates the recording sites from the brain,[83] causing an eventual degradation or disappearance of recordings. All electrode types (i.e., wire, silicon, polyimide) are plagued with this problem to a certain degree and there are currently many efforts focused on improving the tissue interface.

Although telemetry systems are under development that will eliminate output leads and hence improve positional stability and perhaps the tissue interface, most systems in use today are tethered. These assemblies can be classified as fixed, movable, or floating. A *fixed* electrode is inserted into the brain and the proximal end is permanently attached to the skull or chamber above. Microwire arrays, arrays of individually insulated wires bundled together,[27] can be considered to be of the fixed type. While these devices may function well for recording from deep structures and from the cortex of small animals for periods extending over a year, larger animals and humans exhibit greater brain movement with respect to the skull which may lead to increased relative movement of the electrode and more tissue damage. Avoidance of this type of damage is especially critical for very long-term implants such as will be required for a prosthesis. The *movable* electrode, as its name implies, is one that is moved periodically to record from similar or even different brain regions over time.[84–87] Although these electrodes can be quite useful for certain experiments and in fact may alleviate some of the problems associated with electrode failure due to encapsulation, they are not suitable for studies that require chronic recordings from the same cells over time and are probably not the choice for use in a prosthesis. For long-term chronic use, at least until telemetry systems can be fully implemented, the *floating* electrode is most likely the device of choice. In this configuration, a flexible interconnect between the implanted electrode and the percutaneous connector permits the electrode to move with the brain's natural pulsations, minimizing damage caused by relative movement. A classic example of a floating electrode is the "hatpin" electrode developed by Schmidt et al.,[88] which remained functional for 1144 days in monkey cortex.

While there are examples of applications where the Michigan probe has been used in short-term chronic experiments in fixed and movable configurations,[89–91] the chronic assembly that is being developed for long-term use is of the floating type. The flexible interconnect to the probe is achieved using the integrated, multilead silicon ribbon cable described above. These 5-μm-thick cables are extremely flexible; even with multiple leads they are roughly 100 times more flexible than a single 25-μm-diameter gold wire.[73] Because the probe and ribbon cable are monolithic, no further electrical connection is required between the two. The leads are brought out to the external world through a percutaneous connector. An intermediate PC board serves to transfer the ultrasonically bonded leads from the probe to the pins on the connector (Figure 7.11). Currently, 16- and 32-channel versions of this assembly have been developed. Recording periods of over one year have been achieved with this device.[73] The vendors mentioned above are developing

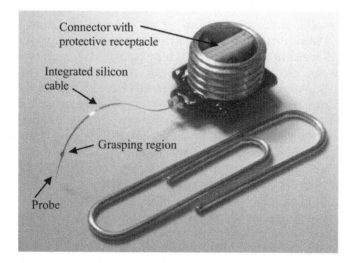

Figure 7.11 Chronic 16-channel Michigan probe assembly. An integrated silicon cable forms the interconnect between the probe and percutaneous connector.

miniature headstages that mate to these assemblies, which will be suitable for use on behaving animals.

Appropriate surgical techniques are the first step in the realization of a successful chronic implant. The probe assembly should first be sterilized using ethylene oxide. As with the acute probes, the dura must be removed at the insertion point to permit penetration. The percutaneous connector is then mounted on the skull using dental acrylic. The probe is next inserted either by hand by grasping the silicon bead at the probe/cable junction with forceps or with a vacuum-pick mounted on a micromanipulator. Quick-cure silicone rubber can then be placed over the implanted electrode to provide support for it and protection for the exiting cable. Dental acrylic is finally used to seal the entire area.

7.5.6   Example applications

The Michigan probes have been utilized in a variety of brain regions for extracellular recording of unit and field activity. Several applications, acute and chronic, will be highlighted here.

7.5.6.1   Mapping using multi-unit recordings

Dr. John Middlebrooks and his colleagues at the University of Michigan are using acute probes to map "cortical images" of activation in guinea pig auditory cortex in response to stimulation through a cochlear implant.[92] The goal of these studies is to identify electrode configurations that optimize information transmission from the cochlear implant to the cortex and to use the results to improve the design of speech processing strategies. In these experiments, a silicon probe was used to record multi-unit auditory cortical activity elicited by cochlear implant stimuli that varied in electrode configuration, location of stimulation within the cochlea, and stimulus level. Cochlear electrode configurations that were investigated were monopolar (MP), bipolar (BP+N) with N active electrodes between the active and return electrodes, tripolar (TP) with one active electrode and two flanking return electrodes, and common ground (CG) with one active electrode and up to five return electrodes. The probe design that was chosen for these experiments was a 16-channel single-shank probe with 100-μm site spacing. This design permitted simultaneous recording of the resulting spatio-temporal pattern of neural activity across the tonotopic axis of the auditory cortex. Figure 7.12 shows an example of cortical images resulting from various cochlear stimulating electrode configurations (rows) and stimulus levels (columns). This figure illustrates several effects. First, it can be seen that the cortical images increase in width in response to increasing current levels. In addition, at the 7-dB level, the images resulting from MP, BP, and CG stimulation spanned nearly the entire 1.5 mm of the cortical recording array, while the image resulting from the TP stimulus remained more restricted even at the higher current levels.

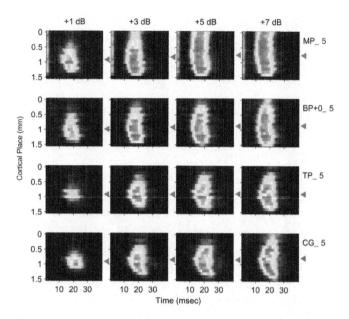

Figure 7.12 Cortical images, derived from Michigan probe recordings for various cochlear-stimulating electrode configurations (rows) and stimulus levels (columns) in the guinea pig. The *y*-axis of each panel represents the silicon probe recording site location relative to the most caudal site, and the *x*-axis represents the time after stimulus onset. The gray scale represents normalized spike probability (see original publication for color version of figure). (From Bierer, J. A. and Middlebrooks, J. C., *J. Neurophysiol.*, 87, 478, 2002. With permission.)

7.5.6.2 Current-source density analysis using field recordings

In addition to their usefulness in studying single- and multi-unit extracellular activity, by broadening the bandwidth of external instrumentation the probes can be used to record field or population potentials. Dr. Gyorgy Buzsaki's group at Rutgers University uses Michigan probes to investigate simultaneous unit and field activity in hippocampus and neocortex. He has published several studies in which he performs current-source density (CSD) analysis to precisely localize current sources and sinks in the awake rat in order to investigate the cellular-synaptic generation of physiologically relevant cortical potentials.[93–96] CSD profiles are derived from field recordings. Changes in the sign of the CSD as a function of depth are interpreted as a change in the local divergence of current flow and correspond to the presence of current sources or sinks.[97]

In one example of Buzsaki's CSD work,[94] he used Michigan probes to study the mechanism of extracellular current generation in the neocortex underlying sleep spindles and spike-and-wave patterns, two low-frequency (<15 Hz) rhythms associated with the thalamocortical system. These spontaneous rhythms were compared to evoked potentials in response to stimulation of various thalamic nuclei. Field potentials and unit activity were

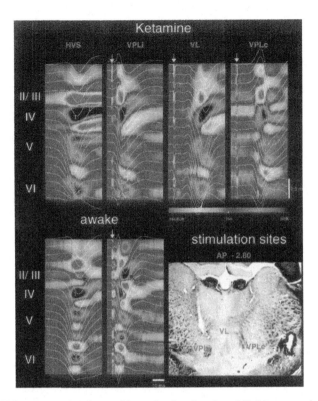

Figure 7.13 Voltage vs. depth profiles, obtained using Michigan probes, superimposed on CSD maps of spike-and-wave (HVS) and thalamic evoked responses (VPLi, VL, VPLc) in awake and anesthetized rats. Thalamic evoked field responses and spontaneously occurring spike-and-wave discharges resulted in spatially similar localized sinks and sources in the various cortical layers (see original publication for color version of figure). (From Kandel, A. and Buzsaki, G., *J. Neurosci.*, 17, 6783, 1997. With permission.)

recorded simultaneously along trajectories perpendicular to the cortical layers using a 16-site silicon probe with a site spacing of 100 μm. For these short-term chronic studies, recordings were obtained in both the anesthetized and the awake rat using a movable probe.

In Figure 7.13, voltage vs. depth profiles is superimposed onto CSD maps of spike-and-wave discharges (HVS) and thalamic evoked responses (VPLi, VL, VPLc) in the awake rat and during Ketamine anesthesia. Thalamic evoked field responses and spontaneously occurring spike-and-wave discharges resulted in spatially similar localized sinks and sources in the various cortical layers. The findings from this study indicate that the major extracellular currents underlying sleep spindles, HVSs, and evoked responses result from activation of intracortical circuitry rather than from thalamocortical afferents alone. The multisite recording technique allowed for the continuous, simultaneous recording of field potentials, current-source density distributions, and unit activity. The Michigan probe has become a powerful tool for CSD analysis. Other examples of CSD studies performed using Michigan probes can be found in References 98 to 102.

7.5.6.3 Chronic recording from behaving animals

Dr. Daryl Kipke of the University of Michigan is working to characterize the electrode-tissue interface with the aim of developing techniques to better control this interface to improve long-term functionality for eventual clinical use. Kipke's group is studying a variety of electrode types including microwires,[27] polyimide,[65] and the Michigan probe. For these studies, the probes are implanted into cerebral cortex of the rat. Long-term recordings (months to over a year) have been obtained in all successful implants. The longest implant to date has been 15 months, at which time the experiment was terminated for complications unrelated to electrode stability. Signal-to-noise ratios have remained high throughout implant durations, as evidenced in Figure 7.14. This figure shows data taken 382 days after the implant.

7.6 Future directions

Solid-state processing technology has permitted the realization of a number of novel devices for extracellular multichannel recording. Advantages over conventional metal electrodes include batch fabrication, highly reproducible geometrical and electrical characteristics, and small feature sizes. In addition, some of the devices described are compatible with on-chip circuitry for signal conditioning. These devices are being used routinely in acute and short-term chronic experiments for recording single units and extracellular field potentials. The future power of these devices, however, will lie in their ability to make long-term chronic connections with the brain to provide control signals

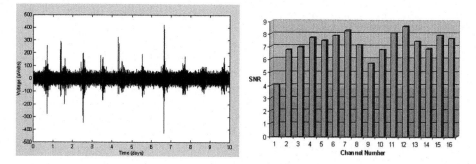

Figure 7.14 Chronic data from a Michigan probe 382 days post-implant. The probe was placed in barrel cortex of the rat. The left panel shows analog data from one channel (sample rate, 40 kHz) that has been evoked using whisker stimulation, and the right panel shows signal-to-noise ratios for all 16 channels of the probe. (Data courtesy of Rio Vetter and Dr. Daryl Kipke.)

for closed-loop prosthetic systems. There are several areas of research focused on improving long-term chronic recordings.

Bioactive coatings hold much promise for improving the electrode-tissue interface and enhancing chronic recording stability.[103–105] Coating the surface of the electrode may be advantageous for several reasons. First, the coating may act as a mechanical buffer between a relatively rigid electrode substrate and the soft neural tissue. In addition to making the electrode more tissue friendly, this may improve positional stability. Second, the coating topology could be made such that it promotes favorable geometrical reactions with cells. Finally, it may be possible to seed the coating with biologically active molecules such as neurotrophins or living cells to positively influence cellular behavior.

Reducing the number of output leads should also improve chronic lifetime by limiting tethering forces on the probe, thereby reducing relative micromotion, migration, and consequent adverse tissue response. Silicon probes are being developed with on-chip CMOS circuitry to amplify, multiplex, and transmit analog neural activity using a total of only three leads for power and bidirectional data transfer. For future prosthetic systems, however, data and power transfer between the implanted unit and the outside world should be achieved without any interconnect leads. Circuits and technologies are being developed for wireless operation and control of implantable recording[106,107] and stimulation[108–110] microsystems.

Acknowledgments

This work is supported by NIH NINDS (N01-NS7-2364) and NIH NCRR (P41-RR09754).

References

1. Hubel, D.H. and Wiesel, T.N., Receptive fields, binocular interaction and functional architecture in the cat's visual cortex, *J. Physiol.*, 160, 106, 1962.
2. Swindale, N.V., Matsubara, J.A., and Cynader, M.S., Surface organization of orientation and direction selectivity in cat area 18, *J. Neurosci.*, 7, 1414, 1987.
3. Lennie, P., Krauskopf, J., and Sclar, G., Chromatic mechanisms in striate cortex of macaque, *J. Neurosci.*, 10, 649, 1990.
4. Chapin, J.K. et al., Real-time control of a robot arm using simultaneously recorded neurons in the motor cortex, *Nature Neurosci.*, 2, 664, 1999.
5. Isaacs, R.E., Weber, D.J., and Schwartz, A.B., Work toward real-time control of a cortical neural prosthesis, *IEEE Trans. Rehab. Eng.*, 8, 196, 2000.
6. Serruya, M.D. et al., Instant neural control of a movement signal, *Nature*, 416, 141, 2002.
7. Taylor, D.M., Tillery, S.I.H., and Schwartz, A.B., Direct cortical control of three-dimensional neuroprosthetic devices, *Science*, 296, 1829, 2002.
8. Schanne, O.F. et al., Electrical properties of glass microelectrodes, *Proc. IEEE*, 56, 1072, 1968.

9. Marg, E. and Adams, J.E., Indwelling multiple microelectrodes in the brain, *Electroencephalogr. Clin. Neurophysiol.*, 23, 277, 1967.
10. Kaltenbach, J.A. and Gerstein, G.L., A rapid method for production of sharp tips on preinsulated microwires, *J. Neurosci. Meth.*, 16, 283, 1986.
11. Kruger, J. and Bach, M., Simultaneous recording with 30 microelectrodes in monkey visual cortex, *Exp. Brain Res.*, 41, 191, 1981.
12. Reitbock, H.J., Fiber microelectrodes for electrophysiological recordings, *J. Neurosci. Meth.*, 8, 249, 1983.
13. Robinson, D.A., The electrical properties of metal microelectrodes, *Proc. IEEE*, 56, 1065, 1968.
14. Abeles, M. and Goldstein, M.H., Multispike train analysis, *Proc. IEEE*, 65, 1977, 762.
15. Gerstein, G.L. and Clark, W.A., Simultaneous studies of firing patterns in several neurons, *Science*, 143, 1325, 1965.
16. Gerstein, G.L. and Perkel, D.H., Simultaneously recorded trains of action potentials: analysis and functional interpretation, *Science*, 164, 828, 1969.
17. Gerstein, G.L. and Perkel, D.H., Mutual temporal relationships among neuronal spike trains, *Biophys.J.*, 12, 453, 1972.
18. Eggermont, J.J., Johannesma, P.M., and Aertsen, A.M., Reverse-correlation methods in auditory research, *Q. Rev. Biophys.*, 16, 341, 1983.
19. Wheeler, B.C. and Heetderks, W.J., A comparison of techniques for classification of multiple neural signals, *IEEE Trans. Biomed. Eng.*, 29, 752, 1982.
20. McNaughton, B.L., O'Keefe, J., and Barnes, C.A., The stereotrode: a new technique for simultaneous isolation of several single units in the central nervous system from multiple unit records, *J. Neurosci. Meth.*, 8, 391, 1983.
21. Drake, K.L. et al., Performance of planar multisite microprobes in recording extracellular activity, *IEEE Trans. Biomed. Eng.*, 35, 719, 1988.
22. Wilson, M.A. and McNaughton, B.L., Dynamics of the hippocampal ensemble code for space, *Science*, 261, 1055, 1993.
23. O'Keefe, J. and Recce, M.L., Phase relationship between hippocampal place units and the EEG theta rhythm, *Hippocampus*, 3, 317, 1993.
24. Gray, C.M. et al., Tetrodes markedly improve the reliability and yield of multiple single-unit isolation from multi-unit recordings in cat striate cortex, *J. Neurosci. Methods*, 63, 43, 1993.
25. Gozani, S.N. and Miller, J.P., Optimal discrimination and classification of neuronal action potential waveforms from multiunit, multichannel recordings using software-based linear filters, *IEEE Trans. Biomed. Eng.*, 41, 358, 1994.
26. Nicolelis, M.A.L. et al., Reconstructing the engram: simultaneous, multisite, many single neuron recordings, *Neuron*, 18, 529, 1997.
27. Williams, J.C., Rennaker, R.L., and Kipke, D.R., Long-term neural recording characteristics of wire microelectrode arrays implanted in cerebral cortex, *Brain Res. Protocols*, 4, 303, 1999.
28. Porada, I. et al., Rabbit and monkey visual cortex: more than a year of recording with up to 64 microelectrodes, *J. Neurosci. Meth.*, 95, 13, 2000.
29. Wise, K.D., Angell, J.B., and Starr, A., An integrated-circuit approach to extracellular microelectrodes, *IEEE Trans. Biomed. Eng.*, 17, 238, 1970.
30. Wise, K. and Angell, J.B., A low-capacitance multielectrode probe for use in extracellular neurophysiology, *IEEE Trans. Biomed. Eng.*, 22, 212, 1975.

31. Starr, A., Wise, K.D., and Csongradi, J., An evaluation of photoengraved microelectrodes for extracellular single-unit recording, *IEEE Trans. Biomed. Eng.*, 20, 291, 1973.

32. Prohaska, O. et al., A multielectrode for intracortical recordings produced by thin-film technology, *Electroencephalogr. Clin. Neurophysiol.*, 42, 421, 1977.

33. Mercer, H.D. and White, R.L., Photolithographic fabrication and physiological performance of microelectrode arrays for neural stimulation, *IEEE Trans. Biomed. Eng.*, 25, 494, 1978.

34. Pickard, R.S., A review of printed circuit microelectrodes and their production, *J. Neurosci. Meth.*, 1, 319, 1979.

35. Pochay, P. et al., A multichannel depth probe fabricated using electron beam lithography, *IEEE Trans. Biomed. Eng.*, 26, 199, 1979.

36. May, G.A., Shamma, S.A., and White, R.L., A tantalum-on-sapphire microelectrode array, *IEEE Trans. Electron. Devices*, 26, 1932, 1979.

37. Prohaska, O. et al., A 16-fold semi-microelectrode for intracortical recording of field potentials, *Electroencephalogr. Clin. Neurophysiol.*, 47, 629, 1979.

38. Kuperstein, M. and Whittington, D.A., A practical 24 channel microelectrode for neural recording *in vivo*, *IEEE Trans. Biomed. Eng.*, 28, 288, 1981.

39. Shamma-Donoghue, S.A. et al., Thin-film multielectrode arrays for a cochlear prosthesis, *IEEE Trans. Electron. Devices*, 29, 136, 1982.

40. White, R.L. et al., Thin film electrodes for an artificial ear, *J. Vac. Sci. Technol.*, 1, 287, 1983.

41. Takahashi, K. and Matsuo, T., Integration of multi-microelectrode and interface circuits by silicon planar and three-dimensional fabrication technology, *Sensors Actuators*, 5, 89, 1984.

42. Barth, P.W., Bernard, S.L., and Angell, J.B., Flexible circuit and sensor arrays fabricated by monolithic silicon technology, *IEEE Trans. Electron. Devices*, 32, 1202, 1985.

43. Harrison, R.V. et al., Technical development of an implantable cochlear prosthesis in Canada, *J. Otolaryngol.*, 16, 311, 1987.

44. van der Puije, P.D., Pon, C.R., and Robillard, H., Cylindrical cochlear electrode array for use in humans, *Ann. Otol. Rhinol. Laryngol.*, 98, 466, 1989.

45. Pickard, R.S., Wall, P., and Ubeid, M., Recording neural activity in the honeybee brain with micromachined silicon sensors, *Sensors and Actuators*, B1, 460, 1990.

46. Blum, N.A. et al., Multisite microprobes for neural recordings, *IEEE Trans. Biomed. Eng.*, 38, 68, 1991.

47. Kovacs, G.T.A., Storment, C.W., and Rosen, J.M., Regeneration microelectrode array for peripheral nerve recording and stimulation, *IEEE Trans. Biomed. Eng.*, 39, 893, 1992.

48. Kovacs, G.T.A. et al., Silicon-substrate microelectrode arrays for parallel recording of neural activity in peripheral and cranial nerves, *IEEE Trans. Biomed. Eng.*, 41, 567, 1994.

49. Kewley, D.T. et al., Plasma-etched neural probes, *Sensors Actuators A*, 58, 27, 1997.

50. Burmeister, J.J., Moxon, K., and Gerhardt, G.A., Ceramic-based multisite microelectrodes for electrochemical recordings, *Anal. Chem.*, 72, 187, 2000.

51. Yoon, T.H. et al., A micromachined silicon depth probe for multichannel neural recording, *IEEE Trans. Biomed. Eng.*, 47, 1082, 2000.

52. Campbell, P.K. et al., A silicon-based, three-dimensional neural interface: manufacturing processes for an intracortical electrode array, *IEEE Trans. Biomed. Eng.*, 38, 758, 1991.

53. Jones, K.E., Campbell, P.K., and Normann, R.A., A glass/silicon composite intracortical electrode array, *Ann. Biomed. Eng.*, 20, 423, 1992.

54. Nordhausen, C.T., Maynard, E.M., and Normann, R.A., Single unit recording capabilities of a 100 microelectrode array, *Brain Res.*, 726, 129, 1996.

55. Rousche, P.J. and Normann, R.A., A method for pneumatically inserting an array of penetrating electrodes into cortical tissue, *Ann. Biomed. Eng.*, 20, 413, 1992.

56. Maynard, E.M., Nordhausen, C.T., and Normann, R.A., The Utah intracortical electrode array: a recording structure for potential brain-computer interfaces, *Electroencephalogr. Clin. Neurophysiol.*, 102, 228, 1997.

57. Rousche, P.J. and Normann, R.A., Chronic recording capability of the Utah intracortical electrode array in cat sensory cortex, *J. Neurosci. Meth.*, 82, 1, 1998.

58. Branner, A. and Normann, R.A., A multielectrode array for intrafascicular recording and stimulation in sciatic nerve of cats, *Brain Res. Bull.*, 51, 293, 2000.

59. Branner, A., Stein, R.B., and Normann, R.A., Selective stimulation of cat sciatic nerve using an array of varying-length microelectrodes, *J. Neurophys.*, 85, 1585, 2001.

60. Cheung, K. et al., A new neural probe using SOI wafers with topological interlocking mechanisms, *Proc. 1st Int. IEEE-EMBS Conf. on Microtechnologies in Medicine and Biology*, Lyon, 2000, p. 507.

61. Norlin, P. et al., A 32-site neural recording probe fabricated by DRIE of SOI substrates, *J. Micromech. Microeng.*, 12, 414, 2002.

62. Ensell, G. et al., Silicon-based microelectrodes for neurophysiology, micromachined from silicon-on-insulator wafers, *Med. Biol. Eng. Comput.*, 38, 175, 2000.

63. Stieglitz, T., Beutel, H., and Meyer, J.-U., A flexible, light-weight multichannel sieve electrode with integrated cables for interfacing regenerating peripheral nerves, *Sensors Actuators A*, 60, 240, 1997.

64. Rodriguez, F.J. et al., Polyimide cuff electrodes for peripheral nerve stimulation, *J. Neurosci. Meth.*, 98, 105, 2000.

65. Rousche, P.J. et al., Flexible polyimide-based intracortical electrode arrays with bioactive capability, *IEEE Trans. Biomed. Eng.*, 48, 361, 2001.

66. Meyer, J.-U. et al., High density interconnects and flexible hybrid assemblies for active biomedical implants, *IEEE Trans. Advanced Packaging*, 24, 366, 2001.

67. Najafi, K. and Wise, K.D., A high-yield IC-compatible multichannel recording array, *IEEE Trans. Electron. Devices*, 32, 1206, 1985.

68. BeMent, S.L. et al., Solid-state electrodes for multichannel multiplexed intracortical neuronal recording, *IEEE Trans. Biomed. Eng.*, 33, 230, 1986.

69. Bai, Q. and Wise, K.D., Single-unit neural recording with active microelectrode arrays, *IEEE Trans. Biomed. Eng.*, 48, 911, 2001.

70. Robblee, L.S., Lefko, J.L., and Brummer, S.B., Activated Ir: an electrode suitable for reversible charge injection in saline solution, *J. Electrochem. Soc.*, 130, 731, 1983.

71. Anderson, D.J. et al., Batch-fabricated thin-film electrodes for stimulation of the central auditory system, *IEEE Trans. Biomed. Eng.*, 36, 673, 1989.

72. Weiland, J.D. and Anderson, D.J., Chronic neural stimulation with thin-film, iridium oxide electrodes, 47, 911, 2000.

73. Hetke, J.F. et al., Silicon ribbon cables for chronically implantable microelectrode arrays, *IEEE Trans. Biomed. Eng.*, 41, 314, 1994.

74. Akin, T. et al., A micromachined silicon sieve electrode for nerve regeneration applications, *IEEE Trans. Biomed. Eng.*, 41, 305, 1994.

75. Kovacs, G.T.A. et al., Silicon-substrate microelectrode arrays for parallel recording of neural activity in peripheral and cranial nerves, *IEEE Trans. Biomed. Eng.*, 41, 567, 1994.

76. Mensinger, A.F. et al., Chronic recording of regenerating VIIIth nerve axons with a sieve electrode, *J. Neurophysiol.*, 83, 611, 2000.

77. Najafi, K. and Hetke, J.F., Strength characterization of silicon microprobes in neurophysiological tissues, *IEEE Trans. Biomed. Eng.*, 37, 474, 1990.

78. Edell, D.J. et al., Factors influencing the biocompatibility of insertable silicon microshafts in cerebral cortex, *IEEE Trans. Biomed. Eng.*, 39, 635, 1992.

79. Najafi, K., Ji, J., and Wise, K.D., Scaling limitations of silicon multichannel recording probes, *IEEE Trans. Biomed. Eng.*, 37, 1, 1990.

80. Chen, J. et al., A multichannel neural probe for selective chemical delivery at the cellular level, *IEEE Trans. Biomed. Eng.*, 44, 760, 1997.

81. Hoogerwerf, A.C. and Wise, K.D., A three-dimensional microelectrode array for chronic neural recording, *IEEE Trans. Biomed. Eng.*, 41, 1136, 1994.

82. Bai, Q., Wise, K.D., and Anderson, D.J., A high-yield microassembly structure for three-dimensional microelectrode arrays, *IEEE Trans. Biomed. Eng.*, 47, 281, 2000.

83. Turner, J.N. et al., Cerebral astrocyte response to micromachined silicon implants, *Exp. Neurol.*, 156, 33, 1999.

84. Sinnamon, H.M. and Woodward, D.J., Microdrive and method for single unit recording in the active rat, *Physiol. Behav.*, 19, 451, 1977.

85. deCharms, R.C., Blake, D.T., and Merzenich, M.M., A multielectrode implant device for the cerebral cortex, *J. Neurosci. Meth.*, 93, 27, 1999.

86. Venkatachalam, S., Fee, M.S., and Kleinfeld, D., Ultra-miniature headstage with 6-channel drive and vacuum-assisted micro-wire implantation for chronic recording from the neocortex, *J. Neurosci. Meth.*, 90, 37, 1999.

87. Szabo, I. et al., The application of printed circuit board technology for fabrication of multi-channel micro-drives, *J. Neurosci. Meth.*, 105, 105, 2001.

88. Schmidt, E.M., McIntosh, J.S., and Bak, M.J., Long-term implants of Parylene-C coated microelectrodes, *Med. Biol. Eng.*, 26, 96, 1988.

89. Bragin, A. et al., Gamma (40–100 Hz) oscillation in the hippocampus of the behaving rat, *J. Neurosci.*, 15, 47, 1995.

90. Bragin, A. et al., Multiple site silicon-based probes for chronic recordings in freely moving rats: implantation, recording and histological verification, *J. Neurosci. Meth.*, 98, 77, 2000.

91. Swadlow, H.A. and Gusev, A.G., The influence of single VB thalamocortical impulses on barrel columns of rabbit somatosensory cortex, *J. Neurophysiol.*, 83, 2802, 2000.

92. Bierer, J.A. and Middlebrooks, J.C., Auditory cortical images of cochlear-implant stimuli: dependence on electrode configuration, *J. Neurophysiol.*, 87, 478, 2002.

93. Bragin, A. et al., Gamma (40–100 Hz) oscillation in the hippocampus of the behaving rat, *J. Neurosci.*, 15, 47, 1995.

94. Kandel, A. and Buzsaki, G., Cellular-synaptic generation of sleep spindles, spike-and-wave discharges, and evoked thalamocortical responses in the neocortex of the rat, *J. Neurosci.*, 17, 6783, 1997.

95. Buzsaki, G. and Kandel, A., Somadendritic backpropagation of action potentials in cortical pyramidal cells of the awake rat, *J. Neurophysiol.*, 79, 1587, 1998.

96. Kamondi et al., Theta oscillations in somata and dendrites of hippocampal pyramidal cells *in vivo*: activity-dependent phase-precession of action potentials, *Hippocampus*, 8, 244, 1998.

97. Nicholson, C. and Freeman, J.A., Theory of current source-density analysis and determination of conductivity tensor for anuran cerebellum, *J. Neurophysiol.*, 38, 356, 1975.

98. Canning, K.J. et al., Physiology of the entorhinal and perirhinal projections to the hippocampus studied by current source density analysis, *Ann. N.Y. Acad. Sci.*, 911, 55, 2000.

99. Castro-Alamancos, M.A., Origin of synchronized oscillations induced by neocortical disinhibition *in vivo*, *J. Neurosci.*, 20, 9195, 2000.

100. Prechtl, J.C., Bullock, T.H., and Kleinfeld, D., Direct evidence for local oscillatory current sources and intracortical phase gradients in turtle visual cortex, *Proc. Natl. Acad. Sci. USA*, 97, 877, 2000.

101. Ahrens, K.F. and Freeman, W.J., Response dynamics of entorhinal cortex in awake, anesthetized, bulbotomized rats, *Brain Res.*, 911, 193, 2001.

102. Biella, G., Uva, L., and de Curtis, M., Network activity evoked by neocortical stimulation in area 36 of the guinea pig perirhinal cortex, *J. Neurophysiol.*, 86, 164, 2001.

103. Huber, M. et al., Modification of glassy carbon surfaces with synthetic laminin-derived peptides for nerve cell attachment and neurite growth, *J. Biomed. Mater. Res.*, 41, 278, 1998.

104. Cui, X. et al., Electrochemical deposition and characterization of conducting polymer polypyrrole/PSS on multichannel neural probes, *Sensors Actuators A*, 93, 8, 2001.

105. Cui, X. et al., Surface modification of neural recording electrodes with conducting polymer/biomolecule blends, *J. Biomed. Mater. Res.*, 56, 261, 2001.

106. Akin, T., Najafi, K., and Bradley, R.M., A wireless implantable multichannel digital neural recording system for a micromachined sieve electrode, *IEEE J. Solid-State Circuits*, 33, 109, 1998.

107. Takeuchi, S. and Shimoyama, I., An RF-telemetry system with shape memory alloy microelectrodes for neural recording of freely moving insects, *Proc. 1st Int. IEEE-EMBS Conf. on Microtechnologies in Medicine and Biology*, Lyon, 2000, p. 491.

108. Rangarajan, R., Von Arx, J.A., and Najafi, K., Fully integrated neural stimulation system (FINESS), *Proc. IEEE Midwest Symp. on Circuits and Systems*, 3, 1082, 2000.

109. Lieu, W. et al., A neuro-stimulus chip with telemetry unit for retinal prosthetic device, *IEEE J. Solid-State Circuits*, 35, 1487, 2000.

110. Suaning, G.J. and Lovell, N.H., CMOS neurostimulation ASIC with 100 channels, scaleable output, and bidirectional radio-frequency telemetry, *IEEE Trans. Biomed. Eng.*, 48, 248, 2001.

section four

Processing neural signals

chapter eight

Wavelet methods in biomedical signal processing

Kevin Englehart, Philip Parker, and Bernard Hudgins

Contents

8.1   Introduction

It has long been recognized that important features of biomedical signals exist in both the time and frequency domains. More recently, a greater understanding of physiological systems has been achieved by articulating the relationship between the time and frequency characteristics of biological signals. The most widely used tool for analyzing signals in the frequency domain is the Fourier transform.[1] Fourier methods assume that the signals of interest consist of a (perhaps infinite) summation of sinusoids. This leaves Fourier analysis poorly suited to model signals with transient components. Alas, this is almost invariably the nature of signals of biological origin, in which the information of interest is often well localized temporally (or spatially). This is certainly the case in the electroencephalograph, the electrocardiogram, the myoelectric signal, and evoked potentials of many forms.

Despite its disadvantages, Fourier methods have been widely used in the analysis of biomedical signals. In an attempt to localize information temporally, the approach taken is to divide long-term signals into shorter windows and perform a Fourier transform on each window. This is the short-time Fourier transform (STFT), which yields a time–frequency representation by expressing the frequency spectrum of each time window. By shortening the windows of the STFT to achieve better time resolution, however, one degrades the resolution of the signal in frequency. Longer data windows improve the spectral resolution but, in degrading the temporal resolution, may violate the assumption of stationarity within each window.

In the past decade, researchers in applied mathematics and signal processing have developed wavelet methods, a powerful new framework for analyzing transient phenomena in signals.[2,15] The fundamental difference between wavelet and Fourier methods is the manner in which they localize information in the time-frequency plane. Whereas the STFT localizes information in time–frequency cells that are of fixed aspect ratio, wavelet methods are capable of trading off time and frequency resolution. Specifically, at low frequencies the representation has good frequency resolution, and at high frequencies good temporal resolution. This tradeoff has been shown to be particularly adept at localizing information in physical signals, particular those of biological origin.

This chapter first provides a brief account of the development of wavelet methods, followed by an introduction to the mathematics of wavelet analysis. In the context of the strengths that wavelet methods have to offer, the application of these techniques in biomedical signal processing is described. The final section of this chapter elaborates on a wavelet-based feature set that has resulted in robust, pattern-recognition-based control of powered artificial limbs.

8.2   A brief history of wavelets

The concept of analyzing signals at different scales or resolutions is not new.[3] Only recently, however, has wavelet theory been unified, mostly due to the work of Grossman (a theoretical physicist) and Morlet (a geophysical engineer) in the late 1970s.[4] Wavelet basis functions were shown to be effective in localizing short, high-frequency sound waves, as well as subtle frequency changes in long, low-frequency tones. In 1985, Meyer (a mathematician at Ecole Polytechnique, near Paris) first demonstrated the construction of an orthonormal wavelet basis with excellent time–frequency localization properties.[5] Shortly thereafter, Battle and LeMarie (colleagues of Meyer) came up with an orthonormal spline wavelet using completely different techniques.[6]

In 1986, Mallat (a graduate student at Penn State University, a specialist in computer vision) conceived of a pyramid algorithm for wavelet decomposition and, in collaboration with Meyer, developed the theory of *multiresolution analysis*.[7] This framework explained all the "miracles" in the wavelet bases constructed up to then and made it very easy to construct new orthonormal wavelet bases. More importantly, multiresolution analysis led to a simple and recursive filtering algorithm to compute the wavelet decomposition of a signal.

A filtering analogy of wavelet decomposition led to the formulation of the subband filtering approach. Here, lowpass and highpass quadrature mirror filters (QMFs) simplified the implementation of wavelet analysis and synthesis. By considering wavelet decomposition as a series of highpass and lowpass filtering operations, wavelet packet bases were conceived[8] that have generalized and extended the capabilities of wavelet bases.

8.3 The theory of wavelets

This section provides a brief introduction to wavelet theory. This is by no means meant to be an exhaustive treatment of wavelet mathematics but simply is intended to provide a basic reference and to examine the properties of the wavelet transform in relation to biomedical applications. Several excellent texts offer a complete background for the interested reader.[5,9,10,15] From the point of view of the practitioner, there are fundamentally two types of wavelet transforms:

1. *Redundant transforms.* These transforms include the continuous wavelet transform and wavelet frames and provide an overcomplete description of the time–frequency plane. They are usually preferable for signal analysis, feature extraction, and detection tasks, as they are shift invariant.*
2. *Non-redundant transforms.* These are orthogonal and biorthogonal wavelet bases. These transforms are desirable when some type of data reduction is preferred, or when orthogonality of the representation is an important factor.

The computational complexity of a decomposition in terms of wavelet bases using Mallat's fast algorithm[7] is typically orders of magnitude less than even the most efficient algorithms for redundant transforms.[11,12] Because of the computational cost of redundant transforms, many researchers have

---

* We may say that a transform is *translation invariant* if, for a simple time shift in the input signal, the transform coefficients experience the same simple shift. Due to the aliasing that occurs in a discrete wavelet decomposition, the transform is not translation invariant.

applied non-redundant transforms to signal analysis, feature extraction, and detection tasks and have obtained satisfactory results.

8.3.1   The continuous wavelet transform

The *continuous wavelet transform* (CWT) provides a variable coverage of the time-frequency plane. The transform is defined as:

$$\text{CWT}_x(\tau, a) \;=\; \int x(t)\, \psi_{a,\tau}^{*}(t)\, dt$$

where

$$\psi_{a,\tau}(t) \;=\; \frac{1}{\sqrt{a}}\, \psi\!\left(\frac{t-\tau}{a}\right)$$

where $\psi(t)$ is a prototype window referred to as the *mother wavelet*. The mother wavelet has the property that the set $\{\psi_{a,\tau}(t)\}_{a,\tau \in Z}$ forms an orthonormal basis in $L^2(\Re)$, the space of square-integrable functions.* The analysis determines the correlation of the signal with *shifted* (by τ) and *scaled* (by a) versions of the mother wavelet.

The parameter scale in the wavelet analysis is similar to the scale used in maps. As in the case of maps, high scales correspond to a non-detailed global view (of the signal), and low scales correspond to a detailed view. Similarly, in terms of frequency, low frequencies (high scales) correspond to a global information of a signal (that usually spans the entire signal), whereas high frequencies (low scales) correspond to a detailed information of a hidden pattern in the signal (that usually lasts a relatively short time). Figure 8.1 demonstrates the nature of a symmetric wavelet at various scales and translations. Note that, at small scales, a temporally localized analysis is done; as scale increases, the breadth of the wavelet function increases, thereby analyzing with less time resolution but greater frequency resolution. Notice, as well, that the wavelet functions are *bandpass* in nature, thus partitioning the frequency axis. In fact, a fundamental property of wavelet functions is that

$$c \;=\; \frac{\Delta f}{f}$$

where Δf is a measure of the bandwidth, f is the center frequency of the passband, and c is a constant. The wavelet functions may therefore be viewed as a bank of analysis filters with a constant relative passband (a "constant-Q" analysis).

---

* A function f is in $L^2(\Re)$ if $L^2(\Re)$ if $\int_{\Re} f^2 < \infty$.

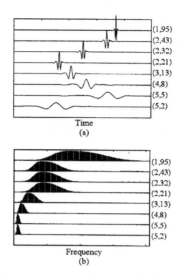

Figure 8.1 Some symmetric wavelets at various scales and locations. The couplet (*a*, *τ*) marking each waveform denotes scale and shift, respectively: (a) time domain and (b) frequency domain of each wavelet.

8.3.2 The discrete wavelet transform

Fast wavelet transforms are obtained through *multiresolution analysis*; a pyramid algorithm with its origins in image processing that was adapted for wavelet analysis by Mallat and Meyer.[7] The fast wavelet transform uses a series of linear filters — lowpass and highpass — to decompose the signal into low- and high-frequency components. The algorithm also combines these filters with *downsampling* operations, that is, steps that decimate the signal at each stage, halving the data each time. This feature accounts for the algorithm's speed, because the downsampling reduces the computations at each iteration geometrically — at *j* iterations, the number of samples being manipulated shrinks by 2^j. This yields very efficient algorithms: most N-point wavelet transforms have complexity on the order of $O(N)$, whereas a Fourier transform is of the order $O(N\log N)$.[13] If the transform's complexity is CN (where C is a constant), then C depends on the wavelet chosen.[14] If C is small, then computing the wavelet transform requires about the same effort as trivial tasks such as copying or rescaling a signal. Wavelets involving only a few terms subtend great efficiency (a small value of C), such as those developed by Daubechies.[15]

In its discrete form, $a = a_0^j$ and $\tau = n \cdot a_0^{-j}$ where *j* and *n* are integers; this is referred to as the *discrete wavelet transform* (DWT). In this context, *j* controls the dilation or compression of the wavelet function, and *n* controls the translation in time. If we choose $a_0 \approx 1$ and $n \approx 0$, we are close to the continuous case. The choice of $a = 2^j$ and $\tau = n \cdot 2^{-j}$ is the most common choice; this is referred to as a *dyadic wavelet basis*. The dyadic form also simplifies the constraints on the mother wavelet to achieve another desirable property: *orthogonality* of the analysis windows, which subtends an efficient representation.

One can interpret the discrete wavelet transform in a number of ways, but the most common are via *subspace analogy* and *multiscale filter banks*.

8.3.2.1 Subspace analogy

In the dyadic case, the mother wavelet is

$$\psi_{j,n}(t) \;=\; 2^{-j/2}\psi(2^{-j}t - n)$$

and the set $\{\psi_{j,n}(t)\}_{j,n \in Z}$ forms a sparse orthonormal basis of $L^2(\mathfrak{R})$.[16] This means that the wavelet basis induces an orthogonal decomposition of any function in $L^2(\mathfrak{R})$:

$$L^2(\mathfrak{R}) \;=\; \bigoplus_{j} \Omega_{j,1}$$

where $\Omega_{j,1}$ is the subspace spanned by $\{\psi_{j,n}(t)\}_{n \in Z}$ and $\oplus$ is the direct sum of the subspaces. Thus, a complete description of the original signal is available from a direct sum of orthogonal subspaces. The subscript 1 is used to differentiate $\Omega_{j,1}$ from its dual space $\Omega_{j,0}$, which is defined as follows. A wavelet function is always associated with a companion: the *scaling function*, $\varphi(t)$, which is also sometimes called the *father wavelet*:

$$\varphi_{j,n}(t) \;=\; 2^{-j/2}\varphi(2^{-j}t - n)$$

and, like the wavelet function, the set $\{\varphi_{j,n}(t)\}_{n \in Z}$ forms a sparse orthonormal basis of $L^2(\mathfrak{R})$. The scaling function induces a chain of nested subspaces:

$$\Omega_{J,0} \subset \Omega_{J-1,0} \subset \cdots \subset \Omega_{1,0} \subset \Omega_{0,0}$$

where $\Omega_{j,0}$ is the subspace spanned by $\{\varphi_{j,n}\}_{n \in Z}$.

What does this really mean? The nature of the scaling function is that a projection of the original signal $x(t)$ onto the space $\Omega_{j,0}$ is a lowpass operation. Specifically, the projection of $x(t)$ onto $\Omega_{j,0}$ is an approximation at scale $a = 2^j$:

$$A_j[n] \;=\; \sum_{n \in Z} a_{j,n}\,\varphi_{j,n}(t)$$

where $a_{j,n} = \langle x(t), \varphi_{j,n}(t)\rangle$ are the *scaling coefficients*. We define $\Omega_{0,0}$ (scale $a = 2^0 = 1$) to be the space of the original signal $x(t)$; that is, $A_0[n] = x[n]$. Thus, $\Omega_{J,0} \subset \Omega_{J-1,0} \subset \cdots \subset \Omega_{1,0} \subset \Omega_{0,0}$ is actually a sequence of successively coarser approximations of $x(t)$ as scale ranges from 0 to J. The subspaces subtended by the wavelet and the scaling functions are related such that:

$$\Omega_{j,0} = \Omega_{j+1,0} \oplus \Omega_{j+1,1} \quad \text{for } j = 0, 1, \ldots, J$$

meaning that $\Omega_{j,1}$ contains the detail needed to go from a coarser to a finer level of approximation. The detail component of $x(t)$ at scale $a = 2^j$ is:

$$D_j[n] = \sum_{n \in Z} d_{j,n} \, \psi_{j,n}(t)$$

where $d_{j,n} = \langle x(t), \psi_{j,n}(t) \rangle$ are the *wavelet coefficients*. This is a bandpass operation, as $D_j[n] = A_{j+1}[n] - A_j[n]$. Therefore, the wavelet transform may be viewed as a way to represent $\Omega_{0,0}$ as a direct sum of mutually orthogonal subspaces:

$$\Omega_{0,0} = \left(\bigoplus_{j=1}^{J} \Omega_{j,1} \right) \oplus \Omega_{J,0}$$

This concept is perhaps more clearly illustrated in Figure 8.2. The wavelet transform processes a signal by decomposing it into successive approximation $A_j[n] \in \Omega_{j,0}$ and detail $D_j[n] \in \Omega_{j,1}$ signals. The approximation signal is resampled at each stage, and the detail coefficients are kept. The aim of the analysis is to arrive, starting from the original sampled signal $x[n] = A_0[n]$, at a decomposition into detail and approximation signals:

$$\left\{ D_1[n], \ldots, D_J[n], A_J[n] \right\}$$

This is the wavelet transform: for a decomposition into J scales, the transform coefficients consist of J scales of detail coefficients and, at the Jth scale, the lowest-level approximation signal.* Equivalently, one may retain the wavelet and scaling coefficients which generate these signals:

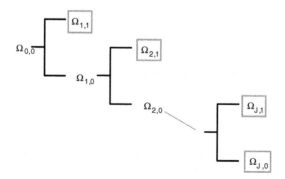

Figure 8.2 Subband decomposition analogy of the wavelet transform. The symbols surrounded by gray represent those subspaces kept intact by the wavelet transform.

* This is illustrated by example in Figure 8.5, in the section "Interpretation of the Wavelet Transform."

$$\left\{ d_{1,n}, \ldots, d_{J,n}, a_{J,n} \right\}$$

8.3.2.2 Multiscale filter banks

For those so inclined, it is instructive to repeat the theory of wavelet decomposition in the language of signal processing. Let the map $\Omega_{j,0} \rightarrow \Omega_{j+1,0}$ be represented by the operator H, and the map $\Omega_{j,0} \rightarrow \Omega_{j+1,1}$ be represented by G. This is illustrated in Figure 8.3 as an operation on an N-sample signal $\mathbf{x} = \left\{ x[n] \right\}_{n=0}^{N-1} \in \Omega_{0,0}$.

As the figure implies, the operators consist of two separate stages: $\mathbf{g} = \left\{ g[n] \right\}_{n=0}^{L-1}$ and $\mathbf{h} = \left\{ h[n] \right\}_{n=0}^{L-1}$ are highpass and lowpass filters of length L, respectively, and the $\downarrow 2$ operator implies a decimation by 2. This may be written as:

$$(H\mathbf{x})_n = \sum_{k=0}^{L-1} h[k] \, x[2n-k], \quad (G\mathbf{x})_n = \sum_{k=0}^{L-1} g[k] \, x[2n-k],$$

for $n = 0, 1, \ldots, N-1$. This makes intuitive sense, that the approximation $\Omega_{j,0} \rightarrow \Omega_{j+1,0}$ should be obtained *via* lowpass filtering, and the detail $\Omega_{j,0} \rightarrow \Omega_{j+1,1}$ by highpass filtering. If $\mathbf{x} = \left\{ x[n] \right\}_{n=0}^{N-1} \in \mathfrak{R}^N$ is a vector to be analyzed, the operators transform the vector $\mathbf{x}$ into two subsequences $G\mathbf{x}$ and $H\mathbf{x}$, of length $N/2$. Next, the same operations are applied to the vector of the lower frequency band $H\mathbf{x}$ to obtain $H^2\mathbf{x}$ and $GH\mathbf{x}$ of lengths $N/4$. If the process is repeated $J \leq \log_2 N$ times, the wavelet decomposition may be then written as:

$$\left\{ G\mathbf{x}, \ GH\mathbf{x}, \ GH^2\mathbf{x}, \ \ldots, \ GH^J\mathbf{x}, \ H^{J+1}\mathbf{x} \right\}$$

of length N. This is demonstrated in Figure 8.4. The wavelet transform thus analyzes the data by partitioning its frequency content dyadically finer and finer toward the low frequency region (and coarser and coarser in the time domain).

The issue to be addressed now is, how are the filters $\mathbf{h}$ and $\mathbf{g}$ determined? The operators H and G are called *perfect reconstruction* or *quadrature mirror filters* (QMFs) if they satisfy the following orthogonality conditions:

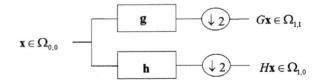

Figure 8.3 Projection operators H and G.

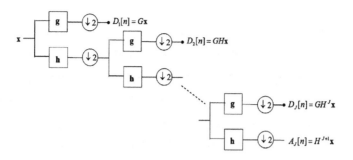

Figure 8.4 The subband coding analogy of the DWT.

$$HG^* = GH^* = 0 \quad \text{and} \quad HH^* = GG^* = \mathbf{I}$$

where * is the adjoint operator and $\mathbf{I}$ is the identity matrix. Various design criteria (concerning regularity, symmetry, etc.) on the lowpass filter coefficients $\mathbf{h}$ can be found in Daubechies.[15] Once the $\mathbf{h}$ are fixed, we can specify the QMFs by setting $g[n] = (-1)^n h[L - 1 - n]$.

The QMFs are related to the wavelet and scale functions by:[16]

$$h[n] = \left\langle \varphi(t), \sqrt{2}\varphi(2t - n) \right\rangle$$

and

$$g[n] = \left\langle \psi(t), \sqrt{2}\varphi(2t - n) \right\rangle$$

If H and G are QMFs, then the perfect reconstruction property allows exact reconstruction of the original signal from the wavelet transform coefficients.[17] Because the wavelet bases are orthogonal, reconstruction takes the form of *upsampling* (inserting zeros at every other sample) and filtering by the QMFs, exactly the reverse operation of the subband decomposition scheme.*

8.3.3   Interpretation of the wavelet transform

After the mathematic details presented above, an example might clarify things a bit. Consider a $N = 1024$ point ramp function, depicted in Figure 8.5a. The ramp is then subjected to a multiresolution analysis using a Symmlet-5 wavelet, decomposed to level 7 (the decomposition could have been carried out to a maximum depth of $J = \log_2 1024 = 10$, but this would have unnec-

---

* For orthogonal wavelet bases, the filters g and h are used for both decomposition and synthesis. An alternative scheme, referred to as *biorthogonal* wavelet bases, allows perfect reconstruction with synthesis filters that are different than the analysis filters.

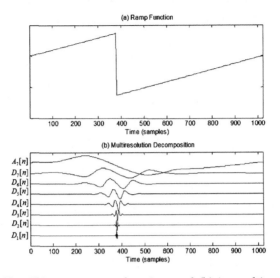

Figure 8.5 (a) A $N = 1024$ point ramp function, and (b) its multiresolution wavelet analysis.

essarily crowded the plot*). Clearly, the detail signal $D_1[n]$ picks up the sharpest detail of the step transition (the highest bandpass); the detail signals from $D_2[n]$ to $D_7[n]$ describe the signal content in bandpass regions of successively lower center frequency. The approximation signal $A_7[n]$ at level 7 provides the lowpass estimation of the signal.

This decomposition can also be represented by the wavelet coefficients themselves, as shown in Figure 8.6. Note the decimation that occurs when progressing from $d_{1,n}$ to $d_{7,n}$. There are $N/2 = 512$ coefficients in $d_{1,n}$, $N/4 = 256$ coefficients in $d_{2,n}$, and so on, until at level 7, $d_{7,n}$ and $a_{7,n}$ have eight coefficients each. This yields a total of 1024 coefficients, the same as the original signal. Due to the perfect regeneration properties of the QMFs, the original signal can be exactly reconstructed from the wavelet coefficients. Note that signal filtering, denoising, or compression is possible by computing the wavelet coefficients and then modifying or eliminating some before reconstructing the signal.

8.3.4 *The wavelet packet transform*

The *wavelet packet transform*[5,18–20] is a generalized version of the wavelet transform; it retains not only the low- but also the high-frequency subband, performing a decomposition upon both at each stage. As a result, the tiling of the time-frequency plane is configurable: the partitioning of the frequency

---

* Although a decomposition can carried out to its maximal depth $J = \log_2 N$, the decomposition may terminate at any level. This means that the process of dividing the frequency axis into finer and finer segments toward the low frequency range stops, and the final detail and approximation signals are of length greater than one. This may be desirable if the subdivision of bands beyond a certain scale does not yield subbands with a significant energy component.

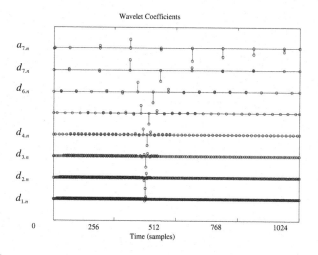

Figure 8.6 The wavelet coefficients of the ramp function shown in Figure 8.5a. Decomposition has been carried out to level 7.

axis may take many forms to suit the needs of the application. This is demonstrated in Figure 8.7.

Starting with a signal $\mathbf{x}$ of length N samples, the first level of the decomposition generates the lowpass and highpass subbands ($H\mathbf{x}$ and $G\mathbf{x}$, respectively) as with the wavelet transform, each of half the length of $\mathbf{x}$. The second level decomposition generates four subsequences: $H^2\mathbf{x}$, $GH\mathbf{x}$, $HG\mathbf{x}$, and $G^2\mathbf{x}$ halved in length again; the decomposition to this level is shown in Figure 8.8.

Figure 8.7 The time-frequency plane tiling of (a) a wavelet basis, and (b) an arbitrary wavelet packet basis.

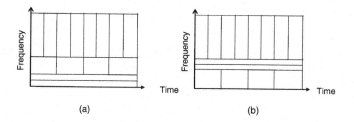

Figure 8.8 The first two levels of decomposition in a wavelet packet transform. The lowpass operations (H) occur to the left, the highpass (G) to the right.

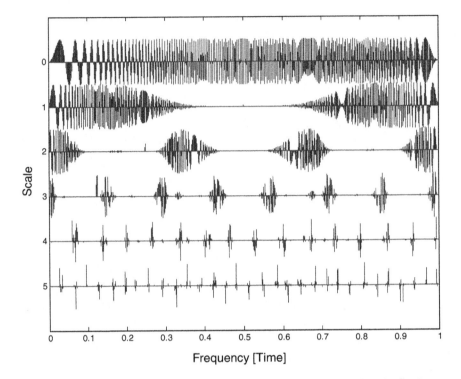

Figure 8.9 The binary wavelet packet tree for a 256-sample chirp signal. A Coiflet-5 wavelet was used, which provides a good tradeoff between time and frequency localization.

This process is repeated J times, where $J \leq \log_2 N$, resulting in JN coefficients. For a subband at scale j, there are $N/2^j$ wavelet packet coefficients sampled at $f_s/2^j$ spanning the entire time record. This iterative process generates a *binary wavelet packet tree* structure where the nodes of the tree represent subspaces with different frequency localization characteristics.

As an example, consider a sinusoid with a frequency that increases linearly with time (a linear chirp). The binary wavelet packet tree of this signal is depicted in Figure 8.9. Each subband in the full decomposition is separated by dashed lines; the wavelet packet coefficients within each subband are shown as solid lines. The vertical axis indicates the depth of decomposition; the zero scale is actually the original signal. The horizontal axis is labeled Frequency [Time], implying that, for a given scale (level), the center frequency localization of each subband increases as one progresses from left to right.

The time-frequency localization of each subband is evident; within each subband, the significant coefficients are temporally localized in the range where the chirp passes through the frequency range corresponding to that subband. The computational cost of this decomposition is on the order of $O(JN) \leq O(N\log_2 N)$.[20]

In subspace notation, the root node of the tree is $\Omega_{0,0}$. The node $\Omega_{j,k}$ is decomposed into two orthogonal subspaces H: $\Omega_{j,k} \to \Omega_{j+1,2k}$ and G: $\Omega_{j,k} \to$

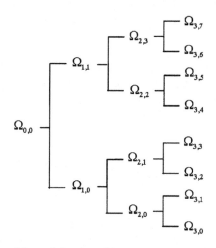

Figure 8.10 A decomposition of $\Omega_{0,0}$ into binary tree-structured subspaces using the wavelet packet transform (with $J = 3$).

$\Omega_{j+1,2k+1}$. Here, j denotes scale, as before, and k indicates the subband index within the scale.* This may be expressed as:

$$\Omega_{j,k} = \Omega_{j+1,2k} \oplus \Omega_{j+1,2k+1}$$

for $j = 0, \ldots, J$ and $k = 0, \ldots, 2^j - 1$, which generates the entire decomposition to scale J. A decomposition to scale $J = 3$ is shown in Figure 8.10.

The set of subspaces in the binary tree is a redundant set. Indeed, there are more than $2^{2^{(J-1)}}$ possible orthonormal bases in this binary tree.[20] Each subspace $\Omega_{j,k}$ is spanned by 2^{n_0-j} basis vectors $\left\{ w_{j,k,n} \right\}_{n=0}^{2^{n_0-j}-1}$. In a given family of wavelet packet bases, the parameter j indicates scale, as before. The parameters k and n roughly indicate frequency band** and the center of the waveform, respectively. The vector $w_{j,k,n}$ is roughly centered at $2^j n$, has length of support $\approx 2^j$, and oscillates $\approx k$ times. For $j = 0$, we have the original signal space (the standard Euclidean basis [$\Re^N$]).

For a J-scale decomposition, the resulting binary tree yields more than $2^{2^{(J-1)}}$ orthonormal bases (or coordinate systems), all of which offer a complete description of the space of the original signal. The power of the wavelet packet transform is that a "best basis" can be chosen for a specific task, if it can be properly identified from the ensemble of possible candidates.

To determine the *best basis*, it is necessary to evaluate and compare the efficacy of many bases. To this end, a *cost function* must be chosen to represent

---

* The wavelet transform has only two subbands per scale, high and low, with $k = 0, 1$.
** The binary tree subband structure, as generated by successive applications of H and G, is called *Paley* or *natural* ordering. In this form, the frequency band of $\Omega_{j,k}$ does not monotonically increase with k. This may be corrected by *Gray-code* permutation.[22]

the goal of the application. The best-basis selection algorithm has its origins in signal compression,[20,21] and the cost functions associated with compression all entail some form of entropy measure. Saito[22] has described a modification to the best-basis algorithm to suit classification problems. Termed the *local discriminant basis* algorithm, it replaces the entropy measure with distance measures between classes of signals.

8.3.5 Time-frequency interpretation

One of the most important ways of interpreting the information content in a signal is through its time–frequency response (TFR). The time–frequency representation tends to be more intuitive than time scale, perhaps because of the historical use of time–frequency methods such as the short-time Fourier transform and the Wigner–Ville distribution. The TFR of the wavelet transform is computed by translating scale into frequency. The generation of the TFR of a wavelet transform is a frequent requirement of many investigators, and it is surprising that the mechanics of this task are overlooked in many texts.

Consider a signal of length $N = 2^{n_0} = 8$ ($n_0 = 3$) sampled at f_s Hz. The tiling of the time-scale grid (with a wavelet decomposition to scale $J = \log_2 N = 3$) will look like that shown in Figure 8.11a. The selection of a dyadic sampling grid ($a = 2^j$, $\tau = n2^j$) means that the resulting TFR will be *critically sampled*; there will be N discrete time–frequency cells.

For a given information cell, the temporal resolution is $\Delta t = 2^j \cdot T_s$, and the scale resolution is $\Delta a = 2^{1-j}$. The subscript 1 or 0 on the scale parameter denotes whether the subband is *detail* (bandpass) or *approximation* (lowpass). As expected, the lowest frequency subband is composed of the approximation coefficients from the highest scale level. At scale level j, there are $((NT_s)/(\Delta t)) = N \cdot 2^{-j} = 2^{n_0-j}$ information cells. The subband at each scale $j < J$ is described by 2^{n_0-j} detail coefficients. At scale J, one bandpass subband is described by 2^{n_0-J} detail coefficients, and one lowpass subband is described by 2^{n_0-J} approximation coefficients.

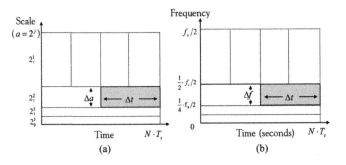

Figure 8.11 The tiling of the time-scale domain (a) and the time-frequency domain (b) of the DWT. In (a), the scale is subscripted by a 1 for detail subbands and 0 for the approximation subband.

Translating this time-scale response into a time–frequency response proceeds as follows. Referring now to Figure 8.11b, the temporal resolution of a given cell is still $\Delta t = 2^j \cdot T_s$. To compute the frequency segmentation corresponding to scale, we may identify the vertical distance from 0 Hz to the bottom of each cell as f, and the height of that cell as Δf. On this grid, $\Delta f = (\Delta a/2^j)(f_s/2) = 2^{-j}(f_s/2)$. The distance f to the lower bound of a cell is given by $f = \Delta f \cdot q$, where $q = 1$ if the subband consists of *detail* coefficients and $q = 0$ if it is an *approximation* subband. Here, we have chosen a definition of Δt and Δf so as to yield a critically sampled time–frequency grid. It is interesting to note that $\Delta t \cdot \Delta f = (2^j \cdot T_s)(2^{-j}f_s/2) = 1/2$, which is somewhat greater than the minimum set by the Heisenberg bound. A chosen wavelet therefore, need not meet the Heisenberg bound but rather must satisfy $\Delta t \cdot \Delta f \leq 1/2$ to avoid overlap amongst adjacent cells in the TFR.* This critically sampled grid also ensures perfect reconstruction of the original signal. The details of constructing the TFR of the wavelet packet transform are somewhat more involved than the wavelet transform. For details, the reader is referred to Simoncelli et al.[23]

8.4 Wavelets in biomedical signal processing

The reader should be cautioned that, as with many tools, there have been inappropriate uses of the wavelet transform:

"When all you have is a hammer, everything is a nail."

For the most part, however, there have been some remarkable advances in biomedical signal processing that exploit the capabilities of the wavelet and wavelet packet transform. An important aspect of biomedical signals is that the information of interest is often a combination of phenomena that are transient (e.g., spikes and action potentials) and diffuse (e.g., small oscillations, and semiperiodic activations). This requires analysis methods that are sufficiently versatile to capture events that present these extremes in time–frequency localization. It is this ability of the wavelet transform that has motivated its use in biomedical applications.

The popularity of wavelets in biomedicine is evidenced by the steadily growing number of publications: two special issues published in *IEEE Engineering in Medicine and Biology Magazine* (March/April 1995 and October 1995), workshops at many biomedical engineering conferences, and some excellent texts[24,25] and review papers[26,27] specifically addressing biomedical applications of wavelets.

The continuous wavelet transform (CWT), in essence, performs a correlation analysis upon the signal of interest. Correspondingly, one can expect

---

* Most wavelets transforms satisfy $\Delta t \cdot \Delta f \leq 1/2$ sufficiently; any overlap is not severe enough to distort the TFR.

its output to be a maximum when the signal most resembles the mother wavelet $\psi_{a,\tau}(t)$. In this sense, the CWT behaves as a *multiscale matched filter*. This property of the CWT has motivated its use as a *detector*, looking for a pattern (or elements of a pattern) that may be scaled by various amounts. The CWT has been used in this context for detecting microcalcifications in mammograms[28] and the QRS complex in the ECG signal.[29]

The ability of the wavelet transform to offer good frequency resolution at low frequencies and good time resolution at high frequencies has made it appealing for the time–frequency analysis of many biological signals. Recent examples of this include the characterization of heart beat sounds,[30,31] the detection of ventricular late potentials in the ECG,[32–35,50] and the analysis of the EEG.[36–38]

The efficiency of representation offered by orthogonal wavelet transforms has made them an attractive choice in data compression and noise removal; this can be achieved by modifying or discarding certain wavelet coefficients that are insignificant.[39] Indeed, wavelet packet based compression has set the standard for fingerprint image compression, offering a 20:1 compression ratio, as compared to the 5:1 ratio offered by JPEG encoding.[14] Orthogonal wavelet decomposition has been found to be very useful in image coding[40] and compression of magnetic resonance images[41] and digital mammograms,[56] the ECG,[42,43] and the myoelectric signal.[44,45] They have also been used for noise removal in evoked response potentials[46] and in the ECG.[47]

Wavelet representations have also been used as feature sets in pattern recognition problems. The problem of characterizing ECG patterns is of enormous interest; it is not surprising that attempts have been made to use wavelet-based feature sets. It has been shown that wavelet features outperform a set of morphological features extracted from a quadratic time–frequency distribution when classifying heart sounds.[48] Others have successfully used Mallat's local extrema representation[49] to recognize cardiac patterns.[50,51] Wavelet coefficients have been used for identification of respiratory-related evoked potentials,[52] image texture segmentation,[53] and classification of magnetic resonance images.[54] Wavelet and wavelet packet coefficients have also been successfully used in myoelectric signal classification for control of prosthetic limbs.[55,56]

Wavelet transforms have also been used for characterization of somatosensory event related potentials for tracking cerebral hypoxia,[57] pitch detection of speech signals,[58] and many applications in biomedical imaging.[26] This list is by no means comprehensive, but rather represents the diversity of wavelet applications in biomedicine.

The next section elaborates upon the use of the wavelet transform in pattern recognition. The control of powered artificial limbs using the myoelectric signal requires a representative feature set to be computed in real time. The computational efficiency of the orthogonal wavelet transform, and its ability to model transient myoelectric signal patterns have resulted in a control system of greater accuracy than its predecessors.

8.5 Myoelectric control applications

8.5.1 Background

The myoelectric signal (MES), recorded at the surface of the skin, has been used for many diverse applications, including clinical diagnosis, and as a source of control of assistive devices and schemes of functional electrical stimulation. This section details recent work that improves the functionality and ease of control of powered upper-limb prostheses using the myoelectric signal.

Many myoelectric control systems are currently available that are capable of controlling a single device in a prosthetic limb, such as a hand, an elbow, or a wrist. These systems extract control information from the MES based on an estimate of the amplitude[59] or the rate of change[60] of the MES. Although these systems have been very successful, they do not provide sufficient information to reliably control more than one function (or device);[61] the extension to controlling multiple functions is a much more difficult problem. Unfortunately, these are the requirements of those with high-level (above the elbow) limb deficiencies, and these are the individuals who could stand to benefit most from a functional replacement of their absent limbs.

In an attempt to increase the information extracted from the MES, investigators have proposed a variety of feature sets and have utilized pattern recognition methods to discriminate amongst desired classes of limb activation. Most work in MES classification has considered the steady-state MES, collected during a maintained (usually constant-force) contraction. Hudgins et al.[62] were the first to consider the information content of the transient bursts of myoelectric activity that accompany the onset of contraction. These data were acquired in a single MES channel, using a widely spaced bipolar electrode pair placed on the biceps and triceps. The data were acquired by triggering on an amplitude threshold of a moving average of the absolute value of the transient waveforms. The structure inherent in the early portion of these transient bursts (roughly the first 100 msec) suggested a promising means of MES classification. Hudgins developed a control scheme based upon a set of simple time domain statistics and a multilayer perceptron artificial neural network classifier, capable of classifying four types of upper limb motion from the MES acquired from the biceps and triceps. This control scheme demonstrated greater discriminant ability than any other at the time and allowed a user to evoke control using muscular contractions that resemble those normally used to produce motion in an intact limb. This system has been implemented as an embedded controller[63] and is currently undergoing clinical trials.

Although the accuracy of Hudgins' controller is good (roughly 10% error, averaged over a set of 10 subjects), there is an obvious motivation to reduce the error as much as possible. This would enhance the usability of the system as perceived by the user and allow greater dexterity of control. A number of approaches have appeared in the literature that have used the transient signal as prescribed by Hudgins, seeking to improve the accuracy of the approach using dynamic artificial neural networks,[64] genetic algorithms,[65] fuzzy logic classifiers,[66] and self-organizing neural networks.[67]

8.5.2   Wavelet based signal representation

Instead of focusing upon the classifier, the authors have demonstrated in previous work that the classification performance is more profoundly affected by the choice of feature set.[56] Specifically, a wavelet-based approach is described that, in direct comparison to Hudgins' time-domain approach, exhibits superior performance. The performance of Hudgins' time-domain (TD) feature set, and those based upon the short-time Fourier transform (STFT), the wavelet transform (WT), and the wavelet packet transform (WPT) were compared using a new dataset. This work is briefly described here.

A roster of 16 healthy subjects participated in the study. Each subject generated six different classes of motion: closing/opening the hand, flexion/extension of the wrist, and ulnar/radial deviation of the wrist. The data were acquired from four channels located on the medial side, top, lateral side, and bottom of the forearm, as shown in Figure 8.12. Each pattern consists of two channels of 256 points, sampled at 1000 Hz. The data were divided into a training set (100 patterns) and a test set (150 patterns).

Each of the STFT, WT, and WPT implementations were empirically optimized to yield the best possible classification performance from the ensemble of 16 normally limbed subjects. For the STFT, it was determined that (from a number of taper windows) a Hamming window of length 64 points with an overlap of 50% gave the best performance. When using the WT, a Coiflet mother wavelet (of order four) yielded better accuracy than a host of other wavelet families of varying order.[15] The WPT experienced the best performance when using a Symmlet mother wavelet (of order five). A number of methods were considered as candidates to determine the best tiling of the WPT. The most common approach to specifying the WPT tiling is by selecting that which minimizes the reconstruction error, using an entropy cost func-

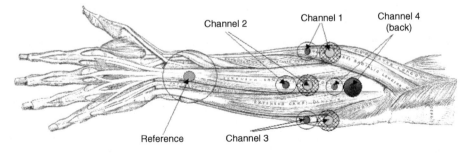

Figure 8.12 The electrode placement used in the four-channel MES acquisition. Four bipolar electrode pairs (Red Dot®; 3M Corp.) were used with a reference at the wrist. Although difficult to show on the figure, the top and bottom electrode pairs are at the same level of the forearm.

tion.[68] This may be considered optimal for signal compression but may be inappropriate for signal classification. A modified form of this algorithm, called the *local discriminant basis algorithm,* has been proposed that seeks to maximize the discriminant ability of the WPT by using a class separability cost function.[22] It is established in Englehart et al.[56] that this discriminant cost function does indeed produce the best classification performance.

From each subject, the TD, STFT, WT, and WPT feature sets were computed. Also considered was the stationary wavelet transform (SWT), which does not decimate the signal at each stage, as does the standard discrete WT.[69] Subsequently, each feature set was subject to dimensionality reduction using principal components analysis (PCA), so as not to over-whelm the classifier with high-dimensional data. It is shown in Englehart et al.[56] that the application of PCA is critical to the success of the time–frequency-based feature sets, and that PCA is clearly superior to other forms of dimensionality reduction. Although the classification performance is not sensitive to the dimensionality of the PCA-reduced feature set, it was demonstrated that at least 5 PCA features are needed, and more than 30 unnecessarily burdens the classifier; 20 PCA coefficients are used in the analyses described here.

Figure 8.13 shows the classification error for each of the candidate feature sets, averaged over all 16 subjects, using a linear discriminant analysis classifier. Clearly, the time-domain feature set performs very well with four channels of MES (as compared to one used by Hudgins), with an error of roughly 4.8%. The STFT does not fare as well, most likely due to its inability to efficiently capture the energy of the signal in the time–frequency plane. The performance of the WT approaches that of the TD feature set, and the WPT surpasses it, due to the adaptive nature of the time–frequency tiling afforded by the local discriminant basis algorithm. The stationary wavelet

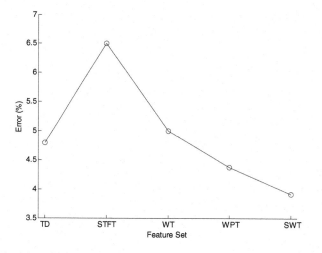

Figure 8.13 The classification error associated with each feature set, averaged across all 16 subjects.

transform performs best of all, due to the fact that it is translation invariant.* Added computational expense is associated with the SWT, but this is not enough to preclude a real-time implementation on modern microprocessors. This improved classification accuracy directly translates into more robust control of a prosthetic limb.

8.6   Conclusions

This introductory view of wavelets provides the practitioner with a brief history, the basic theory, and the fundamental motivation for their use in biomedical signal processing. It should be respected that the use of wavelets should be motivated by their strengths and an understanding of the biology and physics of the problem under consideration, rather than the fact that their use may be considered fashionable. Nonetheless, many advances attributable to wavelets have been made in the past decade, as they are shown to outperform the best techniques otherwise available.

8.7   Further reading

Reviews of wavelet theory as applied to signal processing in general can be found in References 1, 11, 70, and 71. Other information beyond the scope of the discussion here includes wavelet types,[16] biorthogonal transforms,[38,72] translation invariant methods,[73–79] and matching pursuit.[80] By far, the greatest resource for wavelet information is the Internet. It is impossible to summarize the diversity of information available online, but the reader is directed to some "meta" sites that provide direction to literature, software, and discussion groups. These include:

http://www.amara.com/current/wavelet.html
http://www.wavelet.org
http://www.mathtools.net/Applications/DSP/Wavelets/

References

1. Akay, M., *Detection and Estimation of Biomedical Signals*, Academic Press, San Diego, CA, 1995.
2. Vetteri, M. and Kovacevic, J., *Wavelets and Subband Coding*, Prentice-Hall, Englewood Cliffs, NJ, 1995.
3. Haar, A., Zur theorie der orthogonalen funktionensysteme [in German], *Math Anal.*, 69, 331–371, 1910.
4. Grossman, A. and Morlet, J., Decomposition of Hardy functions into square integrable wavelets of a constant shape, *SIAM J. Math Anal.*, 15, 723–736, 1984.
5. Meyer, Y., *Wavelets, Algorithms and Applications*, SIAM, 1993.

---

* The manifestation of a lack of translation invariance in the WT and WPT is a nonlinear distortion of the coefficients with temporal shift, due to the critically sampled nature of the dyadic decomposition.

6. Battle, G., A block spin construction of ondelettes. I. Lemarie functions, *Comm. Math. Phys.*, 110, 601–615, 1987.

7. Mallat, S., A theory for multiresolution signal decomposition: the wavelet representation, *IEEE Trans. Patt. Anal. Mach. Intel.*, 11, 674–693, 1989.

8. Coifman, R.R. and Wickerhauser, M.V., Entropy–based algorithms for best basis selection, *IEEE Trans. Inform. Theory*, 38(2), 713–719, 1992.

9. Kaiser, G., *A Friendly Guide to Wavelets*, Birkhäuser, Boston, MA, 1994.

10. Strang, G. and Nguyen, T., *Wavelets and Filter Banks*, Wellesley-Cambridge Press, Wellesley MA, 1997.

11. Rioul, O. and Duhamel, P., Fast algorithms for discrete and continuous wavelet transforms, *IEEE Trans. Inform. Theory*, 38, 569–586, 1992.

12. Unser, M., Fast Gabor-like windowed Fourier and continuous wavelet transforms, *IEEE Signal Proc. Lett.*, 1, 76–79, 1994.

13. Sweldens, W., Wavelets: what next?, *Proc. IEEE*, 84(4), 680–688, 1996.

14. Bruce, A., Donoho, D., and Gao, H.-Y., Wavelet analysis, *IEEE Spectrum*, October, 26–35, 1996.

15. Daubechies, I., Ten lectures on wavelets, in *CBMS–NSF Regional Conference Series in Applied Mathematics*, Vol. 61, SIAM, Philadelphia, 1992.

16. Meyer, Y., *Wavelets and Operators*, Vol. 37, Cambridge Studies in Advanced Mathematics, Cambridge University Press, Cambridge, MA, 1993 (translated by D.H. Salinger).

17. Cohen, A., Daubechies, I., and Feauveau, J.C., Biorthoginal bases of compactly supported wavelets, *Communications of Pure and Applied Math*, 1990.

18. Coifman, R.R. and Meyer, Y., Nouvelles bases orthonormées de $L^2(\Re)$ ayant la structure du systèm de Walsh, Department of Mathematics, Yale University, New Haven, CT, 1989.

19. Coifman, R.R. and Meyer, Y., Orthonormal Wave Packet Bases, Department of Mathematics, Yale University, New Haven, CT, 1990.

20. Wickerhauser, M.V., *Adapted Wavelet Analysis from Theory to Software*, AK Peters, Ltd., Wellesley, MA, 1994.

21. Coifman, R.R. and Wickerhauser, M.V., Entropy–based algorithms for best basis selection, *IEEE. Trans. Inform. Theory*, 38(2), 713–719, 1992.

22. Saito, N. and Coifman, R.R., Local discriminant bases and their applications, *J. Math. Imaging Vision*, 5(4), 337–358, 1995.

23. Simoncelli, E.P., Freeman, W.T., Adelson, E.H., and Heeger, D.J., Shiftable multiscale transforms, *IEEE Trans. Inform. Theory*, 38(2), 587–607, 1992.

24. Aldroubi, A. and Unser, M., *Wavelets in Medicine and Biology*, CRC Press, Boca Raton, FL, 1996.

25. Akay, M., *Time Frequency and Wavelets in Biomedical Signal Processing*, John Wiley & Sons, New York, 1997.

26. Unser, M. and Aldroubi, A., A review of wavelets in biomedical applications, *Proc. IEEE*, 84(4), 626–238, 1996.

27. Akay, M., Wavelets in biomedical engineering, *Ann. Biomed. Eng.*, 23, 531–542, 1995.

28. Strickland, R.N. and Hahn, H.I., Detection of microcalcifications in mammograms using wavelets, in *Proc. SPIE Conf. Wavelet Applications in Signal and Image Processing II*, San Diego, CA, Vol. 2303, 1994, pp. 430–441.

29. Li, C. and Zheng, C., QRS detection by wavelet transform, *Proc. Annu. Conf. Eng. Med. Biol.*, 15, 330–331, 1993.

30. Khadra, L., Matalgah, M., El–Asir, B., and Mawagdeh, S., The wavelet transform and its applications to phonocardiogram signal analysis, *Med. Informatics*, 16(3), 271–277, 1991.

31. Obaidat, M.S., Phonocardiogram signal analysis: techniques and performance, *J. Med. Eng. Technol.*, 17, 221–227, 1993.

32. Khadra, L. Dickhaus, H., and Lipp, A., Representations of ECG late potentials in the time–frequency plane, *J. Med. Eng. Technol.*, 17(6), 228–231, 1993.

33. Dickhaus, H., Khadra, L., and Brachmann, J., Time–frequency analysis of ventricular late potentials, *Meth. Inform. Med.*, 33(2), 187–195, 1994.

34. Meste, O., Rix, H., Caminal, P., and Thakor, N.V., Ventricular late potentials characterization in time–frequency domain by means of a wavelet transform, *IEEE Trans. Biomed. Eng.*, 41, 625–634, 1994.

35. Taboada-Crispi, A., Improving Ventricular Late Potentials Detection Effectiveness, Ph.D. thesis, Univeristy of New Brunswick, 2002.

36. Schiff, S.J., Aldroubi, A., Unser, M., and Sato, S., Fast wavelet transformation of EEG, *Electroencephalogr. Clin. Neurophysiol.*, 91(6), 442–455, 1994.

37. Kalayci, T. and Ozdamar, O., Wavelet preprocessing for automated neural network detection of spikes, *IEEE Eng. Med. Biol.*, 14(2), 160–166, 1995.

38. Blinowska, K.J., and Durka, P.J., The application of wavelet transform and matching pursuit ot the time-varying EEG signals, in *Intelligent Engineering Systems Through Artificial Neural Networks*, Vol. 4, Dagli, Fernandez, and Gosh, Eds., 1994, pp. 535–540.

39. Donoho, D.L., De-noising by soft thesholding, *IEEE Trans. Inform. Theory*, 41, 613–627, 1995.

40. Lewis, A.S. and Knowles, G., Image compression using the 2-D wavelet transform, *IEEE Trans. Image Process.*, 1, 244–250, 1992.

41. Angelidis, P.A., MR image compression using a wavelet transform coding algorithm, *Magnetic Resonance Imaging*, 1(7), 1111–1120, 1994.

42. Hilton, M., Wavelet and wavelet packet compression of electrocardiograms, *IEEE Trans. Biomed. Eng.*, 44, 394–402, 1997.

43. Shapiro, J.M., Embedded image coding using zerotrees of wavelet coefficients, *IEEE Trans. Signal Process.*, 41(12), 3445–3462, 1993.

44. Wellig, P., Cheng, Z., Semling, M. and Moschytz, G.S., Electromyogram Data Compression Using Single-Tree and Modified Zero-Tree Wavelet Encoding, *Proc. 20th Int. Conf. IEEE Eng. Med. Biol. Soc.*, 20(3), 1303–1306, 1998.

45. Norris, J., Englehart, K., and Lovely, D.F., Myoelectric signal compression using the EZW algorithm, *Conf. of the IEEE Engineering in Medicine and Biology Society*, Istanbul, October, 2001.

46. Carmona, R. and Hudgins, L., Wavelet denoising of EEG signals and identification of evoked response potentials, *Proc. SPIE Conf. Wavelet Appl. Signal Image Process.*, 2303, 91–104, 1994.

47. Tikkanen, P.E,. Nonlinear wavelet and wavelet packet denoising of ECG signal, *Biol. Cybernetics*, 80(4), 259–267, 1999.

48. Bentley, P.M., Grant, P.M., and McDonnell, J.T.E., Time-frequency and time-scale techniques for the classification of bioprosthetic valve sounds, *IEEE Trans. Biomed. Eng.*, 45(1), 1998.

49. Mallat, S. and Zhong, S., Characterization of signals from multiscale edges, *IEEE Trans. Pattern Recog. Machine Intell.*, 14(7), 1992.

50. Senhadji, L., Carrault, G, Ballanger, J.J., and Passariello, G., Comparing wavelet transforms for recognizing cardiac patterns, *IEEE Eng. Med. Biol.*, March/April, 167–172, 1995.

51. Li, C., Zheng, C., and Changfeng, T., Detection of ECG characteristic points using wavelet transforms, *IEEE Trans. Biomed. Eng.*, 42(1), 1992.

52. Lim, L.M., Akay, M., and Daubenspeck, A.J., Identifying respiratory–related evoked potentials, *IEEE Eng. Med. Biol.*, March/April, 174–178, 1995.

53. Bovik, A.C., Gopal, N., Emmoth, T., and Restrepo, A., Localized measurement of emergent image frequencies by Gabor wavelets, special issue on wavelet transforms and multiresolution signal analysis, *IEEE Trans. Inform. Theory*, IT–38(3), 691–712, 1992.

54. Healy, Jr., D.M. and Weaver, J.B., Two applications of the wavelet transform in magnetic resonance imaging, *IEEE Trans. Inform. Theory*, 38, 840–860, 1992.

55. Englehart, K., Hudgins, B., and Parker, P.A, A Wavelet Based Continuous Classification Scheme for Multifunction Myoelectric Control, *IEEE Trans. Biomed. Eng.*, 48(3), 302–311, 2001.

56. Englehart, K., Hudgins, B., Parker, P.A., and Stevenson, M., Classification of the myoelectric signal using time-frequency based representations, special issue on intelligent data analysis in electromyography and electroneurography, *Med. Eng. Phys.*, 21, 431–438, 1999.

57. Thakor, N.V., Gou, X., Sun, Y.-C., and Hanley, D.F., Multiresolution wavelet analysis of evoked potentials, *IEEE Trans. Biomed. Eng.*, 1085–1094, 1995.

58. Kadambe, S. and Boudreaux–Bartels, G.F., Applications of the wavelet transform for pitch detection of speech signals, *IEEE Trans. Inform. Theory*, 38, 917–924, 1992.

59. Dorcas, D. and Scott, R.N., A three state myoelectric control, *Med. Biol. Eng.*, 4, 367–372, 1966.

60. Childress, D.A., A myoelectric three state controller using rate sensitivity, in *Proc. 8th ICMBE*, Chicago, IL, 1969, pp. S4–S5.

61. Vodovnik, L., Kreifeldt, J., Caldwell, R., Green, L., Silgalis, E., and Craig, P., *Some Topics on Myoelectric Control of Orthotic/Prosthetic Systems*, Report No. EDC 4–67–17, Case Western Reserve University, Cleveland, OH, 1967.

62. Hudgins, B., Parker, P.A., and Scott, R.N., A new strategy for multifunction myoelectric control, *IEEE Trans. Biomed. Eng.*, 40(1), 82–94, 1993.

63. Hudgins, B., Englehart, K., Parker, P.A., and Scott, R.N., A microprocessor-based multifunction myoelectric control system, in *Proc. 23rd Canadian Medical and Biological Engineering Society Conf.*, Toronto, Canada, May, 1997.

64. Englehart, K., Hudgins, B., Stevenson, M., and Parker, P.A., Classification of transient myoelectric signals using a dynamic feedforward neural network, *World Congress of Neural Networks*, Washington, D.C., 1995.

65. Farry, K.A., Fernandez, J.J., Abramczyk, R., Novy, M., and Atkins, D., Applying genetic programming to control of an artificial arm, in *Proc. Myoelectric Control '97 (MEC '97) Conference*, Fredericton, New Brunswick, Canada, July 23–25, 1997, pp. 50–55.

66. Leowinata, S., Hudgins, B, and Parker, P.A., A multifunction myoelectric control strategy using an array of electrodes, in *Proc. 16th Annual Congress of the International Society Electrophysiology and Kinesiology*, Montreal, Canada, 1998.

67. Gallant, P.J., An Approach to Myoelectric Control Using a Self–Organizing Neural Network for Feature Extraction, Master's thesis, Queens University, Kingston, Ontario, 1993.
68. Coifman, R.R. and Wickerhauser, M.V., Entropy-based algorithms for best basis selection, *IEEE. Trans. Inform. Theory*, 38(2), 713–719, 1992.
69. Nason, G.P. and Silverman, B.W., The stationary wavelet transform and some statistical applications, *Lect. Notes Statistics*, 103, 281–299, 1995.
70. Rioul, O. and Vetterli, M., Wavelets and signal processing, *IEEE Signal Process. Mag.*, 8, 14–38, 1991.
71. Shensa, M.J., Affine wavelets: wedding the *à trous* algorithm and Mallat lags, *IEEE Trans. Signal Proces.*, 40, 2464–2482, 1992.
72. Cohen, A., Daubechies, I., and Feauveau, J.C., Biorthoginal bases of compactly supported wavelets, *Communications of Pure and Applied Math*, 1990.
73. Pesquet, J.C., Krim, H., and Carfantan, H., Time invariant orthonormal wavelet representations, *IEEE Trans. Signal Processing*, 44(8), 1964–1970, 1996.
74. Cohen, I., Raz, S., and Malah, D., Shift invariant wavelet packet bases, *Proc. of ICASSP*, 2, 1081–1084, 1995.
75. Coifman, R.R. and Donoho, D.L., Translation-invariant de-noising, in *Wavelets and Statistics*, Antoniadis, A. and Oppenheim, G., Eds., Lecture Notes in Statistics, Springer-Verlag, Berlin, 1995, pp. 125–150.
76. DelMarco, S. and Weiss, J., Improved transient signal detection using a wave-packet-based detector with an extended translation-invariant wavelet transform, *Optical Eng.*, 35(1), 131–137, 1996.
77. Liang, J., and Parks, T.W., A translation–invariant wavelet representation algorithm with application, *IEEE Trans. Signal Process.*, 44(2), 225–232, 1996.
78. Simoncelli, E.P., Freeman, W.T., Adelson, E.H., and Heeger, D.J., Shiftable multiscale transforms, *IEEE Trans. Inform. Theory*, 38(2), 587–607, 1992.
79. Nason, G.P. and Silverman, B.W., The stationary wavelet transform and statistical applications, in *Wavelets and Statistics*, Antoniadis, A. and Oppenheim, G., Eds., Lecture Notes in Statistics, Springer-Verlag, Berlin, 1995, pp. 281–299.
80. Mallat, S. and Zhang, Z., Matching pursuit with time-frequency dictionaries, *IEEE Trans. Signal Process.*, 41, 3397–3415, 1993.

chapter nine

Neuroprosthetic device design

Donald L. Russell

Contents

0-8493-1100-4/03/$0.00+$1.50
© 2003 by CRC Press LLC

9.1   Introduction

The goal of this chapter is to provide the researcher with a general context for the design of neuroprosthetic devices. Specifically, the practical issues surrounding the design and development of devices that will actually be used are presented. Concepts are given and exemplified by referring to the development of powered prosthetic elbows. In short, this chapter attempts to describe the overall design challenges (the "forest") rather than the details of any specific area (the "trees").

Neuroprosthetic devices are objects that interact with the nervous system to perform a function that maintains or improves the health and quality of life of an individual. They can be broken down into three classes: those that make use of neural measurements as an *input* to control a mechanical device that replaces a function of the body (for example, a prosthetic hand that uses myoelectric based control); those that, based on some action or situation, create an *output* signal that is transmitted to a nerve to have a desired action (for example, neuroprosthetic bladders); and those that have both neural *inputs* and neural *outputs* (for example, devices aimed at bypassing spinal cord damage).

Even when neural signals can be reliably measured, interpreted, and transmitted, a device must be constructed to perform the required function. The technical demands on such devices may be significant, as they typically require their own power sources and cannot take advantage of the ability of the body to repair damage and compensate for wear. Neuroprosthetic organs that process material may require supplies of material and the ability to discharge waste material. Further, internal devices must work within strict temperature limits or else surrounding tissues will be destroyed.

To be useful (in the eyes of the user) the device must improve both health and quality of life, it must be reliable, the performance of the device must be predictable, the device must be safe and not create other health hazards as a result of its operation, and the effort and concentration required in training and day-to-day use must be reasonable. These are significant challenges. The use of the device will be compared to the ability of the user to "make-do" and adapt to life without using the device. The ability of people to adapt and find innovative solutions to challenging situations after the loss of function of a limb or organ is significant. This is, however, the level of function that must be surpassed if the device is to be used. Such mundane features as the appearance, operating life, or time between recharging batteries of the device often have a significant impact on the very practical decision that a user makes about using a device.

This chapter will present a general classification of different types of devices and the challenges faced in their practical design. A review is presented of current approaches to the design of a number of current neuroprosthetic devices: limbs, the bladder, and the ear. Finally, the impact of current technologies on the design of practical devices is briefly discussed. The primary example used in this chapter will be the powered, myoelectrically controlled prosthetic elbow.

9.2 The device classes

9.2.1 Devices with neural input

This class of neuroprosthetic devices consists of those that base their operation, either directly or indirectly, on measurements of neural activity (see Figure 9.1). The measurements are then interpreted to create control signals appropriate for the device. Those signals are then used to control the action of the device, be it movement, the release of materials, or some other action. In most cases, some sort of feedback exists. This feedback is critical for the body or device to regulate its action. In some instances the feedback may be entirely confined to the device; in other cases, the body's natural senses (such as vision) and nervous system may provide the feedback.

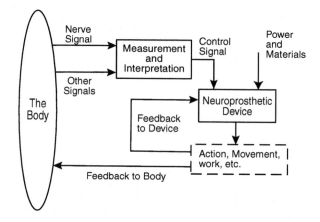

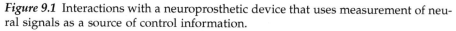

Figure 9.1 Interactions with a neuroprosthetic device that uses measurement of neural signals as a source of control information.

9.2.2 Devices with neural output

In this class of devices the stimulus for action does not arise from measurement (direct or indirect) of neural activity but from some other action (for example, pushing a button). This input signal is generally processed and manipulated to create a signal that is designed for transmission to a nerve (see Figure 9.2). Because this signal is to be transmitted back into a nerve, the design must take into account knowledge of other associated neural connections. Devices, such as pacemakers, which respond to the needs of the body but not to direct measurement of neural signals, fall into this category as they generate signals that interact with the nervous system. A group of devices that is placed in this class (rather than the second class) are those that are used to implement functional neuromuscular stimulation. These devices measure a nervous signal or some other input signal and transmit this signal, via electrodes, to a muscle. In this way external devices can control aspects of the body over which the subject has lost control.

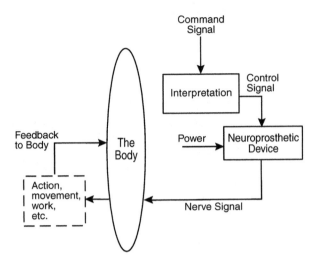

Figure 9.2 Interactions with a neuroprosthetic device that creates a signal that is transmitted to a nerve to achieve a desired function.

9.2.3 Devices with both neural input and neural output

This third class of device essentially replaces the transmission function of the nervous system or modifies a nerve signal to achieve desired goals (Figure 9.3). Both its output and input are neural signals so that the challenges of accurate interpretation of a measured nerve signal and of creating an appropriate, interpretable output signal are significant. In general, devices in this category fulfill the role of transmission of the nervous signal and as such do not have significant material or power requirements (relative to devices that actually move or manipulate material).

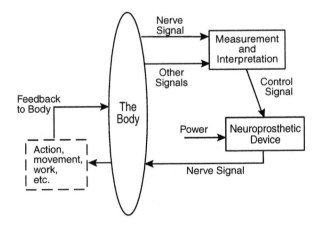

Figure 9.3 Interactions with a neuroprosthetic device that measures neural activity and transmits it (or a modification of it) back to the nervous system.

9.3   The design challenges

The challenges in designing practical neuroprosthetic devices are many. They generally include signal measurement, signal interpretation, performing the desired action, intent interpretation, signal generation, and signal transmission. These challenges are outlined here from the point of view of the designer of a practical device. A specific device may require consideration of some or all of these challenges.

9.3.1   Signal measurement

For a device to be useful and to be used it must be reliable. The ability to measure a signal from neural inputs over a span of time without the need for significant medical intervention is critical. Components that gather the signals must do so in a way that is not subject to a response from the body that inhibits their function, either directly or through a side effect. Conquering these challenges requires a range of efforts and solutions, including the creation of implantable electrodes that use materials and designs that avoid these problems, the careful placement of surface electrodes to avoid skin abrasion or local pain, and the development of other sensors to measure various quantities the represent the state of the body.

9.3.2   Signal interpretation

The neural signal, once measured, must be interpreted; the designer must have the answer to the question, "What does the signal mean?" The body's network of nerves is extremely complex, offering many opportunities for interaction of neural signals with related (or sometimes unrelated) actions and significant feedback (usually to the central nervous system). This network is also nonlinear and the signals measured may not be directly interpretable in terms of the quantities and variables engineers and scientists may choose to describe the resulting actions. The parts of the body that the measured signals would normally control and interact with are, in general, far more complex and nonlinear than the devices used to replace them. In many cases, the state of the art of a device represents a first-order solution to a highly complex problem.

Artificial neural networks often can achieve high levels of robust performance, yet the internal signals within those networks are not meaningful. In most cases where neural measurement is used to control a device, the measured signals are those from within the network of nerves. By analogy with artificial neural networks, these signals may not have any apparent or simple meaning. Significant amounts of experimentation on the system concerned must be performed in order to arrive at an interpretation of a neural signal that can be used to infer a desired action (for example, to flex the limb). This correlation between a measured neural signal and the intended action represents a significant challenge, especially as the details almost certainly vary from individual to individual and because injury or disease may alter "normal" patterns of activity.

9.3.3   Intent interpretation

This challenge is analogous to the challenge of signal interpretation but applies to inputs that indicate the state of the body and not the action intended by the body. For example, future artificial hearts may respond to measurements of oxygen or carbon dioxide levels within the blood or an input from a physician indicating a lack of oxygen in some organ. Having such an input then requires the decisions to be made – what should be done? In this example, should the heart rate be increased or lowered? Are other aspects of the body that are unmeasured (such as, the presence of internal bleeding) important in making this decision? The body is a complex system with many possible solutions to a given problem but also with many possible outcomes to a single action. Designers must focus on the whole body and in creating responses to specific problems.

9.3.4   Performing the desired action

Once an action is required, the device itself must be able to adequately perform the action. In the case of prosthetic limbs, significant energy is required to manipulate objects. Power and material are often required. Pacemaker batteries last for years and performance of the pacemaker itself degrades "gracefully" as power levels drop. Replacement of the battery requires minor surgery but this is infrequent enough that it is not a significant deterrent to the use of pacemakers. Prosthetic limbs also work on batteries. In this application, however, limbs must be designed with removable battery packs so that a single day of practical use is possible. This has a significant impact on limb design. Challenges surrounding the realistic performance of a device are often those that limit the development of practical devices. The amazing capabilities of the body make the challenges of designing replacement parts significant.

9.3.5   Signal generation

When a neural signal is to be created, enough knowledge of the system must exist to understand what signal is required to achieve a given goal or requirement. This is the inverse of the signal interpretation problem. A designer may be able to create an input that indicates a desired action. However, inserting a neural signal into the midst of a complex array of neurons depends on knowing the body's neural representation corresponding to the desired action and a careful and complete understanding of other possible results of the neural signal. Examples are the muscle fatigue and muscle cramps that can result when the output signal of a functional neuromuscular controller is not carefully controlled. If the neuroprosthetic device functions to gather sensory information, a large quantity of detailed information may have to be processed and transmitted to nerves.

9.3.6   Signal transmission

Even when a signal can be created that will do the appropriate thing without inducing other problems, the practical issues of transmitting the signal back to a nerve with an approach that is reliable and robust over the long term can be challenging. Other more practical issues often can have a great impact on the usefulness of the neuroprosthetic devices. For example, the devices must by mounted and supported so that their operation is not affected by movement of the user. The techniques used to locate the device must not cause irritation or other problems for the user.

9.4   An example: prosthetic arms

A limb is a complex set of tissues. Limbs are the primary way in which we interact with the physical world. The loss of a limb through disease or accident is a traumatic occurrence for which physicians and engineers have few solutions. In terms of the variety and complexity of tasks few neuro-prosthetic devices offer as many challenges to be overcome. One of the simplest joints in the human body is the elbow and, because of this, the powered prosthetic elbow will be used to illustrate examples of the types of challenges outlined in the previous section.

The human arm has many uses, and an arm amputee using a neuro-prosthetic device has an expectation that the prosthesis will be capable of performance approaching that of the human arm. In fact, discussions with amputees show that many amputees have expectations that their robotic arms will be much more powerful based on various science-fiction stories. This expectation that an artificial device should be in some way better than the biological tissues it attempts to replace is a non-technical challenge faced by all those who assist patients in the training and use of these devices.

9.4.1   State-of-the-art prosthetic elbows

The ideal prosthetic arm would replace all of the functions of the natural arm. It must be strong, but also gentle, and fast, but also capable of precision, it must be able to operate with a range of stiffness. It must be able to sense all of those senses that the arm possesses (e.g., heat, pressure); it must look and move in a manner that gives the appearance of the natural limb. Further, the arm must respond correctly and naturally to the neural commands that are sent to it. All sensing capabilities of the arm must be translated back into neural signals and transmitted back into the body. These are challenging requirements that are discussed in more detail below. A complete historical review of pros-thetic arms is presented in Mann.[1] Two of the more common, commercially available neuroprosthetic limbs are the Boston arm[2] and the Utah arm.[3]

The performance of commercial prosthetic elbows does not approach that of the intact arm. They are capable of lifting a maximum load that is typically much less than 10 lb; for example, the Boston elbow can lift 9 lb at the hand.[4] While most prosthetic elbows are fitted with some type of mechanical clutch to hold static loads, these clutches do not mimic natural behavior. Elbow speed varies slightly with the limb's orientation relative to gravity, but typical limbs move with a maximum movement time from one limit of motion to the other (roughly 135 degrees) of about 1 sec; for example, the Utah arm will flex against gravity in 1.1 sec.[4]

Several limbs also incorporate other modes of movement. For example, keeping the limb rigid during walking causes the amputee to appear unnatural. In order to overcome this, the Utah arm incorporated a powered "free-swing" mode, and the Boston arm incorporates a mechanical "free-swing" mode. The goal of these modes is to allow the arm to swing naturally during walking. In order to change the behavior of the limb from a movement-controlled mode to a free-swing mode (or any other mode), some sort of switching is required. This can be accomplished by the use of switches or by the identification of special features in the neural inputs (for example, the Utah arm uses a rapid simultaneous contraction of two neural inputs to switch between modes).

Commercial powered limbs work on removable, rechargeable batteries. These batteries must be removable because they are not capable of providing the limb with enough energy for a full day of use. Amputees requiring a prosthetic elbow also require terminal devices to allow them to interact with environment. While beyond the scope of this discussion, the need to control further degrees of freedom, such as a hook or hand, requires either a greater number of operation modes (and the complexity of switching between them) or more inputs.

9.4.2   *Signal measurement and interpretation*

The human arm responds to a large number of nerves that can control a wide variety of movements. Even restricting the discussion to prosthetic elbows, many muscles cross the elbow joint and each muscle has many motor units which are controlled by different neural inputs. Questions and hypotheses about how these neural commands work together to produce the coordinated, smooth motions of a limb abound and present many challenges.[5] Variations from individual to individual make data analysis difficult. Possible changes in neural behavior as a result of the amputation or perhaps the body's adaptation to the amputation further complicate the need to understand the relation between neural signal and intended action.

Neuroprosthetic limb designers have developed a range of solutions to the challenge of measuring signals that can be used to control the limb. In general, commercial limbs use measurements of myoelectric activity by using surface electrodes. These electrodes (typically mounted inside the socket of the limb so that in use they sit on the surface of the stump over appropriate muscles) sense the neural signals as they are transmitted throughout a muscle or group of muscles. The detected signals contain superimposed information from active motor units of any muscles close enough to the electrode to make their signal measurable. All of the signals are further modified due to the nature of transmission through the heterogeneous material in the arm between the active muscles and the electrodes. The raw signals measured in this way do not appear to differ significantly from white noise. The use of surface measurement of these signals has proven to be a practical, reliable, and robust technique of gaining information on the neural input to the limb.

Other inputs are possible, including the use of cables arranged so that rounding of the shoulders will flex the elbow, the use of switches that convert some other body movement into a command for the elbow, and the use of signals measured from other muscles. The typical drawback encountered with these inputs that couple another action of the body to the motion of the elbow is the loss of a degree of freedom; however, directly coupling the motion of a part of the body to the motion of the limb can allow a kind of feedback. This approach, incorporating extended physiological proprioception, has a number of benefits and has been a subject of recent research.[6]

Two basic approaches exist to extract useful information from this form of neural activity measurement. The differences in approach reflect decisions made about the level of physiological meaning assumed in use of the measured signals. The first approach is fairly commonly used, with variations in the details of implementation, in most myoelectric prosthetic situations.[2,3] Signals are measured using surface electrodes. Typically, the signals are processed by first rectifying the signal and then calculating a mean squared amplitude estimate. No attempt is made in this type of approach to differentiate between different muscles or motor units that may be contributing to the myoelectric signal. The second approach uses an array of electrodes to collect information on the spatial distribution of the signal over the surface of the arm.[7] This information is processed to determine more accurate estimates of the signals in the muscles. This second approach requires significantly more computational effort and is far more complex than the first approach.

The use of surface electrodes also can cause problems ranging from skin irritation with prolonged use to the presence of extraneous components in the measured signals (such as the so-called movement artifacts, which result from the movement of the electrode as the muscle contracts and changes the limb geometry).

9.4.3   Intent interpretation

After the measurements are taken they must be interpreted. A significant amount of effort has been, and continues to be, put into establishing the meaning of these signals. In fact it is not necessary to be able to determine the physiological meaning or the intent behind the signals in order to use them to control a prosthesis. Several prosthetic control schemes[8] base the action of the prosthesis on the magnitude of a single measured signal. For example, a low signal may correspond to a closed hand, and a higher signal may cause a hand to open. In the case of a joint, a low signal below an established threshold would cause the arm to remain stationary, a signal above a certain higher established threshold may cause the arm to flex, and a signal between the two thresholds would cause the arm to extend. By extrapolation, higher levels of discretization of the magnitude can achieve more actions. The limit on this control approach is the amputee's ability to gain a facility at precisely regulating the signal amplitude. This is complicated by the lack of any close relationship between the original meaning of the signal and its use. Also, amputees often have difficulty regulating these signals without much practice and focused concentration. These difficulties limit the number of levels of signal magnitude that can be differentiated. Recent work has identified characteristics of the myoelectric signal during the onset of activity that can be used to switch between different devices or control actions.[9]

The measured signal can be interpreted in other ways. One of the common ways of interpreting the signal relies on having signals from two muscles in an agonist–antagonist configuration. For example, in the case of an above-elbow amputation, signals from the biceps muscle remnant and from a triceps muscle remnant can be used together to infer something about the intended movement of the limb. In particular, it is common practice to first rectify and average the individual signals and then use the difference between the signal amplitudes as the control input. Specifically, a large difference in the signals implies a large joint velocity, and a small difference implies a small joint velocity. Care is taken to ensure that a "dead zone" exists so that neural activity below a certain threshold does not result in any motion.

While this approach is more physiological than the switching control, in reality the relationship between measured neural or myoelectric activity and limb movement is far more complex. It also results in limb motion that appears to be very robotic. That is, prostheses using this control scheme move in a manner that is easily distinguishable from the movement of a healthy limb. As will be discussed more below, appearance is a critical factor in designing a prosthetic limb that will be used. In fact, many prosthetic arms now incorporate separate modes of operation (e.g., the free-swing mode in the Utah arm) that ensure that the limb has a more natural appearance during walking. In order to detect when a switch between modes of operation (controlled movement and free swing) is desired, sudden co-contraction events are used. When the amputee suddenly contracts both muscles, the prosthesis will switch modes.

Research efforts are extending the approach of using two antagonistic muscles to generate control inputs. Efforts by Hogan[10] suggest using the sum of the two signals as an indication of limb stiffness, thus mimicking the natural arm in which co-activation of antagonist muscles results in a stiffness increase of the limb. Preliminary experiments on a prosthesis emulator used for the evaluation of new control schemes indicated significant promise in this approach. However, significant changes in limb design are necessary before such an approach can be used in a practical commercial limb.

9.4.4 Performing the desired action

If the intent of the user is known the limb must be able to execute this movement. Knowing what to do is only part of the problem. In many cases, the significant challenges are actually accomplishing the desired tasks. In the case of the prosthetic elbow, determining what the desired action should be is a significant challenge but most commercial limbs are severely limited in their mechanical performance.

An actual limb has very complex mechanical properties. Commercial limbs respond to inputs by moving at specific velocities, without regard for any contact between the limb and the environment. Muscles, as actuators, have very different properties from electric motors and mechanical transmissions. One of the fundamental differences between these actuators is the stiffness (or more generally, the impedance), the response of the actuator when it is forced to move from its commanded equilibrium position.

Muscles have relatively low stiffness when compared to motors and gear sets.[11] Typical robotic control schemes have the goal of ensuring that the robot is moving as commanded. Many commercial prosthetic controllers share this feature. Designers must consider what would happen if the limb is forced to move by an external force, as might occur when bumping an object or in attempting to carry an object for which the weight exceeds the lift capacity of the arm. Only two possibilities are possible: either the motor and transmission prevent the motion from occurring or the device is designed so that it is possible for an external force to move the limb. If the motion is prevented, both the limb and the amputee must support the applied load. As impact loads can be significant this would generally result in damage to the prosthesis and pain for the amputee as the load is transmitted from the prosthesis to the residual limb. This is why commercial limbs will move when a sufficient load is applied to them (they are "back-drivable").

For the second possibility, consider a limb that is moving as directed, at the speed corresponding to the measured muscle signals, but bumps into an object. The object may be moved or damaged. More commonly, the object is too heavy to be moved by a prosthetic limb and too strong to be damaged by it. The result is that the controller causes the arm to apply as much force as it can to achieve the desired movement. When that force surpasses the maximum force that can be generated by the motor, the limb is forced to move from its commanded position, overpowering the motors and creating a high-pitched squealing sound. As one of the main goals is to provide the amputee with the ability to participate in society and to feel acceptable, this high-pitched sound is a significant drawback for many users as it brings attention to their missing limb.

The general idea of the appearance (or cosmesis) of the limb is significant to amputees.[12] Approximately 75% of arm amputees who choose not to wear a prosthesis have looked for a device that appeared the same as their sound arm.[13] For many individuals, the need to have a limb that looks like it is real, both when stationary and when in motion, is one of the major factors dictating the level of use of the prosthesis. Also of significance is the manner in which the amputee supports the limb; the fit of the socket is critical for the long-term comfort and to prevent irritation of the residual limb during use.

One feature of the way that the healthy body responds to neural controls is the lack of a requirement for concentration. For most people, activities such as listening, keeping the heart beating, or using a limb to walk or move a sandwich from the plate to the mouth do not require significant focus and concentration. Amputees learning to use a new prosthetic limb can require very significant levels of concentrated mental effort. Training periods that are too long with too little indication of progress and a return to normal function can reduce the morale and interest of the amputee to a point where interest in gaining a facility in using the device disappears. The most advanced neuroprosthetic limbs will be unused if the training requirement is too long and difficult. Each individual user may have a different level of tolerance for extensive training and a different level of desire to master the device.

Finally, for many prosthetic devices power use is an issue. Pacemakers use power at a rate that allows them to function for years before new batteries are needed. However, for prosthetic limbs significant power is necessary; in fact, one view of the interaction of a limb and the world is to describe it as the controlled distribution and absorption of kinetic energy. An arm provides energy to an object as its velocity increases and absorbs energy as its velocity is reduced. The need for removable and/or rechargeable batteries in prosthetic arms has a significant impact on the geometry of the limb and the efficiency of space usage. An increase in the efficiency of the limb or its use will lengthen the time before it is necessary to replace or recharge the batteries. When a design is achieved that allows limbs to be used for a complete day of use, many other aspects of the designs can change. Under such conditions, batteries could be distributed in the device and charging occur overnight when the limb is not in use.

Given the limited performance of current technology, many advances in artificial limbs are possible. One promising approach is to construct the prosthetic limb using biomimetic principles. That is, the limb can be designed so that its mechanics mimic an intact limb. With this approach, muscle signals might be fed directly to actuators without the need to interpret the signal or perform extensive manipulations on the signal.

9.4.5 Signal generation and transmission

Commercial prosthetic limbs do not provide direct feedback to the user to replace the sensations that were present in the amputated limb. The volume of sensory information to be transmitted is such that researchers concentrate on provision of very limited sensing back to the user.

9.5 Other neuroprosthetic devices

While entire books or chapters could be written on each of these devices, this section summarizes the current state of a number of these devices with reference to the design challenges given above. Specifically, the ear (cochlear implant) and the bladder are parts of the body for which commercial neuroprosthetic devices have been developed. These two areas are discussed briefly here as examples of where the technology can lead. The large body of work on pacemakers and similar cardiovascular devices that interact with the nervous tissue surrounding the heart is not discussed here.

9.5.1 Cochlear implants

These neuroprosthetic devices are intended to provide hearing to the deaf.[14] They function by converting sounds picked up by a microphone into signals that are transmitted with an array of small, implanted electrodes to the auditory nerve. A microphone worn by the user collects sound information that is processed, commonly with digital signal processing (DSP)-based hardware to process and extract features from the measured sound. A signal is then created that is transmitted to the implanted portion of the device (often using radiofrequency [RF] technology). The signal is then used to stimulate the auditory nerve with electrodes. These electrodes must be carefully placed to generate useful signals in the auditory nerves. While no claim is made that hearing is restored, companies such as Advanced Bionics (www.bionicear.com) and AllHear, Inc. (www.allhear.com) suggest that their devices provide enough sensation and information about the sound environment for many to recognize speech, especially in conjunction with experience in other techniques such as lip reading.

The challenges confronting the design or design improvements to cochlear implants follow the general outline presented here. Collecting the input signal (sound) using a microphone should be done in a way that allows the user to participate in society without significant, visible signs of the use of the device. Perhaps the biggest challenge in the design of these neuroprosthetic devices is dealing with the large volume of information in the measured sound. Current solutions to the problem involve extracting features from the sound or processing the sound in a fashion that reduces the information content to a level that can be transmitted into the nervous system. By improving the level of understanding of the way that the healthy ear functions these solutions can be improved so that a larger amount of more accurate information can be transmitted to the auditory nerves.

The components that process the signals and generate the signal to be transmitted to the implants are external units that must be carried by the user. Further development of signal processing electronics to reduce the size and power requirements of this part of the technology will provide a far more user-friendly product that is less intrusive into other aspects of the users life. Even when the signal is properly generated, it must be transmitted to the nervous system. Careful development of finer arrays of electrodes (or the equivalent) and development of techniques for implanting, locating, and calibrating these electrodes are two other important steps in moving these devices forward.

9.5.2 Bladder

The loss of bladder control presents an individual with a medical problem that also has serious social implications. Regaining control of the bladder allows these individuals to improve their health and participate and contribute to society without fear of embarrassment. Commercial neuroprosthetic bladders (for example, the Vocare system made by Neurocontrol Corp.; remote-ability.com) provide the patient with an external unit to control when the bladder will empty. These units generate signals that are transmitted to an implanted receiver, which uses pacemaker-like technology to activate appropriate nerves. Future challenges in neuroprosthetic bladder development center on device design to make the devices smaller and less obtrusive and to perhaps develop approaches that allow the user control of the device without the need for an external triggering unit.

9.6 Conclusions

The design of actual, usable neuroprosthetic devices requires an understanding of the complex details underlying the function of the body the device is intended to replace. For the devices that are developed to be practical and used by the client group they were designed for, a holistic approach must be taken to their design. The chapter has outlined the basic nature of the challenges facing neuroprosthetic device designers. Ideally, the devices must be able to perform the required actions in a way that does not impact the users' health in another way. Muscles and nerves have fundamentally the same principles of operation as their technological equivalents — motors and wires.

References

1. Mann, R.W., Cybernetic limb prosthesis: the ALZA distinguished lecture, *Ann. Biomed. Eng.*, 9, 1–43, 1981.
2. Williams, III, T.W., The Boston elbow: the path to a mature myoelectric prosthesis, *SOMA*, 29–32, 1986.
3. Jacobsen, S.C., Knutti, D., Johnson, R., and Sears, H., Development of the Utah artificial arm, *IEEE Trans. Biomed. Eng.*, BME-29(4), 249–269, 1982.
4. Liberty Technology, Prosthetics and Orthotics Group, *The Liberty Boston Elbow: A Complete Prosthetic Arm System*, Liberty Technology, Hopkinton, MA, 1997.

5. Winters, J.M. and Woo, S.L.-Y., Eds., *Multiple Muscle Systems: Biomechanics and Movement Organization,* Springer-Verlag, New York, 1990.
6. Richard, F., Heckathorne, C.W., and Childress, D.S., Cineplasty as a control input for externally powered prosthetic components, *J. Rehab. Res. Dev.,* 38(4), 2001.
7. Clancy, T.E., Morin, E.L., and Merletti, R., Sampling, noise-reduction and amplitude estimation issues in surface electromyography, *J. Electromyogr. Kinesiol.,* 12(1), 2002.
8. Scott, R.N., Parker, P.A., O'Neill, P.A., and Morin, E.L., Criteria for setting switching levels in myoelectric prostheses, *J. Assoc. Children's Prosthetic-Orthotic Clin.,* 25(1), 1990.
9. Hudgins, B., Englehart, K., Parker P., and Scott, R.N., A microcomputer-based multifunction myoelectric control system, *CMBEC '97,* Toronto, 1997.
10. Abul-Haj, C.J. and Hogan, N., Functional assessment of control systems for cybernetic elbow prostheses. I: Description of the technique, II: Application of the Technique, *IEEE Trans. Biomech. Eng.,* 37(11), 1990.
11. Burke, T., Modelling and Evaluation of Nonbackdrivable Transmissions for Variable Stiffness Prostheses, Master's thesis, Carleton University, Ottawa, Canada, 2000.
12. Kaczkowski, M. and Jeffries, G.E., Cosmesis is much more than appearance ... it's function, *Motion,* 9(3), 1999.
13. Melendez, D. and Leblanc, M., Survey of arm amputees not wearing prostheses: implications for research and service, *J. Assoc. Children's Prosthetic-Orthotic Clin.,* 23(3), 1998.
14. NIH, *Cochlear Implants,* Health Information Bulletin, NIH Doc. No. 00–4798, National Institutes of Health, Washington, D.C., 2000.

section five

Prosthetic systems

chapter ten

Implantable electronic otologic devices for hearing rehabilitation

Kenneth J. Dormer

Contents

10.1 Introduction

Beginning in the 1960s, interest was generated among auditory physiologists, engineers, and otologists in devising implantable electronic hearing devices for the restoration of hearing. The idea perhaps first was conceived around 1800 when Alesandro de Volta inserted two metal electrodes, from the electrolytic cell he discovered, into his ear canals and reported a bubbling sensation. Nevertheless, attempts over the next 50 years to understand the phenomenon were sporadic. In the mid-1800s, alternating current was tested, to no avail. Wever and Bray, in 1930, reported that the electrical response recorded in the vicinity of the auditory nerve of a cat was similar in frequency and amplitude to the sounds to which the ear had been exposed.[9,48] One of the first attempts to stimulate the VIIIth nerve was by Lundberg in 1950, who did so during a neurosurgical operation.[9,48] Djourno and Eyries, in 1957, placed electrodes on the auditory nerve and stimulated them at different pulse rates, and the patient was able to distinguish such words as *papa* and *allo*.[9,48] These early observations were the forerunners of today's cochlear implant device, the world's first successful neural prosthesis. This technology also spun off the development of three other types of implantable hearing devices to address the specific types of hearing losses prevalent worldwide. Each of the four technologies seeking to improve the auditory input to the brain over the VIIIth cranial nerve will be reviewed in this chapter. Virtually all types of hearing loss, except for supramedullary brain damage have been ameliorated by implantable electronic otologic prosthetic devices: bone conduction devices, middle ear implantable hearing devices (MEIHDs), cochlear implants (CIs), and auditory brainstem implants (ABIs).

10.2   Review of auditory mechanisms

The ear, of course, provides one of the five senses of the human body. The outer ear, consisting of the pinna and outer ear canal, is basically the sound gathering portion of anatomy. The input impedance at the entrance to the ear canal would be important for the acoustic load of earphones but is of little consequence for the consideration of, say, implantable middle ear devices. Our sense of directionality is dependent on differential arrival times at the outer ear, and the binaural difference in loudness and arrival time at the auditory cortex of the brain is the neurophysiological basis of determining directionality of sound.

The outer ear is connected to the middle ear by the tympanic membrane through the ear canal. The middle ear consists of the tympanic membrane, ear ossicles (malleus, incus, and stapes), an attic air space dorsal to the middle ear space, the middle ear space containing the ossicles, the antrum with air-containing air cells, and the aditis ad antrum, which connects the antrum and middle ear space. As the impedance-matching portion of the auditory pathway, the middle ear converts variations in air pressure into variations in fluid pressure (perilymph in the cochlea), where the footplate of the stapes enters the cochlea or inner ear. This impedance mismatch from air to fluid mechanics is accomplished by the middle ear anatomy and mechanics, with the primary basis of middle ear impedance matching being the surface area ratio of the tympanic membrane and the footplate of the stapes. For normal sound transmission (but not excessively high frequencies) the ossicles act as a unit, a solid body, and the footplate moves as a piston. Hence, the middle ear transfer function would involve the acoustic input pressure at the tympanic membrane and the middle ear output being the same as the input pressure in the scala vestibuli of the cochlea (through the stapes footplate). Transfer function and impedance characteristics are important for middle ear hearing devices when the ossicular chain is preserved, residual hearing is unimpaired and the ability of high fidelity of amplification is provided by a MEIHD.

The inner ear is the cochlea, a fluid-filled chamber where displacement of the fluids by a traveling wave is converted into a nerve action potential by hair cells. The 30,000 hair cells act as miniature displacement transducers that respond to deformations in the perilymph (cochlear fluid). They elicit action potentials that are tonotopically arranged (in an ordered frequency array) both in the cochlea and the brain. Cochlear outer hair cells respond to the traveling wave in the cochlea, have a contractile function (actin filaments), and serve as controllable amplifiers for the inner hair cells, which send action potentials to the auditory cortex. Loss of outer hair cells results in about a 60-dB loss of hearing. The inner ear normally encodes frequencies by responding to the cochlear fluid movements where different frequencies cause maximum vibration amplitude at different points along the basilar membrane in the cochlea. The threshold for hearing is 10^{-11} M of movement by the tympanic membrane with a dynamic range of 120 dB (sound pressure level, or SPL). Higher frequencies cause traveling waves in the basal portions of the basilar membrane, while lower frequencies affect apical regions of the membrane. This is called *tonotopic organization* (frequency selectivity) and the cochlear acts as a spectrum analyzer.

In reality, the "sensorineural" hearing loss that is treated by hearing aids and MEIHDs is the same pathology as is indicated for cochlear implants: loss of inner ear hair cells. In moderate to severe hearing loss, the reduction in hair cell numbers is such that sensitivity to sounds has been reduced; amplified sounds or ossicular movements recruit a greater portion of the remaining hair cells and socially adequate hearing is restored. In profound hearing loss, the hair cell population is so low that neither acoustic amplification by hearing aids nor mechanical overdrive by MEIHDs can provide adequate restoration of hearing. Direct stimulation of spiral ganglion cells by passing electrical current across these nerves effectively bypasses the hair cell transducers and initiates action potentials to the first relay in the auditory pathway to the brain, the cochlear nucleus. In yet a third form of neural hearing loss, the VIIIth cranial nerve, the statoacoustic nerve, has been damaged or diseased and the auditory pathway interrupted. This most often occurs during surgical removal of acoustic neuromas, tumors that invade the VIIIth nerve. In this instance, for patients with no VIIIth nerve, neural prostheses have been designed for provision of an auditory signal by directly stimulating the cochlear nucleus in the brain stem by electrical means. The auditory brainstem implant is the most recent of neural prostheses to be implanted for a form of neural deafness. Still, it may be appropriate to refer to all of the above forms of deafness simply as "sensory" impairments, as hearing is one of the five senses.

10.3 Hearing losses

Approximately 10% of any nation's population has a hearing loss of one type or another. It is estimated that 21.9 million U.S. adults have hearing loss, 15.2 million of those having moderate to severe loss. About 4.7 million use hearing aids and 2.3 million are unsatisfied with their hearing aids. There are two basic mechanisms for hearing loss. In one, *conductive hearing loss*, the mechanically conductive pathway to the inner ear is impaired, although the inner ear works fine. The second loss mechanism, *sensorineural hearing loss*, involves the nervous system, meaning that the conduction to the inner ear is operative but the transduction mechanism to a nerve action potential by hair cells is impaired; the sound signal does not get conveyed to the auditory cortex of the brain. When devices are implanted to amplify or restore hearing, they deal with mechanics or neural stimulation, depending upon the hearing loss type. In some individuals, there is a *mixed hearing loss*, meaning both conductive and neural impairments exist.

The etiology of hearing loss is varied. Chronic infections of the middle ear can lead to erosion of the ossicles (conductive loss) or bacterial or viral invasion of the inner ear, causing loss of hair cells (sensorineural loss). Some antibiotics containing aminoglycosides (used to treat infections) are ototoxic and kill hair cells. Today, five genes are known to be related to hearing, and genetic causes of hearing loss are just being understood. For example, a hereditary defect on the gene *connexin*-26 creates a protein necessary for the conveyance pathway of potassium within the inner ear. Potassium contributes to the large generator potential of the hair cells and the absence of that gene results in hearing loss. Hair cell damage due to excessive noise (noise pollution) is a more recent phenomenon in our younger populations because of portable audio systems and high-fidelity, powerful, stereophonic amplifiers. Other chemical pollutants or oxygen radicals from air, water, foods, or smoking are suspect in hearing loss, as they may adversely affect hair cells. Outright damage to the ear from trauma is another cause of hearing loss. Finally, aging is correlated with hearing loss, presbycusis possibly being related to long-term exposure to any or all of the above or simply the normal loss of hair cells with age.

Hence, approaches to restoring hearing by implanted devices and direct intervention into the sensory nervous system require an understanding of the mechanism of loss and whether the implantable hearing device amplifies sound to a reduced population of hair cells, restores the mechanical connection to a normal population of hair cells, or restores the ability to initiate action potentials to the brain in an otherwise normal cochlea (auditory nerve is present), except for the absence of hair cells.

10.4 Types of implantable hearing devices

The four categories of implantable hearing devices (IHD) are classified according to the types of hearing losses in individuals. These categories are *bone conductors, middle ear implants, cochlear implants,* and *auditory brainstem implants.* When conductive hearing loss is present, a bone conduction IHD such as the bone-anchored hearing apparatus (BAHA®; Entific Medical Systems, Gothenberg, Sweden) seeks to restore hearing by vibratory conduction through the skull to the inner ear. Bone conduction devices are intended for use with conductive hearing loss, whereas the remainder of the IHDs seek to restore hearing due to sensorineural hearing loss. Sensory hearing impairment is addressed by MEIHDs and the cochlear implant. MEIHDs treat moderate, moderate-to-severe, and severe hearing losses by amplification of sound signals through implanted transducers. These amplification mechanisms utilize either piezoelectric crystals or electromagnetic transducers where greater displacement of the footplate of the stapes will activate a greater number of the (reduced) hair cell population and thus amplify sounds. Cochlear implants treat profound sensory hearing loss and such patients essentially hear nothing without their implant being activated. The hair cell population is essentially decimated with profound hearing loss and only the fibers of the spiral ganglion cells that project to the auditory nerve remain. The CI provides direct bipolar or monopolar electrical stimulation to the spiral ganglion cells according to their tonotopic (frequency) arrangement within the cochlea. Finally, total deafness due to the loss of the VIIIth nerve during tumor removal is treated by direct electrical stimulation of the brainstem cochlear nucleus, the first relay of the VIIIth nerve en route to the auditory cortex. The auditory brainstem implant (available from Cochlear Corp.; Englewood, CO) provides multielectrode stimulation of the surface of the cochlear nucleus.

10.4.1 Bone conduction devices

Currently, one implantable bone conduction device is commercially available, the bone-anchored hearing apparatus (BAHA®) (Figure 10.1). It has been widely used in Europe since 1977 and is FDA-approved in the United States.[31,49,50] This percutaneous titanium system is implanted in more than 7000 patients as a hearing prosthesis and for the attachment of prosthetic noses, ears, eyes, etc. The attachment is based on the physiological phenomenon of osseointegration (or osteofixation), where the titanium interface becomes attached to living bone tissue through the cell matrix ground substances such as glucose aminoglycans. The principle of bone conduction amplification is that an external sound-processing device converts auditory signals into mechanical vibratory signals and conveys them to the percutaneous post that is osseointegrated into the post-auricular region of the temporal bone. Basically, the output speaker of a hearing aid circuit is physically coupled to the titanium post. There is no airborne sound, but the vibrations in the skull are conveyed to the inner ear and hair cells are bent by the normal physiological mechanism. The middle ear is bypassed. Patients with conductive hearing loss up to an average bone threshold of 70 dB or who are unable to wear an air-conduction hearing aid can benefit from the BAHA.

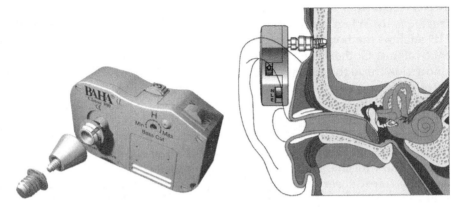

Figure 10.1 Bone-anchored hearing aid (Entific Medical Systems; Gothenberg, Sweden). The external sound processor and percutaneous abutment and titanium implant screw are shown on the left; the abutment screws into an osseointegrated fixture in the temporal bone behind the ear. The diagram on the right shows the sound transfer pathway to the inner ear through the temporal bone. Vibration of the cochlear capsule activates the hair cells.

It was recently approved by the FDA for bilateral implantation and attempts are being made to amplify sounds using more powerful systems (BAHA Cordelle II) so as to partially restore both conductive and sensorineural hearing losses.[46]

10.4.2 Middle ear devices

Ten middle ear implantable hearing devices (MEIHDs) are in active stages of research, development or commercialization at this time globally: University of Dundee, in Dundee, Scotland; IMPLEX® GmbH Hearing Technology, in Ismaning/Munich, Germany; Otologics LLC, in Boulder, CO; Rion Corporation, in Tokyo, Japan; St. Croix Medical, in Minneapolis, MN; University of Sheffield, in England; SOUNDTEC, Inc., in Oklahoma City, OK; Symphonix Devices, Inc., in San Jose, CA; Technical University, in Dresden, Germany; and University of Virginia, in Charlottesville, VA. Five devices use electromagnetic or motor actuators[19] for mechanical micromovements of the ossicular chain and five devices use piezoelectric ceramic or piezoelectric–hydroacoustic actuators.[47] Electromagnetic devices are characterized by high efficiency and variable placement of the implant magnets. Piezoelectric MEIHDs are characteristically small and simple, electronically speaking, but are also less efficient than their magnetic counterparts. An advantage is that piezoelectric devices have been made totally implantable, as early as the 1970s.

Middle ear implants are one of the newer technologies in hearing restoration but the concepts behind the technology are the oldest, the first electromagnetic hearing amplification having been tested in 1935. The first clinically wearable devices were made in the early 1970s in Japan and late 1990s in Germany. The first FDA-approved device was implanted in 2000[44] followed by a second in 2001,[24] with two other devices late into clinical trials at this time. The current MEIHD application is for sensorineural hearing loss in ears with normal middle ear mechanics, which is characteristic of the greatest number of hearing-impaired people worldwide.[32]

Functional gain is the most important rehabilitation parameter for all IHDs. For MEIHDs, the position of the FDA is that comparisons for efficacy and safety should be made with the nominal or best alternative hearing technology available to a hearing-impaired patient, in this case a hearing aid. That is, the gain above that of the (control) optimally fit hearing aid (or other device) is the functional gain criterion for device evaluation of efficacy. Today's high-performance, digital, programmable hearing aids provide a rigorous standard of comparison for substantial equivalency. MEIHD comparisons against either unilateral or bilateral hearing aids for equivalency in performance are still a matter of investigation. Additionally, a new standard is emerging for comparison of different IHD performances: laser Doppler interferometry. This contactless means of measuring displacements is being used on human temporal bones for middle ear mechanical evaluations and functional gain *ex vivo.*[16]

IMPLEX GmbH, Munich, Germany, has been commercializing the totally integrated cochlear amplifier (TICA®) (Figure 10.2).[29,54] Having received the CE mark with its first implantation in Europe (June 1998), the TICA consists of a totally implantable system with a microphone, rechargeable lithium battery, sound processor, and piezoelectric driver with a titanium-coupling rod that vibrates the head of the stapes or incus. Because of the necessary osseointegration of the coupling rod to produce the push–pull mechanical transduction that sets the fluid in the cochlea in motion, the TICA design is being modified for a hook-type attachment to the incus, stabilized with ionomeric cement, for more assurance of mechanical connection between actuator and ossicle. The microphone is located under the skin in the ear

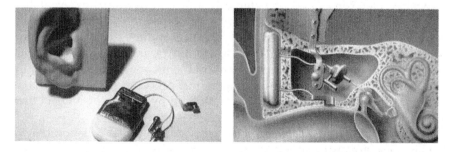

Figure 10.2 Totally integrated cochlear amplifier MEIHD (IMPLEX GmbH; Munich, Germany). On the left is shown the implant portion with implantable microphone and driver. Shown on the right is the implanted device with the microphone above the ear canal and piezoelectric driver on the incus.

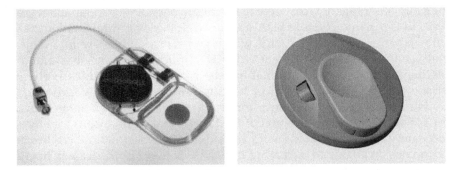

Figure 10.3 Middle ear transducer ossicular stimulator MEIHD (Otologics, Inc.; Denver, CO). The implant portion (left) contains a receiving antenna, demodulating circuitry, and electromagnetic driver on the end of the cable. The external Button sound processor (right) contains microphone, battery, transmitting antenna, and magnet for transcutaneous attachment.

canal for detection of sounds gathered by the outer ear and is especially designed with a silicone rubber collar to prevent tissue overgrowth onto the diaphragm of the microphone. Some occurrences of acoustic feedback (the eardrum driven by the transducer and becoming a speaker) due to the microphone pickup in the ear canal have necessitated partial ossicular disconnection in some patients. The rechargeable battery allows 60 hours of continues use by patients and requires 90 minutes to recharge the battery, which has a projected lifetime *in vivo* of 5 years. Such a direct-drive MEIHD has a maximum power output of 90 dB at 4000 Hz and a maximum functional gain of 50 dB at 3000 Hz. Functional gain is the improvement over their preexisting condition or control comparison.

At this time, at ten sites in the United States, Otologics, Inc., is conducting Phase II clinical trials on the middle ear transducer ossicular simulator (MET™), currently a partially implantable device (Figure 10.3).[14] An implantable transducer designed around an electromagnetic motor with associated electronics and an externally attached Button™ (external audio processor) drive the ossicular chain by a thin probe that is surgically placed onto the incus in a laser-drilled hole. The aluminum-oxide vibrating probe is connected to the incus by a fibrous union (scar tissue). The Button is attached to the patient's head by transcutaneous magnetic coupling. Like the TICA, the MET contains a digital sound processor (two channels and 12 digital filter bands) that is programmable by means of NOAH-based fitting software.

Rion Co., Ltd., and Sanyo Electric Co., Ltd., in Japan collaborated with the Japanese government and pioneering otologists Suzuki and Yanagihara beginning in 1978.[20,28,47] In 1983, they began clinical investigations, producing the first clinically wearable MEIHD (Figure 10.4). Using piezoelectric technology, they developed two types of implantable hearing aids (IHAs), a partially implantable hearing aid (PIHA) and a totally implantable hearing

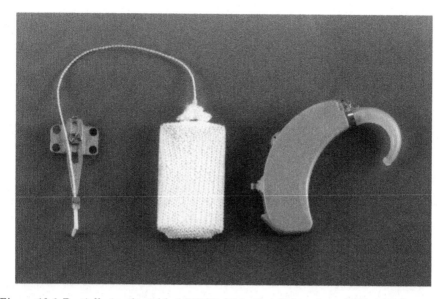

Figure 10.4 Partially implantable MEIHD (Rion Co.; Tokyo, Japan). The implant portion is shown on the left with a mounting fixture to aid placement of the piezoelectric driver on to the head of the stapes. The external sound processor (right) contains a microphone, power supply, and radiofrequency transmitting circuitry for conveying the sound signal to the implantable portion.

aid (TIHA). Two types of implantable batteries were developed, a rechargeable and non-rechargeable type. Noteworthy was the implantable rechargeable battery, which became a forerunner of IHD technology. The IHAs were designed for patients with chronic middle ear dysfunction and, interestingly, all the recent MEIHD designs are made keeping in mind those patients for whom middle ear function is normal. A piezoelectric bimorph is attached to the head of the stapes and driven by an amplifier. In the PIHA, an external behind-the-ear receiver contains microphone, battery, sound processor, and transmitting antenna. The received environmental sounds are processed and transmitted transcutaneously by radio frequency to the receiver demodulator unit under the skin. This internal receiver drives the piezoelectric bimorph as an ossicular vibrator. In Japan, approximately 100 patients have been implanted with either the TIHA or PIHA, and U.S. FDA approval is being sought.

St. Croix Medical recently completed Phase I clinical trials and was approved in 2001 by the FDA to begin Phase II studies on its Envoy™ system, a totally implantable MEIHD (Figure 10.5). The Envoy uses piezoelectric ceramic transducers placed on the malleus or incus and stapes. A first piezoelectric transducer (the sensor) on the malleus or incus detects tympanic membrane motion in response to sound. The sound signal is processed by a programmable digital circuit and is used to drive the second transducer (the driver) that is attached to the stapes by ionomeric cement. Programming

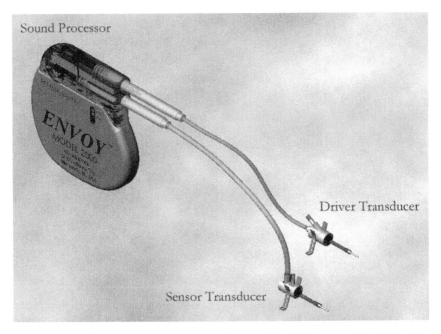

Figure 10.5 Envoy system MEIHD (St. Croix Medical; Minneapolis, MN). Totally implantable, the sound processor is powered by a pacemaker-type battery. The sensor transducer is a piezoelectric pickup for vibrations from the malleus or incus. Sounds are processed and conveyed to the piezoelectric driver transducer (also piezoelectric), attached by probe to the stapes.

of the implanted digital chip by an audiologist is accomplished with bidirectional telemetry. Personal programming also utilizes the telemetry link to allow the patient to change volume, select a desired environmental setting, or place the Envoy in standby mode. Normally, the incus is interposed between the malleus and stapes, but, to prevent mechanical feedback to the sensor, the incus is surgically resected, interrupting the normal conductive pathway to the cochlea. Analogous to acoustic feedback in hearing aids, the output transducer could drive the input sensor and create oscillatory feedback unless the pathway was interrupted.

The Direct System™ is manufactured by SOUNDTEC, Inc., and is an electromagnetic, partially implantable system, the only implant component being a rare earth, permanent magnet in a titanium canister and attached by fibrous tissue around the incudo–stapedial joint (Figure 10.6).[16,24] The external sound processor consists of analog or digital sound processor, battery, microphone, and electromagnetic coil, packaged for behind-the-hear or in-the-canal wear by the patient. The external electromagnetic coil, currently worn in the ear canal, produces a magnetic field that interacts with the permanent magnet implant across the intact tympanic membrane. The Direct System, approved by the FDA for implantation in 2001, is the only contactless MEIHD with an intact ossicular chain, and the principle

Figure 10.6 Direct System MEIHD (SOUNDTEC, Inc.; Oklahoma City, OK). This semi-implantable electromagnetic device is shown on the left; the sound processor and electromagnetic coil are worn in the ear canal. The magnet implant portion is surgically placed through the ear canal and attached around the incudostapedial joint. The implant portion (right) is a permanent magnet hermetically sealed in a titanium canister.

of operation is like that of an acoustic speaker, with a driving coil and magnetic driver directly driving the ossicular chain mechanically. Middle ear mechanics demonstrate that the closer a transducer is located to the footplate of the stapes, the higher the frequency response. This is due to the rocking motion of the stapes at higher frequencies. The magnetic implant of the Direct System is the closest to the stapes footplate of all MEIHDs, with an intact ossicular chain producing the high-frequency amplification sought after with this technology. Phase II clinical trials at ten U.S. sites were completed on 103 patients, and Canadian regulatory agency approval is also anticipated late 2001.

The Vibrant Soundbridge™, manufactured by Symphonix Devices, Inc., was the first MEIHD to attain FDA approval for marketing in the United States (2000) and Europe (CE mark) (Figure 10.7).[43,44] Approximately 667 patients worldwide, 110 in the United States, have received the Vibrant Soundbridge, which has a unique electromagnetic configuration. The floating mass transducer is an integrated magnet-coil unit that produces the auditory frequency vibrations on the incus where it is attached. Semi-implantable, the Vibrant Soundbridge system has an implantable electronics package that receives both power and the auditory signal through a transcutaneous radiofrequency link. The signal transmitted across the skin is via an amplitude-modulated signal to the internal receiver in the vibrating ossicular prosthesis (VORP; the electronics package). The VORP contains a magnet for transcutaneous attachment to the skin and antennae alignment, a demodulator, and a cable connecting to the FMT. The auditory processor contains a transmitting coil and microphone, with programmable multiple-band digital signal processor, modulator circuit, and battery.

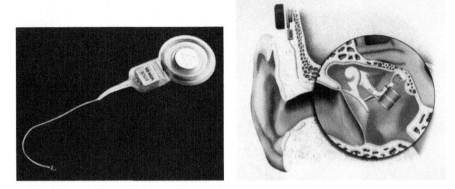

Figure 10.7 Vibrant Soundbridge MEIHD (Symphonix Devices, Inc.; San Jose, CA). This semi-implantable electromagnetic device is shown on the left, where a transducer (magnet and encircling coil) is attached to the incus. The internal receiver demodulates an radiofrequency signal containing frequency and amplitude information. The middle ear placement is depicted on the right; the hardwired, hermetically sealed transducer is attached to the incus.

Technical University in Desden has been exploring a hydroacoustic transmission principle where vibrations from a piezoelectric driver are conveyed to a fluid-filled conductor that is in contact with the perilymphatic fluid of the cochlea.[26] The ossicular chain is intact and, presumably, because the air-to-fluid impedance matching function of the middle ear is bypassed and compensated for by this hydroacoustic transducer, the MEIHD demonstrates a high-frequency response.

Using guinea pigs and human temporal bones, Pellegrin Hospital at University of Bordeaux[12] has been investigating a piezoelectric vibrator application where the vibrator is placed onto a round window and the ossicular chain is left intact. A platinum ball tethered on the end of the piezoelectric bender bimorph is (slightly) spring loaded onto the round window; thus, no attachment device or polymer is necessary.

The Case Western Reserve University[32,33] researched a semi-implantable, middle ear, electromagnetic hearing device (SIMEHD) during 2 years of Phase I clinical testing. The functional gain observed was not superior to the control acoustic hearing aid and the project was discontinued. In the SIMEHD, an implanted electromagnetic coil interacted with a magnet attached to the incus using dental cement. In more recent investigations, the magnet, coupled to the tympanic membrane, is being used as a mechanical sound detector. It interacts with the electromagnetic coil, the changing magnetic environment of the coil becoming a sensor. This detected signal can then be processed, converted into electrical stimuli, and presented to intracochlear electrodes for cochlear implant stimulation. Thus, their original MEIHD technology is now being investigated for inclusion in a totally implantable cochlear implant device.

The University of Virginia, Charlottesville,[45] has also been investigating a middle ear electromagnetic device; in this instance, a magnet is placed on the round window membrane of the cochlea. This membrane moves in the opposite direction of the oval window membrane under the footplate of the stapes. Thus, it too is a reflection of the cochlear fluid movement that activates hair cells and produces the transduction from fluid movement into afferent action potentials toward the cochlear nucleus. Mechanically sound in principle, the biomedical hurdle remains one of permanent attachment of the magnet to a constantly growing epithelium of the round window membrane.

The University of Dundee[1] MEIHD research is directed toward the design of hearing implants and prostheses for ear surgery using a multilayered piezoelectric actuator for directly driving the ossicles. This work has been carried out in collaboration with the University of Edinburgh, forming the collaborative United Kingdom Middle Ear Research Group.

Xomed-Treace, Inc.,[23] was the first commercial venture in the United States to design a MEIHD, and the company received the first FDA investigative device exemption to clinically test the electromagnetic device in the early 1980s. In collaboration with the Hough Ear Institute (Oklahoma City, OK), this venture has employed the magnetic coupling technology that is now in use worldwide for positioning cochlear implants, as well as MEIHDs and auditory brainstem implants. This approach first used rare earth magnet formulations placed onto the ossicular chain. Early failures with hermeticity and excessive mass loading on the ossicular chain led to corporate abandonment of the project by Xomed-Treace. This beginning, however, became the forerunner of the SOUNDTEC Direct System now entering commercialization almost two decades later. The pioneer use by Xomed-Treace and Hough Ear Institute of reduced-mass and increased-strength rare earth magnets for sound transduction at the incudo–stapedial joint and their determination of ossicular mass loading limits became critical constituents in the success of today's MEIHD technology.

In the 1980s, Gyrus-ENT[22,27,35,51] worked on a partially implantable electromagnetic design that was directed toward patients with chronic middle ear disease and, as such, had partial destruction of the ossicular chain. Because there was a conductive loss as well, their design incorporated a permanent magnet (samarium cobalt) into either a partial ossicular replacement prosthesis (PORP) or total ossicular replacement prosthesis (TORP). Hence, the ossicular chain was reconstructed with a prosthesis that was magnetically responsive to an electromagnetic field emanating from the ear canal and crossing the tympanic membrane. Difficulties encountered during the limited clinical trials for this device included alignment of the magnet in the prosthesis with the electromagnetic coil and electromagnetic coupling, sizing of the implant to allow for vibratory movements, and preloading of the ossicular replacement. A corporate decision was reached to discontinue the research.

NanoBioMagnetics, Inc., in collaboration with the University of Oklahoma Health Sciences Center, is exploring applications of nanotechnology in MEIHDs. Additional miniaturization of implantable device components can lead to increased safety, performance, and implant life. No results have been published to date.

Stanford University[19] was the site for some of the pioneering electromagnetic MEIHD concepts. Investigators glued an AlNiCo magnet to the umbo of the malleus on the outside of the tympanic membrane and placed an electromagnetic coil either in the ear canal or outside of the ear on the post-auricular skin. This same basic concept of driving a movable permanent magnet with a fixed electromagnet was tested acutely on patients undergoing other procedures in the operating room where the magnet was placed on the incus and round window membrane. No mechanical advantage over the umbo site was observed. However, when the magnet implant was clipped to the malleus, one subject wore an 85-mg implant for 22 months without adverse effects or changes in air conduction hearing thresholds up to 400 Hz. This work led to important discussions regarding mass loading of the ossicular chain. Early data were related to the limitations of loading that would impair residual hearing of a person with an intact ossicular chain and a MEIHD.

Resound Corporation[41] investigated the placement of a disc-shaped magnet onto the outside of the tympanic membrane. This magnet was then driven using the same principle used in research at Stanford University and Case Western Reserve University and for the SOUNDTEC Direct System. For the Ear Lens, as it was called, an electromagnetic coil was placed externally at the ear level or around the neck as a collar. This non-implanted IHD held the magnet onto the tympanic membrane by surface tension of a drop of oil between a custom-molded silicone rubber "lens" and the tympanic membrane. Shortcomings in the available battery power necessary to drive the large electromagnetic coils caused this commercial venture to be discontinued.

Sheffield University[2] is investigating the direct drive approach using an integrated electromechanical device where the electromagnetic coil and a central magnet (like an acoustic speaker diaphragm with central magnet and surrounding coil) are in direct contact with the head of the stapes. In this MEIHD, the coil is linked to an external amplifier and microphone either by direct connection outside of the ear or by a radiofrequency transmission link. The required driving forces for mechanical stapes vibration are in the range of 0.16 to 16 μN and must produce amplitudes of the stapes between 0.1 and 10 nm.

Based on auditory physiology, MEIHD technology has several limitations:

1. One cannot regain frequency sensitivity by pushing the basilar membrane harder.
2. When the cochlear amplifier (outer hair cells) is lost, internal noise is generated by external amplification.
3. Prolonged overdrive of the diminished population of hair cells may exacerbate the underlying hearing loss.

For long-term restoration of hearing, then, cochlear implant technology may be the optimal methodology.

10.4.3   Cochlear implants

The technology underlying the 30 or more cochlear implant devices investigated over the last 20 years sought to stimulate the available neurons of the auditory nerve (spiral ganglion cells) by passing current in the vicinity of those neurons.[30,39] The challenge included not only safely electrically depolarizing auditory neurons but also conveying meaningful information about speech to the auditory cortex. Preclinical research on cochlear physiology, electrode design, current delivery, stimulus paradigms, speech encoding, and biomedical engineering has taken place in the United States,[25,37,42,53] Australia,[8] France,[7] Austria, United Kingdom,[52] Switzerland,[10] Denmark, West Germany,[3] and Spain.[4]

The two physiological hypotheses for auditory frequency discrimination have been modeled in CI technology. The *place principle* states that the basilar membrane of the cochlea has the ability to separate out different frequencies from complex sounds. It states that neurons have characteristic frequencies and that the VIIIth nerve has a tonotopic organization. The *temporal principle* states that a time pattern in sounds can convey information about the sound spectrum (frequency distribution of energy) within that sound to the auditory cortex. How the spectra of sounds are distributed in the auditory cortex is the basis of speech discrimination. Spectral analysis by the normal cochlea is also dependent on the intensity of sounds. Both principles are involved in hearing and were considered in the design of speech processing algorithms for the CI. For example, localized, controlled, charge-balanced, biphasic waveforms are used to depolarize selected pools of neurons according to their frequency specificity along the cochlea. Stimulating different electrodes depending upon the frequency of the signal restores the filtering function of the cochlea and presents electrical stimulation coded for interpretation of loudness and pitch information.

Common features of a contemporary CI device include a microphone, external sound processor and power supply, transmitting circuitry, receiver/stimulator package, and electrode array. The microphone picks up sounds and the sound processor filters and selects information that is converted into electrical signals that are transmitted to the intracochlear array of electrodes. For example, the voicing frequency may be coded as the electrode pulse rate.[18] Both the encoded signal and power are transmitted transcutaneously, using radiofrequency antennae, to a demodulator that assigns the speech information to the electrode array. Single-channel systems use only one electrode, while multichannel systems employ up to 31 electrodes. Variations occur in the packaging and location of external sound processors, microphones, and power supplies, but an attracting pair of magnets on the surface of and under the skin are usually used to align the antennae and hold the external devices onto the head. Cochlear implant devices differ in electrode design, type of stimulation, and mode of signal processing. Contemporary CI systems include AllHear, Inc., in Aurora, OR; Clarion®, Advanced Bionics, Inc., in Sylmar, CA (Figure 10.8); Nucleus®, Cochlear Corporation, in

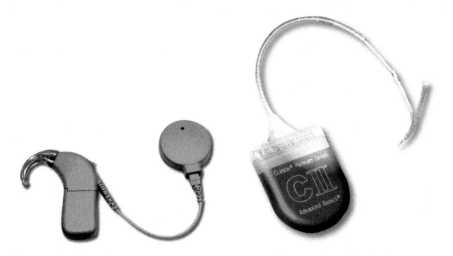

Figure 10.8 Clarion II Bionic Ear system CI (Advanced Bionics, Inc.; Sylmar, CA). The ear-level sound processor (left) contains microphone, battery, sound processing circuitry, transmitting antenna, and magnet for attachment to the head of a patient. The implant (right) contains a receiving antenna, demodulating circuit, and planar contact electrode array (31) at the end of the cable. Both Clarion and Nucleus electrode arrays are curved to fit inside the cochlear canal.

Englewood, CO (Figure 10.9); Digisonic®, MXM Co., in Vallauris, France; Ineraid, Symbion, Inc., in Provo, UT; Laura Flex, Antwerp Bionic Systems, in Belgium; and COMBI™-40+, MED-EL Corp., in Insbruck, Austria (Figure 10.10). In the research phase is the University College London implantable device (UCLID), employing a percutaneous connector link between the speech processor and electrode array. The Clarion, Nucleus and COMBI-40+ are approved for patient use in the United States.

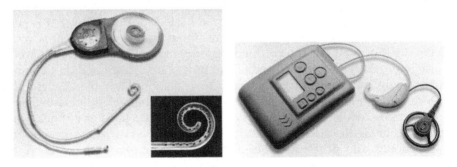

Figure 10.9 Nucleus 24 Contour CI (Cochlear Corp.; Englewood, CO). The ear-level sound processor (right) contains a microphone, battery, sound processing circuitry, and transmitting antenna with magnet in its center for attachment to the head of a patient. The implant (left) contains a receiving antenna, demodulating circuit, ground (ball) electrode, and semibanded electrode array (22) at the end of the second cable.

Figure 10.10 COMBI-40+ CI (MED-EL Corp.; Insbruck, Austria). The ear-level sound processor (upper) contains a microphone, battery, sound processing circuitry with ear hook for attachment to the ear, and transmitting antenna with magnet in its center for attachment to head of a patient. The implant (below) contains a receiving antenna, demodulating circuit, ground (ball) electrode, and split electrode array (12 pairs) at the end of the second cable.

10.5 Signal processing

Signal processing strategies, the conversion of the speech signals into electrical stimuli, vary from device to device. Processing algorithms over 30 years sought either to preserve waveforms or preserve speech envelope information or preserve spectral features from speech (such as formants). Development of single-channel CI devices began in the 1970s (House/3M and its later iteration, Vienna/3M single-channel implants[25]); they are relatively simple in design and cost, using a single electrode, but they became a clinical–engineering controversy based on the principle of channel theory and additional information transfer that became possible through multichannel CI systems. Loudness and temporal variations in speech but limited spectral information were conveyed to deaf patients but their open set speech recognition scores were extremely limited. Important information is contained in the speech signal up to 4000 Hz.

10.5.1 Cochlear electrode types

Electrode design for the CI critically evaluates their placement, number and spacing, orientation with respect to spiral gangion cells, and configuration.[53] Usually placed within the scala tympani (intracochlear), electrode stimulus paradigms preserve the "place" mechanism of the cochlear for coding frequencies. The multichannel electrode array of a CI consists of a flexible silicone rubber carrier, in some instances, shaped to fit inside of the scala tympani, with noble metal (usually platinum) electrodes spaced within the bony modiolus of the cochlea. Because high-frequency neurons are at the basal end of the cochlea, the further the electrode insertion, the lower the frequency response of the stimulation. Most electrode arrays do not extend more than 30 mm into the scala tympani so as to minimize damage to cochlear neurons from insertion trauma. The number of electrodes and spacing between them affects the resolution for coding those frequencies, within the constraints of the number of surviving neurons and the spread of electrical excitation. Electrode design seeks to discretely depolarize a local population of neurons.[15] Approximately 30,000 branches (normal cochlea) of the auditory nerve theoretically can be divided and assigned frequencies by the receiver– stimulator–electrode design. In reality, current spreads symmetrically in the cochlea, and an isolated set of neurons is difficult to stimulate. This is especially so with monopolar stimulation but constrained to a degree with bipolar stimulation. Channel theory predicts that separation of neuronal depolarizations within the cochlea will facilitate frequency analysis and speech discrimination by the cortex.[36]

Currently, CI devices offer either monopolar or bipolar electrode arays or both. The AllHear is a single monopolar CI. The Digisonic has 15 channels for intracochlear implantation and a multiarray with three insertion leads for ossified cochlea where spiral ganglion cells are positioned between electrode arrays for depolarization. Ineraid, the only percutaneous cochlear implant device, is no longer manufactured; it used six electrodes spaced 4 mm apart, with the four apical electrodes in monopolar configuration. The Nucleus device uses up to 22 electrodes spaced 0.75 mm apart; electrodes 1.5 mm apart are used as bipolar pairs. The Clarion, CII Bionic Ear® also provides both monopolar and bipolar configurations and uses up to 31 electrodes. The COMBI-40+ uses 12 pairs of electrodes equally dispersed over 26.4 mm (C40+S) or a split compressed electrode array of five and seven pairs, to be used in an ossified cochlear where drilling and electrode insertion are performed in a monopolar configuration. The UCLID employs 8 to 22 active electrodes. Recent CI developments place the electrode array as close to the inner wall or spiral ganglion of the VIIIth nerve as possible.

10.5.2   Types of cochlear stimulation

The two types of stimulation, analog and pulsatile, depend on how information is presented to the electrodes. In analog stimulation, the acoustic waveform is replicated and presented to the electrode. For multichannel systems, the acoustic waveform is bandpass filtered and presented to all the electrodes simultaneously in analog form.[38] The neural plasticity of the brain is allowed to sort out the information and formulate a meaningful auditory percept.[40] Neurophysiologically, a disadvantage of this simultaneous analog stimulus paradigm is channel interaction, or a lack of discretization of information. In pulsatile stimulation, electrodes deliver a narrow set of biphasic and charge-balanced pulses, the amplitude being related to the height of the envelope of the filtered waveforms. Rate of stimulation affects speech recognition performance and this is likely due to the pulses not overlapping (or not being simultaneous). Both analog and pulsatile stimuli can be sent transcutaneously by radiofrequency transmission or percutaneously by hardwire as with the Ineraid.

Multichannel CI devices began to appear in the 1980s and immediately questions arose as to how many electrodes were best and what kind of information should be transmitted to each electrode. The Ineraid first used the compressed analog (CA) speech-processing algorithm. In CA, the signal is first compressed with automatic gain control, then filtered into four contiguous frequency bands with center frequencies. The filtered waveforms have individual gain controls, then are fed to the four intracochlear electrodes. The CA algorithm uses analog stimulation, delivering continuous analog waveforms to four electrodes simultaneously. Channel separation has not been optimally demonstrated with CA processing.

The next iteration of speech processing algorithms in multichannel CI devices is a continuous interleaved sampling (CIS) strategy, using non-simultaneous, interleaved pulses.[17,55] The pulse amplitudes were derived from the speech envelopes, which were derived by full-wave rectification and low-pass filtering. Several CIS parameters have been optimized over the years that have increased speech recognition in quiet from the initial single-channel results of 5 to 15% to over 80% today. Pulse rate and pulse duration at each electrode vary from patient to patient so the programmable receiver–stimulator is mapped for each patient. Pulses between 100 and 2500 pulses/sec are being used at approximately 33 µsec/pulse. Sound delivery (non-simultaneous) by speech processors can be 18,200 pulses/sec (COMBI-40+), 14,000 pulses/sec (Nucleus 24 Contour™), or 250,000 pulses/sec (Clarion CII Bionic Ear). Stimulation order of the electrodes has been used to optimize channel separation,[5] sometimes from apex to base of the cochlea and sometimes staggered to maximize spatial separation between stimulated electrodes. The compression of the speech envelope outputs is the important determinant of pulse electrical amplitudes. The speech-processing algorithm must ensure that the range of acoustic envelope amplitudes conforms to the auditory dynamic range of the patient. Dynamic range is the range between barely audible speech and uncomfortably loud sounds. Implant patients may have dynamic ranges of only 5 dB, hence CIS strategies use nonlinear compression functions to fit the patient's comfort. Such logarithmic functions are programmed into the patient's map in the erasable programmable read-only memory of the speech processor.

In addition to continuously improving speech-processing algorithms, additional improvements have occurred in CI technology. For example, reverse telemetry was first incorporated into the multichannel CI by the Nucleus Neural Response Telemetry (NRT™). Here, interrogation by the digital processor of the cochlear environment can report selective electrode viability in stimulating the spiral ganglion cells. A brief set of current pulses is delivered to an electrode while a second electrode nearby records the neural response. This response is essentially an intracochlear recording of wave I of the electrically evoked auditory response (EABR) that allows an audiologist to optimally assign electrode stimulations. Thresholds of auditory nerve stimulation can also be accurately detected and better establish the dynamic range of a patient's hearing.

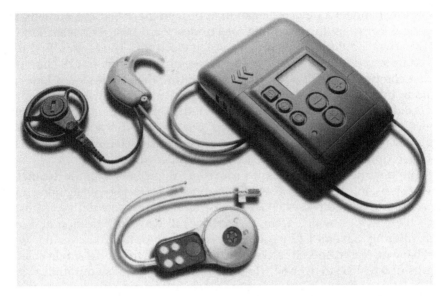

Figure 10.11 Auditory Brainstem Implant (Cochlear Corp.; Englewood, CO). The wearable speech processor (upper) has an ear-level microphone and associated transmitting antenna, and the magnet for attachment contains the batteries and sound processing circuitry. The implant portion (below) is comparable to a cochlear implant with a ball and ground electrode but with a flat surface array of electrodes at the end of the second cable that is placed on the medullary surface, over the cochlear nucleus. The implant also contains a receiving antenna, attaching magnet, and demodulating circuit.

10.5.3   Auditory brainstem implants

If the auditory nerve has been damaged, a modified CI device can accomplish direct stimulation of the cochlear nucleus when electrodes are placed in a carrier for placement onto the surface of the medulla, over the cochlear nucleus.[11,34] Auditory brainstem implants by Cochlear Corp. and Digisonic have been approved for use in humans, and patient data are just being compiled at this time (Figure 10.11).

10.6   Summary

The types of hearing losses experienced today by every population range from mild to severe, most commonly involving loss of inner ear hair cells and the transduction method for initiating the auditory percept. This sensory loss is being ameliorated today, not only by classical acoustic hearing aids but by implantable, sophisticated electronic devices. Middle ear, direct drive mechanical systems (applicable for 19 million Americans) are an emerging technology after electrical neuroprosthetic cochlear implants, which have limited application (about 1 million in the United States). Nevertheless, over

40 research endeavors worldwide have resulted in three leading cochlear implant devices and six promising MEIHDs. The future technological directions of these neuroprosthetic devices include further miniaturization, total implantation, and greater lifetimes in *vivo*. If recombinant or other genetic means of restoring (hair cells) hearing do not emerge, then IHDs will continue as a viable bioengineering approach to hearing loss.

References

1. Abel, E.W., Wang, Z.G., Mills, R.P., and Liu, Y., Performance and power consumption of a multi-layer piezoelectric actuator for use in middle ear implants, *Proc. 3rd Int. Symp. on Electronic Implants in Otology and Conventional Hearing Aids*, May 31–June 2, Birmingham, 2000.
2. Affane, W. and Birch, T.S., A microminiature electromagnetic middle-ear implant hearing device, *Sensors Actuators A*, 46–47, 584–587, 1995.
3. Banfai, P., Karczag, A., Kubik, S., Luers, P., and Surth, W., Extracochlear sixteen channel electrode system, *Otolaryngol. Clin. North Am.*, 19(2), 371–408, 1986.
4. Bosch, J., Prades, J., Colomina, R., and Monferre, A., A model multichannel cochlear implant, *Ear Hear.*, 1(4), 226–228, 1980.
5. Carlyon, R.P., Geurts, L., and Wouters, J., Detection of small across-channel timing differences by cochlear implantees, *Hear. Res.*, 1(1–2), 140–154, 2000.
6. Chasin, M. and Spindel, J., Middle ear implants: a new technology, *Hearing J.*, 54(8), 2001.
7. Chouard, C.H., The surgical rehabilitation of total deafness with the multichannel cochlear implant: indications and results, *Audiology*, 19(2), 137–145.
8. Clark, G.M., Kranz, H.G., Minas, H., and Nathar, J.M., Histopathological findings in cochlear implants in cats, *J. Laryngol. Otol.*, 89(5), 495–504, 1975.
9. Clark, G.M., Tong, Y.C., Patrick, J.F., *Cochlear Prostheses*, Churchhill-Livingstone, Melbourne, 1990.
10. Dillier, N., Spillmann, T., Fisch, U.P., and Leifer, L.J., Encoding and decoding of auditory signals in relation to human speech and its application to human cochler implants, *Audiology*, 19(2), 146–163, 1980.
11. Di Nardo, W., Fetoni, A., Buldrini, S., and Di Girolamo, S., Auditory brainstem and cochlear implants, functional results obtained after one year of rehabilitation, *Eur. Arch. Otorhinolaryngol.*, 258, 5–8, 2000.
12. Dumon, T., Zennaro, O., Aran, J.M., and Bébéar, J.P., Piezoelectric middle ear implant preserving the ossicular chain, *Otolaryngol. Clin. North Am.*, 28(1), 173–187, 1995.
13. Fredrickson, J.M., Coticchia, J.M., and Khosla, S., Current status in the development of implantable middle ear hearing aids, *Adv. Otolaryngol.*, 10, 189–203, 1996.
14. Fredrickson, J.M., Coticchia, J.M., and Khosla, S., Ongoing investigations into an implantable electromagnetic to severe sensorineural hearing loss, *Otolaryngol. Clin. North Am.*, 28 (1), 107–120, 1995.
15. Friesen, L.M., Shannon, R.V., and Slattery, III, W.H., Effects of electrode location on speech recognition with the Nucleus-22 cochlear implant, *J. Am. Acad. Audiol.*, 11, 418–428, 2000.

16. Gan, R.Z., Wood, M.W., Ball, G.R., Dietz, T.G., Dormer, K.J., Implantable hearing device performance measured by laser doppler interferometry, *Ear Nose Throat J.*, 76(5), 297–309, 1997.

17. Geurts, L. and Wouters, J., Coding of the fundamental frequency in continuous interleaved sampling processors for cochlear implants, *J. Acoustic. Soc. Am.*, 109(2), 713–726, 2001.

18. Geurts, L. and Wouters, J., Coding of the fundamental frequency in contiuuous interleaved sampling processors for cochlear implants, *J. Acoust. Soc. Am.*, 109(2) 713–726, 2001.

19. Goode, R.L., Current status of electromagnetic implantable hearing aids, *Otolaryngol. Clin. North Am.*, 22, 201–209, 1989.

20. Gyo, K., Saiki, T., and Yanagihara, N., Implantable hearing aid using a piezoelectric ossicular vibrator: a speech audiometic study, *Audiology*, 35, 271–276, 1996.

21. Håkansson, B., Lidén, G., Tjellström, A., Ringdahl, A., Jacobsson, M., Carlsson, P., and Erlandson, B.E., Ten years of experience with the Swedish bone-anchored hearing system, *Ann. Otol. Rhinol. Laryngol.*, 99(10, suppl. 151, pt. 2), 1–16, 1990.

22. Heide, J., Tatge, G., Sander, T. et al., Development of a semi-implantable hearing device, in *Advances in Audiology*, Vol. 4, *Middle Ear Implant, Implantable Hearing Aids*, Hoke, M., Ed., Karger, Basel, 1988.

23. Hough, J., Vernon, J., Dormer, K., Johnson, B., and Himelick, T., Experiences with implantable hearing devices and a presentation of a new device, *Ann. Otol. Rhinol. Laryngol.*, 95(1), 60–65, 1986.

24. Hough, J.V., Dyer, R.K., Matthews, P., and Wood, M.W., Early clinical results, SOUNDTEC implantable hearing device Phase II study, *Laryngoscope*, 111(1), 1–8, 2001

25. House, W.F., Cochlear implants, *Ann. Otol. Rhonol. Laryngol.*, 85(3, suppl. 27), 1–93, 1976.

26. Huttenbrink, K.-B., Zahnert, Th., Bornitz, M., and Hoffmann, G., Biomechanical aspects in implantable microphones and hearing aids and development of a concept with a hydro dynamic acoustical transmission, *Acta Otorhinolaryngol.*, (121), 185–189, 2001.

27. Kartush, J.M. and Tos, M., Electromagnetic ossicular augmentation device, *Otolaryng. Clin. North Am.*, 28, 155–172, 1995.

28. Kodera, K., Suzuki, J.I., Nagai, K., and Yabe, T., Sound evaluation of partially implantable piezoelectric middle ear implant: comparative study of frequency responses, *Ear Nose Throat J.*, 73(2), 108–111, 1994.

29. Leysieffer, H., Baumann, J.W., Müller, G., and Zenner, H.P., An implantable piezoelectric hearing aid transducer, *HNO*, 45, 792–800, 1997.

30. Loizou, P.C., Introduction to cochlear implants, *IEEE Signal Process. Mag*, 101–130, 1998.

31. Lustig, L.R., Arts, A.H., Brackman, D.E. et al., Hearing rehabilitation using the BAHA bone-anchored hearing aid: results in 40 patients, *Otol. Neurotol.*, 22(3), 328–334, 2001.

32. Maniglia, A.J. and Proops, D.W., Eds., Implantable electronic otologic devices: state of the art, *Otolaryngol. Clin. North Am.*, 34(2), 1–522, 2001.

33. Maniglia, A.J., Ko, W.H., Rosenbaum, M., Falk, T., Zhu, W.L., Frenz, N.W., Werning, J., Masin, J., Stein, A., and Sabri, A., Contactless semi-implantable electromagnetic middle ear device for the treatment of sensorineural hearing loss, *Ear Nose Throat J.*, 73, 78–90, 1994.

34. Marangos, N., Stecker, M., Sollman, W.P., and Laszig, R., Stimulation of the cochlear nucleus with multichannel auditory brainstem implants and long-term results: Freiburg patients, *J. Laryngol. Otol.*, (27, suppl.), 27–31, 2000.
35. McGee, T.M., Kartush, J.M., Haide, J.C. et al., Electromagnetic semi-implantable hearing device: Phase I clinical trials, *Larynogoscope*, 101, 355–360, 1991.
36. Mehr, M.A., Turner, C.W., and Parkinson, A., Channel weights for speech recognition in cochlear implant users, *J. Acoust. Soc. Am.*, 109(1), 359–366, 2001.
37. Michaelson, R.P., Merzenich, M.M., Schindler, R.A., and Schindler, D.N., Present status and future developments of the cochlear prosthesis, *Ann. Otol. Rhinol. Laryngol.*, 84(4, pt. 1), 494–498, 1975.
38. Morse, R.P. and Roper, P., Enhanced coding in a cochlear-implant model using additive noise: aperiodic stochastic resonance with tuning, *Phys. Rev.*, 61(5), 61–66, 2000.
39. Niparko, J.K., Kirk, K.I., Mellon, N.K., Robbins, A.M., Tucci, D.L., and Wilson, B.S., Eds., *Cochlear Implants: Principles & Practices*, Lippincott, Williams & Wilkins, Philadelphia, 2000.
40. Nishimura, H., Doi, K., Iwaki, T., Hashiwawa, K., Oku, N. et al., Neural plasticity detected in short- and long term cochlear implant users using PET, *NeuroReport*, 11(4), 811–815, 2000.
41. Perkins, R. and Pluvinage, V., The ear lens: a new method of sound transmission to the middle ear-current status [abstract 34], presented at the Int. Symp. on Electronic Implants in Otology, November, Orlando, FL, 1993.
42. Rabinowitz, W.M. and Eddington, D.K., Effects of channel-to-electrode mappings on speech perception with the Ineraid cochlear implant, *Ear Hear.*, 16(5), 450–458, 1995.
43. Snik, A.F.M. and Cremers C.W.R.J., First audiometic results with the vibrant soundbridge: a semi-implantable hearing device for sensorineural hearing loss, *Audiology*, 38, 355–338, 1999.
44. Snik, A.F.M. and Cremers C.W.R.J., The effect of the "floating mass transducer" in the middle ear on hearing sensitivity, *Am. J. Otol.*, 21, 42–48, 2000.
45. Spindel, J.H., Corwin, J.T., Ruth, R.A. et al., The basis for around window electromagnetic implantable hearing aid, in *Proc. of the 13th Annual Int. Conf. of IEEE Eng. in Med. Biol. Society*, October, Orlando, FL, 1991.
46. Stenfelt, S., Håkansson, B., Jönsson, R., and Granström, G., A bone-anchored hearing aid for patients with pure sensorineural hearing impairment, *Scand. Audiol.*, 29, 175–185, 2000.
47. Suzuki, J.I., Ed., Middle ear implant: implantable hearing aids, *Adv. Audiol.*, 4, 1–174, 1988.
48. Syms, III, C.A. and House, W.F., Surgical rehabilitation of deafness, *Otolaryngol. Clin. North Am.*, 30(5), 777–782, 1997.
49. Tietze, L. and Papsin, B., Utilization of bone-anchored hearing aids in children, *Int. J. Pediatr. Otorhinolaryngol.*, 58(1), 75–80, 2001.
50. Tjellström, A. and Håkansson, B., The bone-anchored hearing aid, *Otolaryngol. Clin. North Am.*, 28(1), 53–72, 1995.
51. Tos, M., Salomon, G., and Bonding, P., Implantation of electromagnetic ossicular replacement device, *Ear Nose Throat J.*, 73(2), 92–103, 1994.
52. Walliker, J.R., Rosen, S., Douek, E.E., Fourcin, A.J., and Moore, B.C., Speech signal presentation to the totally deaf, *J. Biomed. Eng.*, 5(4), 316–320, 1983.

53. Wilson, B.S., Cochlear implant technology, in *Cochlear Implants-Principles & Practices*, Nikarko, J.K., Kirk, K.I., Mellon, N.K., Robbins, A.M., Tucci, D.L., and Wilson, B.S., Eds., Lippincott, William & Wilkins, Philadelphia, 2000, pp. 109–128.

54. Zenner, H.P., TICA totally implantable system for treatment of high frequency sensorineural hearing loss, *Ear Nose Throat J.*, 79(10), 770–777, 2000.

55. Ziese, M., Stützel, A., von Specht, H., Begall, K., Freigang, B., Sroka, S., and Nopp, P., Speech understanding with the CIS and the N-of-M strategy in the MED-EL COMBI-40+ system, *Otorhinolaryngology*, 62, 321–329, 2000.

chapter eleven

Visual neuroprostheses

David J. Warren and Richard A. Normann

Contents

0-8493-1100-4/03/$0.00+$1.50
© 2003 by CRC Press LLC

11.1   Introduction

In the eighteenth century, Luigi Galvani, an Italian physician, observed that dissimilar metals, attached to a frog's leg and connected together, can cause the frog's skeletal muscles to contract. Subsequently, natural philosophers, and later physiologists, have come to appreciate that electricity is one of the key features associated with the control of the body. Based upon the pioneering work of Einthoven in studying the role of electricity in the sensory and motor parts of the body, it has become clear that the contraction of all muscle is associated with local electrical fields and that, by the discriminate application of externally applied electrical fields, one can intervene with damaged or diseased parts of the nervous system and thereby restore lost functions. The first, and most lifesaving, example of this electrical intervention is the cardiac pacemaker, which was developed in the middle of the last century and has become a standard therapeutic approach to a variety of cardiac arrhythmias and pathologies.

The extension of this successful therapy to other sensory and motor systems has fostered the neuroprosthetic field. This field is in its infancy but is experiencing great interest in both academic and commercial sectors. With the widespread availability of VLSI (very-large-scale integration) circuitry and with the emergence of microfabrication technologies, researchers are now able to build interfaces to the nervous system with greater complexity than could have been imagined only a few decades ago. These interfaces are being built of highly biocompatible materials that contain feature sizes comparable to the neurons that the devices are intended to interact with. Further, our understanding of the function and dysfunction of the nervous system has improved to the point that, in many cases, researchers have suggested plausible interventional routes whereby neuroprosthetic systems could expect to be efficacious. It is clearly the right time for this technology.

About a half century ago, Brindley, at Cambridge University in England, conducted a bold set of experiments wherein he tried to restore a visual sense to a few individuals who had become profoundly blind. These pioneering experiments did not result in the restoration of useful vision in these subjects, but they did indicate that the concept could be entertained seriously. Thus, the visual neuroprosthetic field was born. Because the technological innovations described above had yet to be developed, little progress was made in the field up until the last decade. Over this last 15 years, many individuals have begun working in this field, and, while clinical systems are still probably a decade away, it is becoming clear that such systems will likely find their way into the operating rooms of major hospitals around the world.

This chapter provides an overview of some of the progress that has been made in this field. We will describe the anatomy and physiology of the visual pathways and human-engineered systems that have been designed to interface with neuronal ensembles at the levels of the retina, the optic nerve, and the visual cortex. We describe animal experiments that support the safety of these systems, and preliminary human experiments that are beginning to demonstrate the efficacy of the approach. It is stressed that the human experimentation is still in its infancy, and as a result the findings are mixed and inconclusive, but encouraging. However, the increased numbers of experimenters working in the area, the acceleration of technological innovation, and the commercial impetus to develop successful clinical systems make it clear that visual neuroprosthetic systems will provide, perhaps, the first interventional systems that could restore useful vision to those with profound blindness. Useful vision in this context will not be vision as enjoyed by normally sighted individuals. First-generation systems will be pixelized, will contain small numbers of pixels, and will recreate narrow field views of the visual world in front of the blind subject. However, it is hoped that such systems will allow the subject independent mobility without requiring a guide dog or a family member or friend, at least in familiar environments, and perhaps even in unfamiliar visual environments.

11.2 Anatomy and physiology of the visual pathway

Despite almost a half century of experimentation, the best site for implementing a visual neuroprosthesis has yet to be resolved. As will be seen in subsequent sections, numerous sites have been investigated, each with its relative merits and shortcomings. To provide a better understanding of each neuroprosthetic approach and to give grounds for comparing the competing approaches, an awareness of the underlying neural system is necessary. As details of the visual pathway are readily available in any elementary anatomy and physiology text, the discussion here is limited to a short review highlighting the potential sites for a neuroprosthesis. In particular, we have chosen to highlight the neural organization of these sites and how this organization relates to a neuroprosthetic application. The majority of the information provided in this section comes from either well-known neuroscience texts[1,2] or the excellent Web site http://www.webvision.med.utah.edu.

The visual pathway, highlighting the points where vision neuroprostheses have been or could be implemented is shown schematically in Figure 11.1. Light entering the eye falls upon the *retina*, located at the back surface of the eye. Here, photoreceptor neurons convert the electromagnetic energy of the light into electrochemical signals. This is the first stage of a series of

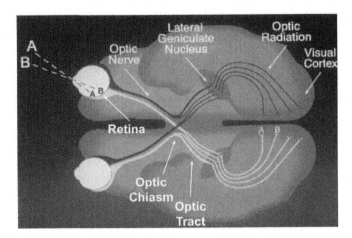

Figure 11.1 Schematic view of the visual pathway, highlighting points where a visual neuroprosthesis has been or could be investigated. This illustration also shows the concept of visuotopy with the letters A and B. The visuotopic organization of the pathway results in nearby objects in visual space being represented in nearby neurons throughout the visual pathway.

retinal neurons that process features of the visual scene. The axons of the last of the series, the retinal ganglion cells, are collected together into the *optic nerve* (ON). These axonal fibers are reorganized at the *optic chiasm* (OC) and then project to a number of subcortical structures through the *optic tract* (OT). The majority of these axons form synapses in the *lateral geniculate nucleus* (LGN) of the thalamus, where another series of neurons further process the visual scene. In turn, the axons of the LGN neurons project through the *optic radiation* (OR) to the cerebral cortex. Here, a hierarchical scheme of visual processing occurs over much of the posterior region of the occipital cerebral cortex. The regions subserving this processing are collectively called *visual cortex* (VC).

A number of themes are common to the neural organization at all sites along the visual pathway. The first of these is that, for any neuron along the pathway, its *receptive field* describes the type of the visual stimuli that causes the neuron to respond. The receptive field characterization of a neuron details the nature of the visual stimulus (location in visual space, shape, size, intensity, color, etc.) that optimally drives the neuron and how the neuron's activity changes for other than the optimal stimulus. For example, a particular photoreceptor neuron may reserve its greatest response for a small blue spot of light at a particular location relative to central vision. Most probably, external electrical stimulation of the same neuron could evoke in a similar percept. A related and very critical theme for a neuroprosthesis is that the map from visual space to neural space is *visuotopic*. That is, the neurons at any site along the visual pathway are arranged so that their receptive field locations form an organized and approximately linear map of visual space. This implies that nearby objects in visual space evoke activity in nearby neurons. Hence, the outline of a rectangle presented in visual space will result in activity in a similarly shaped arrangement of neurons. The arrangement of neurons may be differentially stretched in each axis, rotated, and evenly slightly warped but still would appear as the outline of a rectangle. Presumably, external electrical stimulation of the same ensemble of neurons will result in the perception of the outline of a rectangle. A last theme is that the visual pathway represents a massively parallel method of signal processing. Associated with this parallelization are a number of distinct processing pathways found in primates, including humans. The principal two pathways, the M pathway (for *magno*, or large) and the P pathway (for *parvo*, or small), begin at the retina and are segregated throughout much of the visual pathway. It is thought that the M and P pathways represent two broad features of an object in visual space: where the object is located and what the object is, respectively. At this point, it is unclear whether a neuroprosthesis should (or could) preferentially evoke activity in the M pathway, the P pathway, or both pathways in order to provide the best percept. Nevertheless, due to the segregation of the two pathways, there could be a preference for stimulating one pathway over the other at various points along the visual pathway.

11.2.1   Retinal anatomy and physiology

If one were to observe a normal human retina through an ophthalmoscope, one would see the gross morphological features illustrated in Figure 11.2. On the right side of the figure is the head of the optic nerve, also called the *optic disk*. A number of blood vessels originate from approximately the center of the optic disk and spread to cover much of the inner retina (the side of the retina closest to the lens). These vessels are supplied by the central artery and vein of the retina, which pass through the optic nerve. The outer retina (the side of the retina farthest from the lens and not visible in the figure) is perfused by the blood vessels running behind the retina in the choroid. The blood vessels on the inner retina and in the choroid form the retinal and choroidal circulations, respectively. To the left of the optic disk, one observes a small (~1.5-mm diameter) darker region of the retina devoid of major blood vessels. This region is called the fovea and it subserves the approximately 6 degrees of central vision, equivalent to a 10-cm-diameter circular region at a distance of 1 m. The fovea and the closely surrounding region are called the macula lutea or, more commonly, the macula. As the macular region is not readily seen in the figure, a dashed line indicates the border of the macula. This region, having a diameter of approximately 5 mm, subserves the approximately 20 degrees of the central visual field. This is equivalent to a 35-cm-diameter circular region at a distance of 1 m, or approximately the entire screen on a standard 15-inch monitor observed at 1 m. As the optics of the eye are relatively linear, the magnification factor (millimeters of neural tissue per degree of visual space) is approximately 0.25 throughout the retina.

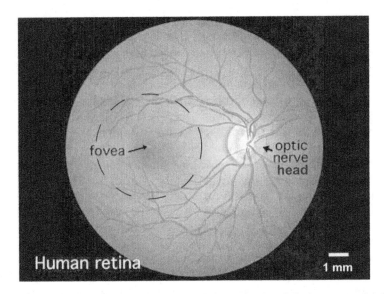

Figure 11.2 Normal human retina as viewed through an ophthalmoscope, highlighting the regional specialization of the retina. The optic nerve head, also known as the optic disk, is where the axons of the retinal ganglion cells are collected together into the optic nerve as these axons course towards their subcortical targets. This region is devoid of photoreceptors. Additionally, the blood vessel of the retinal circulation originates from the optic nerve head. The fovea, the region of darker pigmentation, subserves the central 6 degrees of visual space and is the region of greatest visual acuity. The photoreceptors in this region are primarily cones. The macula lutea, outlined with a dashed line, subserves the central 20 degrees of visual space. Both rod and cone photoreceptors are found outside the fovea, with rod photoreceptors becoming more prevalent the farther one is from the fovea. The two principal sources of untreatable blindness, age-related macular degeneration and retinitis pigmentosa, are characterized by loss of photoreceptors in the macular region and outside the macular region, respectively. (Adapted from Webvision. With permission.)

The retina has a laminar neural organization, a highly simplified view of which is presented in Figure 11.3. This organization consists of three nuclear layers (layers of neuronal cell bodies) separated by two plexiform layers (layers of synaptic connections). An inner and outer limiting membrane as well as an outer epithelial layer surround the neural retina. As the photoreceptors are located in the outermost nuclear layer in humans, the description logically starts with the outermost layer. The *retinal pigmented epithelium* (RPE) is not properly part of the neural retina but bears discussion due to its association with pathologies of the retina. This layer provides critical nutrients to the photoreceptors as well as phagocytosis of the exhausted components of the photoreceptors that are shed on a daily basis. The *outer nuclear layer* (ONL) consists of the photoreceptor neurons that convert light's electromagnetic energy into electrochemical energy. In humans, there are two morphological classes of photoreceptor neurons, rods and cones. Cone pho-

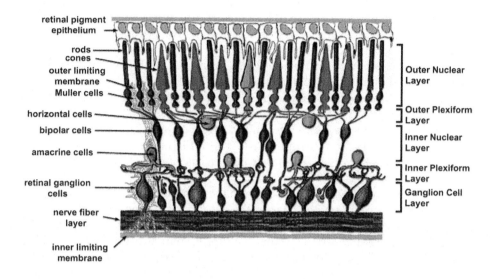

Figure 11.3 Schematic diagram of the organization of the retina, highlighting the principal cells of the retina and the nomenclature of the neural retina. Light entering the eye passes through the entire neural retina before being transduced into electrical signals by the rod and cone photoreceptors of the outer neural layer. These signals are processed in a columnar fashion by the bipolar cells of the inner neural layer and the retinal ganglion cells of the ganglion cell layer as well as being processed horizontally by the horizontal and amacrine cells of the inner nuclear layer. The axons of the retinal ganglion cells are collected together into the nerve fiber layer, which lies between the ganglion cells and the inner surface of the retina, or the epiretinal surface. (Adapted from Webvision. With permission.)

toreceptors are functional in typical daytime viewing conditions and mediate color vision. Humans have three subclasses of cone receptors (S, M, and L) that are most sensitive to blue, green, and red colors, respectively. Rod receptors become functional only at low light levels and, hence, are associated with night vision. As all rod receptors are most sensitive to a single color (blue-green) in humans, at night we can only distinguish relative intensity. Both rod and cone photoreceptors utilize graded voltage potentials to signal a change in stimulus intensity. That is, a change in luminance results in a change in the membrane potential, not an action potential. The densities of rod and cone photoreceptors vary with distance from the fovea (or with retinal eccentricity). Within the fovea, the photoreceptors are exclusively cones. For the first 10-degree radius about the fovea, covering approximately the macular region, the density of cone photoreceptors drops off rapidly to approximately 1/20th the density at the center of the fovea. An inverse relationship occurs with the density of rod photoreceptors, starting with a nil density at the fovea that grows rapidly for the first 20-degree radius about the fovea. The *outer plexiform layer* (OPL) separates the ONL and the *inner nuclear layer* (INL) and is the region where the photoreceptors of the ONL form synapses with the two classes of neurons of the INL (bipolar and horizontal neurons). Like photoreceptors, both bipolar and horizontal neurons utilize graded voltage potentials. The third class of neurons of the INL, amacrine cells, fire action potentials. The *inner plexiform layer* (IPL) separates the INL and the *ganglion cell layer* (GCL) and is the region where the bipolar and amacrine neurons of INL form synapses with GCL neurons. The neurons of the GCL consist almost exclusively of action potential firing *retinal ganglion cells* (RGC). The axons of the RGC form the *nerve fiber layer* (NFL) as they cross the inner surface of the retina, heading towards the optic disk. These fibers are unmyelinated until they reach the optic disk. The *inner limiting membrane* (ILM) separates the retina from the vitreous humor.

Visual information is processed from the outer to inner layers in a columnar functional organization. That is, the activity of a ganglion cell is modulated by the photoreceptor and bipolar neurons along a column from the inner to outer retina. This way the visuotopic map is preserved across the layers of the retina. Additionally, the horizontal and amacrine neurons integrate information across the neighboring regions of visual space. The degree of convergence of information across visual space depends on the retinal eccentricity of the ganglion cell. At the fovea, there is a one-to-two relationship between photoreceptors and ganglion cells; hence, the fovea is the region of greatest visual acuity. As one moves to the periphery, a large number of photoreceptors drive a single ganglion cell.

The degree of convergence is reflected in the receptive field properties of the RGC. At the fovea, the receptive field size is on the order of a few minutes of arc. At the outer edge of the macula, the receptive field size is as large as 3 to 5 degrees. The majority of the ganglion cells are optimally driven by a center-surround stimulus, or small circular of light at one intensity (or color) surrounded by an annulus of light at another intensity (or color). Cells of the M pathway, representing only 8% of the RGC, are best driven by a luminance difference between the center and surround. Two possible luminance differences are possible: a well-lit circle surrounded by a dark annulus (ON ganglion cell) or a dark circle surrounded by a well-lit annulus (OFF ganglion cell). Cells of the P pathway, representing 80% of the RGC, are best driven by a color difference in the center and surround. The two types of optimal differences are green vs. red and blue vs. yellow. The remaining 12% of the retinal ganglion cells do not fall into the M or P classification and have uncharacterized function.

The receptive field characteristics of bipolar cells are very similar to those of the RGC. In response to the reduction of neurotransmitter release by photoreceptors with increased illumination, bipolar cells either depolarize or hyperpolarize; hence, the two classes of bipolar cells are ON center and OFF center. The opposing surround is thought to result from lateral inhibition by horizontal cells. The similarity between bipolar cell and RGC receptive field properties is thought to be the result of RGC receiving exclusively excitatory input from a small number of bipolar cells.

11.2.2   Organization of the optic nerve and tract

Together, the optic nerve and tract cross along the ventral surface of the brain. From the retina to the optic chiasm (the intersection of the nerves from the left and right eyes), the nerve is called the *optic nerve* (ON) and, from the chiasm to the subcortical targets, it is called the *optic tract* (OT). In humans, both the nerve and tract have a diameter of approximately 3 mm. The length of the optic nerve is approximately 50 mm, and the tract is approximately 30 mm. Neither the nerve nor the tract has neuronal cell bodies. Instead, each carries the approximately 1,200,000 axons from the retinal ganglion cells of each eye.[3] (In comparison, the auditory nerve has around 30,000 fibers.) The optic nerve contains the axons of the visual fields nasal and temporal to the fovea for a single eye. At the optic chiasm, the fibers are reorganized so that the optic tract contains almost exclusively the axons representing the contralateral visual hemifield. In addition to the RGC fibers, the central artery and vein of the retina pass through the optic nerves, approximately in the middle of the nerve.

Indications suggest that the fibers of the optic nerve are visuotopically organized with the upper retina (lower visual field) represented along the dorsal side of the nerve, the central retina along the lateral side, and nasal visual field along the medial size[,4,5] however, this visuotopic organization appears to vary along the length of the nerve. The distribution of P and M fibers within the nerve and tract is unclear at this time, but, as the M fibers have twice the conduction velocity, they likely have a larger diameter. As the optic nerve and tract are axonal extensions of the retinal ganglion cells, they have the same receptive field characteristics.

11.2.3   Anatomy and physiology of subcortical structures

The axons of the RGC target three subcortical structures, the superior colliculus, the pretectum, and the *lateral geniculate nucleus* (LGN), with the majority (90%) targeting the LGN. The superior colliculus and the pretectum are located on the roof of the midbrain and are associated with saccadic eye movements and pupillary reflexes, respectively. As blindness results when the LGN is damaged but these two structures are preserved, neither structure is well suited for a vision neuroprosthesis.[2] The LGN is a small structure (approximately 7×7 mm by 2 mm deep) located on the ventral side of the thalamus. The neurons of the LGN are often considered relay neurons, merely receiving input from the RGC and passing the same on to cortex, but this is an overly simplistic view of the LGN. The LGN has six independently acting lamina arranged as six upside down U's stacked in the ventral–dorsal axis. Each lamina receives input from only one of the P or M pathways and only one of the contralateral or ipsilateral eyes. Due to the much higher proportion of P neurons, four laminae are dedicated to this pathway. Each lamina is visuotopically organized and the visuotopic maps are registered between lamina so that all neurons in a ventral–dorsal path have receptive fields in the same region of visual space. Almost half of the neurons of the LGN (representing nearly half of the area of the LGN) have receptive fields in the foveal or surrounding region. Although a neuroprosthesis could take advantage of this unique neural organization, the LGN generally has not been considered a potential site due to the surgical difficulties in its exposure.

11.2.4   *Visual cortex anatomy and physiology*

The final stage of the visual pathway is the visual cortex. Even within the cortex, visual processing continues to be a hierarchical operation with the M and the P pathways segregated. Of the LGN axons projecting to visual cortex, almost all project to the gyri lining the calcarine fissure, a large sulcus running anterior–posterior along the medial surface, just anterior to the occipital pole. This 2500- to 3200-mm^2 region is known as primary visual cortex or *visual area 1* (V1) to indicate that it is the first region of visual processing at the cortical level. The region is also called *striate cortex*, due to its unique laminar appearance with certain histological stains, and Area 17, from Brodmann's classification of the regions of cerebral cortex. The regions of cortex subserving visual processing that surround V1, including the lateral surface of cortex anterior to the occipital pole, are called visual areas 2, 3, 4, and 5 (V2, V3, V4, and V5), indicating somewhat the hierarchy of visual processing.

All of these cortical areas have a similar laminar structure, consisting of six neural laminae arranged in planes from the pia mater to white matter, with a total thickness of approximately 2 mm in humans. Visual processing is performed both in a columnar manner, with information passing between neurons in a column from pia mater to white matter, and in a horizontal method, where information is integrated across a number of columns. Generally, the fourth lamina from the pia mater (layer 4) receives input from the preceding stage in visual processing, layers 2 and 3 project to the next stage in processing, and layer 6 projects back to the preceding stage in processing. In V1, layer 4 is further subdivided into four sublaminae, 4A, 4B, 4Cα, and 4Cβ. The latter two sublaminae receive LGN input from the M and P pathways, respectively. Consistent with the general architecture of cortical lamina, layer 6 in V1 projects back to the preceding stage in processing, the LGN in this case. Each of the areas of visual cortex is visuotopically organized. Consistent with the columnar processing of information, the visuotopic maps are registered between lamina so that all neurons in a column from pia mater to white matter have receptive fields in the same region of visual space. Almost half of the neurons in any visual area (representing nearly half the size of the region) have receptive fields in the foveal or surrounding region. The foveal region of visual space is represented at the posterior part of V1.

With the exception of some neurons of layers 4Cα, and 4Cβ, which have LGN-like receptive fields, the receptive field characteristics of V1 neurons are much richer than those of the preceding stages of processing. Here, the receptive fields continue to have subregions preferring illumination (ON regions) and lack of illumination (OFF regions) but the subregions are elongated, more of a hotdog in a bun shape than circular with an annular surround. The elongation of the subregions leads to a preference for bars of particular orientation, resulting in the new receptive field characteristic of orientation preference. The segregation of the retinal input from each eye, so carefully preserved in the laminar structure of the LGN, becomes less distinct at the cortical level. Here, a neuron may be best driven by the contralateral eye or ipsilateral eye, or be equally driven by both eyes. This leads to the receptive field characteristic of ocular dominance, the measure of which eye is the dominant driving source. Similar to visuotopy, nearby neurons tend to prefer the same orientation and receive input from the same eye, observed both when examining a column from pia to white matter and when traversing across cortex. Further, when traversing across cortex, one tends to observe slow changes in either the preferred orientation or preferred eye, or both. A complete sequence of orientations (0 to 180 degrees) and eye preferences (contralateral to ipsilateral and back to contralateral) occurs over approximately a 1-mm² area of cortex. Although it is tempting to map the changes in orientation and eye preferences as a square 1 × 1-mm grid overlaying the visuotopic map, recent studies have suggested that the prior two maps have a much more complex organization.[6] These studies have shown the existence of regions where the preferred orientation (or eye dominance) rapidly changes and other regions where the same orientation (or eye dominance) is preferred for a few millimeters.

After V1, the organization of the visual pathway becomes more complex and the optimal visual stimuli become less clear. Throughout the visual pathway to V1, the M and P pathways pass through the same neural structures but target different lamina in order to remain distinct. In V2, the pathways begin to diverge with the pathways targeting neighboring horizontal subdivisions of V2. Past V2, the pathways completely diverge. The P pathway heads toward posterior parietal cortex and principally targets visual areas on the dorsal aspect of occipital cortex. The M pathway heads toward inferior temporal cortex and principally targets visual areas on the ventral aspect of occipital cortex. Further, the simple hierarchical organization is replaced with a more complex connectivity as can be seen in the block diagram of the visual area interconnections shown in Figure 11.4.[7,8] Finally, concurrent with higher level feature extraction, the receptive field characteristics V2 neurons and those of subsequent visual areas become much more intricate. For example, some neurons of inferior temporal cortex are most responsive to features that appear like a face.[9] Due to complexity of the interconnectivity and the receptive field characteristics, visual areas beyond V1 have not been proposed as sites for vision prostheses.

11.2.5   Pathologies leading to blindness

The leading causes of untreatable blindness are *age-related macular degeneration* (AMD), *retinitis pigmentosa* (RP), accidents, and cancers. Additionally,

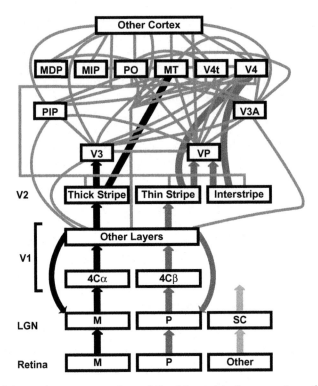

Figure 11.4 Schematic representation of the hierarchical processing of information along the visual pathway. The two principal pathways (M for magnocelluar pathway and P for parvocellular pathway) are represented by thick black and thick dark-gray arrows, respectively. These pathways are thought to represent the location (where) and the type (what) of objects in visual space, respectively. Beyond V1, the hierarchical organization becomes somewhat less clear, suggesting that these regions would not be prime candidates for visual neuroprostheses. (Adapted from Van Essen.[7])

blindness can be caused by glaucoma and diabetic retinopathy, but these two pathologies currently are treatable and blindness can be prevented if treated at an early stage.

Age-related macular degeneration is pathology of the retina characterized by the progressive loss of central vision, generally occurring in the elderly. It is the leading cause of blindness in developed countries, with an estimated 2,000,000 individuals having some form of AMD in the United States. For unknown reasons, the pigmented epithelium of the retina degenerates, leading to a subsequent degeneration of the photoreceptors and fluid leakage into the neural retina. As this occurs in the macular region, central vision is compromised, leading to the loss of the ability to distinguish fine detail. Retinitis pigmentosa is an inherited pathology of the retina characterized by the loss of peripheral vision and night vision. An estimated 100,000 individuals are affected by RP in the United States. For unknown reasons, the rod photoreceptors degenerate. As these are the principal photoreceptors outside the macula, peripheral vision is lost. Further, as these photoreceptors are functional only in dim light, night vision is compromised. Individuals with advanced RP have tunnel vision, only seeing objects with the centrally located cone photoreceptors. Despite the degradation of the outer nuclear layer in both of these retinal forms of blindness, there are reports that inner nuclear and retinal ganglion layers are partially preserved, having 80 and 30% of the neurons found in a normal eye, respectively.[10–12] On the contrary, another group has reported more significant cell loss in transgenetic animal models of RP.[13] Further, this group reports that the remnant neurons become so disorganized and dislocated within the neural retina that the laminar structure of the retina is not apparent in histological sections.

For other sources of blindness, typically accidents and cancer, either the eye is lost or the visual pathway is disrupted downstream from the retina. In either case, the earliest site in the visual pathway that vision prostheses would be possible is the LGN.

11.2.6 Concluding remarks

Of the neural structures described previously, only the photoreceptor layer of the retina, ganglion cell layer of the retina, the optic nerve, and the cerebral cortical region V1 have been proposed as potential sites for a vision neuroprosthesis for restoration of function in the blind. These sites principally have been proposed due to the relative ease of surgical access, with perhaps the exception of the optic nerve, and the suggestion that the pathology leading to blindness leaves the remnant neurons somewhat functional.[13] As will be seen in the subsequent sections, some success has been achieved for all four potential sites. However, due to the need to use "heavy-handed" methods in order to evoke a useful percept, it is likely that only a crude sense of vision will be possible in the near term. Nevertheless, using the progress in cochlear neuroprostheses as an example, once an even partially useful vision neuroprosthesis becomes available steady improvements will occur and a more useful sense of artificial vision will result. When these second and subsequent generation vision prostheses appear, it is likely they will be specifically designed to take better advantage of the characteristics of the underlying neural system.

11.3 Elements of a visual neuroprosthesis

In order for a visual neuroprosthesis to become an accepted therapeutic approach to sight restoration, it is clear that those who will utilize this technology must realize that they will not be receiving sight like those with normal vision. It is also clear that a clinically acceptable system must be virtually invisible. This means that the components must be integrated into normal systems typically worn by individuals such as eyeglasses and an external package not much larger than a pocket organizer. While such invisibility would be the eventual goal of a commercial system, first-generation experimental systems are not expected to be so constrained.

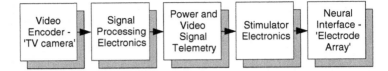

Figure 11.5 Schematic representation of the elements of a visual neuroprosthesis. With the exception of retinal neuroprostheses, all of the devices reviewed in this chapter consist of the items illustrated. Retinal devices, due to their location, might not require telemetry and use very simplistic signal-processing and stimulator electronics. Further, these devices derive their power from ambient light. The video encoder essentially replicates the function of the photoreceptors, that is, transforming the visual signal into an electrical signal. The signal-processing electronics perform necessary filtering and remapping. As external electronics cannot be as efficient as the biology, which has been tuned through millions of years of evolution, external power may be necessary. The stimulator electronics convert the video signals into electrical signals that more effectively stimulate neurons. The interface to the neurons is provided by an array of electrodes.

A visual neuroprosthesis will be a complex system containing a number of interconnected elements. A block diagram of the elements that a visual prosthesis will likely contain is shown in Figure 11.5. It must contain the following components: a video encoder to capture the visual field in front of the user of the system, signal-processing electronics to transform the video image into a set of discrete signals that can be used to control the injection of current through the neural interface, some mechanism by which these processed signals and electrical power can be delivered in a wireless fashion to the implanted neural interface, stimulator electronic circuitry to control the currents injected through each electrode in the neural interface, and the neural interface that evokes the neuronal excitation. As this minimum set of components will be present in some form in a retinal, optic nerve, or a cortical visual prosthesis, we will expand upon these elements in the following sections.

11.3.1 Video encoder

The electrical currents that will be injected at some site into the visual pathways must be related to the visual scene in front of the blind subject using the visual prosthesis. Thus, the first key element in the prosthesis is the video encoder (this element would mimic the lost function of the photoreceptors in the retina and transform visible images into electrical images). For cortical or optic-nerve-based visual neuroprostheses, this system could be a photodiode array, a dedicated CCD (charge-coupled device) array, a conventional video camera, or a miniaturized video camera, mounted within a pair of eyeglasses worn by the subject (to achieve invisibility of this device). For a retinal-based visual neuroprosthesis, the encoder could be integrated into the neural interface and, therefore, reside within the plane of the retina. This would have the obvious advantage that the optics of the eye could be

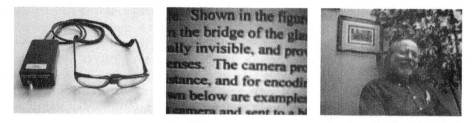

Figure 11.6 Example of video encoder. The first panel illustrates glasses that contain a miniaturized camera in the bridge of the glasses and the associated electronics that easily could fit into a pocket. The electronics provide a standard color NTSC signal. The second panel illustrates printed text as viewed through the camera, and the third panel shows one of the authors of this chapter as viewed through the camera. Both panels have been converted to gray scale.

used to project an image of the world onto the encoder. Thus, the acquisition of visual information could function more naturally, without the use of head movements for image acquisition (normal eye movements could provide part of this function).

The spatial resolution of early generation retinal, optic nerve, or cortical visual neuroprosthetic systems is expected to be quite low because of the limited numbers of electrodes in the neural interface. Thus, just about any conventional video camera could provide adequate spatial resolution for a laboratory-based system. As the temporal resolution of the human visual system is relatively low (on the order of 30 Hz), conventional, inexpensive video cameras can easily provide adequate spatial and temporal resolution for a laboratory experimental system. For a clinical device, miniaturized video cameras are already commercially available. Shown in Figure 11.6 are glasses that contain a miniaturized camera in the bridge of the glasses. The very small aperture of the camera (1 mm) makes it virtually invisible and provides for a very large depth of field without the use of additional lenses. The camera provides a color NTSC signal and allows for encoding images at distance, and for encoding of text when the text is brought close to the glasses. Also shown in the figure are examples of images of text (font size 12) and an office encoded with this camera and sent to a black and white monitor (the photographs were taken off the monitor screen).

11.3.2   Signal Processing

As discussed earlier in this chapter, the foundation upon which a visual prosthesis will be built is the notion that the visual system is organized into a hierarchical sequence of maps. One of the most relevant maps is called *visuotopy*: points in space that are close to each other excite neurons at various levels in the visual pathways that are also close to each other. If this visuo-topic mapping were perfectly conformal and linear, then a high-quality neural "image" of the visual space would be found at each level in the visual pathways. This is not the case, however. The visuotopic map is only conformal when considered from a low-resolution perspective. If one studies the visuotopic organization with high spatial resolution techniques, it becomes clear that the mapping is somewhat locally random. Recent work by Warren et al.[14] have shown that one can use the relative location of a neuron in the cortex to predict where spots of light projected in visual space will excite the neuron. However, this can only be done with an accuracy of about 1/2 degree of visual angle. This uncertainty reflects the nonconformal nature of this visuotopic organization.

Signal-processing electronics must accomplish a number of tasks. First, the electronics must transform the signal coming out of the video encoder into a set of discrete signals, one for each electrode in the neural interface. Next, the encoder must be able to adapt the incoming light signals into a range of stimulus levels that is appropriate for the neurons that are being stimulated; that is, the range of stimulus levels must be the same regardless of the level of ambient illumination (a bright, sunny afternoon or a dimly lit restaurant). This will require an automatic gain control circuit that will duplicate the adaptation properties of the human photoreceptors. The signal-processing electronics could also perform image compression, somewhat like that achieved with the MPEG techniques used to compress audio and visual images in digital cameras and DVD video disks. Finally, the signal-processing electronics may have to accomplish remapping of the visual image such that a vertical line in visual space evokes the percept of a vertical line. The degree to which this remapping may be required will depend upon the degree of randomness in the visuotopic organization at the implant site and the degree of plasticity that remains in the subject's visual system. This problem of remapping of visual space into appropriate signals is being pursued by the Eckmiller laboratory in Germany.[15] While this level of signal processing appears highly complex, recent developments in integrated circuit technology will make this a complex but achievable task.

11.3.3   Telemetry and power interface

In order to achieve the challenge of developing an invisible clinical visual neuroprosthesis, it will be necessary to use a wireless link by which power and video signals are delivered to the implanted neural interface. The design constraints of such a wireless link are such that they are only recently becoming possible due to recent advances in integrated circuit technology. The wireless link could be mediated by radiofrequency or by light, and recent work on integrated optical systems and cellphone technologies are making these approaches tenable. The telemetry link should be bidirectional so the implanted electronics can inform the external electronics of the need for more or less power. In order to minimize excessive power requirements of the transmitting electronics, it is expected that the distance between the transmitter and receiver will be on the order of a centimeter. This suggests that the receiving coil will likely be implanted under the scalp (for a cortical prosthesis) or within the eye for a retinal based system. The coupling of power between transmitters and receivers can be achieved with very high efficiencies over such short distances. One possible design would be the use of transmitting coils built into the image-encoding eyeglasses. Such a design will have the advantage that registering the transmitting coil over the implanted receiving coil can be readily accomplished and will not vary significantly during the course of normal daily activities of the wearer. Further, the external electronics would consist only of the eyeglasses, signal-processing electronics, and the telemetry circuitry.

In order to ensure that the telemetry system does not deliver damaging levels of heat and radiation to the tissues near the implanted electronics, the frequency of the radiated power must be limited and the efficiencies of the implanted electronics must be very high. Further, the implanted receiver must be very small and have ultra-high reliability, considering that the implanted electronics will be immersed in the corrosive environment of the human body, possibly for decades. Finally, the bandwidth of the telemetered video signals will depend upon a number of factors. As the number of electrodes in the implanted neural interface increases, the bandwidth of the transmitted signals increases. If the signal-processing electronics use clever encoding strategies, the bandwidth of the video signals can be significantly reduced. It is clear that all these design considerations present the design engineer with very significant challenges.

An experimental wireless power and video signal telemetry link has been designed and fabricated by Dr. Phillip Troyk at the Indiana Institute of Technology. This system is intended to be used in a cortical visual prosthesis. The system has a modular design so that failure of any individual components will not cause failure of other modules. The system has also been designed to consume little power and to facilitate the interconnection of neural interfaces.

11.3.4 Neural stimulator

The video signals will have to be processed to make them compatible with, and appropriate for, stimulation of neurons at some level in the visual pathways. Thus, the next major element in the visual prosthesis is the neural stimulator. As we have discussed earlier, an experimental visual prosthesis system may use external stimulator electronics, which would offer the advantage that commercially available electronic stimulators could be used. However, clinical systems will likely require that the stimulator electronics be implanted near to the neural interface. The stimulator will most likely be a digital device, which will facilitate its development and minimize power dissipation by the device. Thus, the stimulator will likely contain on-chip memory locations, with one location dedicated to each electrode in the neural interface. These memory locations will be updated via the telemetry link as the image being encoded changes. The considerations associated with the design of a neural stimulator have been discussed in a paper by Jones et al.,[16] and this design has been implemented in a VLSI (very-large-scale integration) chip. The problems of hermetically sealing the electronics, however, represent one of the more formidable challenges for the design engineer.

11.3.5   Neural interface

Any visual prostheses will require an interface between the nervous system and internal and external electronics, a so-called *brain–computer interface*. This interface should be thought of as an active element that transduces electronic currents that flow in the human-engineered devices that are connected to the device into ionic currents that flow in the human body. While most materials accomplish this transduction to a limited degree, certain materials do this relatively well. However, some of these materials, such as silver, are toxic to the neurons in the immediate vicinity. Platinum is a highly biocompatible material but it is not particularly effective in transducing electrons into ions. Oxidized iridium, on the other hand, is not only biocompatible but is also an excellent electronic-to-ionic transducer. It is the material of choice today and is generally being used in most implantable neural interfaces.[17]

In other human-engineered implant systems, the interface between the living and non-living has generally been difficult to develop. The human body has developed numerous defense systems that have specifically evolved to ward off intrusions into the body. These biological defense systems are so effective that most devices that are implanted in the body eventually either become degraded into harmless molecular species or become sealed off by a fibrotic tissue capsule. The challenge of the bioengineer is to thwart this biological process with the use of materials that the body recognizes as benign. While this seems like a straightforward task, the more one delves into the problem, the more complex it becomes.

11.3.5.1   Physical biocompatibility

11.3.5.1.1 Density of the implant system.   Ideally, the implant system must appear invisible to the host tissue. The invisibility applies not only to the chemical nature of the surfaces that come in contact with the host tissue but also to the physical aspects of the implant. Specifically, if the density of the implant is not identical with that of the host tissues, gravity or kinematic accelerations associated with normal body movements can produce shear forces between the implant and the host tissue. If these shear forces are sufficiently large, the implant can migrate through the tissue, or at least experience micromovements with respect to the host tissue. These micromovements can produce a chronic inflammatory response to the host tissue due to chronic mechanical irritation of the neurons near the implant. This problem is exacerbated in the visual pathway, where saccadic eye movements can produce large accelerations of the eye. Any neural interface that is implanted within the eye could be subjected to particularly large shear forces if the density of the implant differs significantly from the host tissues.

11.3.5.1.2 Physical barriers. The cells of the nervous system are in a homeostatic equilibrium with their local environment. Each cell exchanges nutrients and waste products with the extracellular fluid surrounding the cell, and these molecules eventually become transported into or out of the capillaries that support the tissues. Implanting large structures into the nervous system, even if their surfaces are biochemically invisible, could disrupt this exchange of nutrients and wastes. This aspect of physical biocompatibility is particularly relevant in various tissues of the visual system such as the retina. The oxygen that nourishes the retina comes from two sources: the choroidal and retinal circulations. Similarly, the carbon dioxide waste is also removed from the retina via these two circulations. The choroidal circulation is a highly vascularized bed that ramifies throughout the choroid (located between the pigment epithelium and the sclera) and, because of its proximity to the photoreceptors, provides the major nourishment of the outer retina (photoreceptors, horizontal cells and bipolar cells). It also nourishes the pigment epithelium. Any implant system placed at any point between the choroid and the retina must not act as a significant barrier to the exchange of oxygen and carbon dioxide between these tissues, or retinal degeneration could be exacerbated by the implant system. The retinal circulation nourishes the cells in the inner retina (the ganglion cells, amacrine cells, and, to some degree, the bipolar cells). As the retinal circulation is integrated with the retina itself, it is difficult to imagine how an implant system could be interposed between the retinal circulation and the cells it supports. However, the existence of any molecular exchange between the vitreous humor and the inner retina could be compromised by implant systems placed on the retinal surface. The degree to which molecular exchange between the pigment epithelium and the retina takes place and its importance in retinal support are not clear.

11.3.5.1.3 Mechanical compliance of the implant system. The neural tissues into which a visual prosthesis will be implanted are relatively compliant, compared with the materials with which the implant system is typically constructed. Further, the tissues into which (or onto which) the implant is inserted typically experience relative movements due to gross movements of the body or more subtle displacements due to blood pulsations or to respiration. Because of the compliant nature of the host tissues, the tissues can be slightly compressed or distended as a result of these mechanical displacements. Clearly, if the implant materials are virtually noncompliant, a mechanical compliance mismatch can develop at the interface between the host tissues and the implant. Polymeric materials are generally more compliant than the metals and silicon used in many implant systems. It is likely that future generation implant systems may profitably utilize new generations of polymeric materials to mitigate this potential compliance mismatch problem.

11.3.5.1.4 Tethering produced by lead wires. While clinical visual prostheses will eventually be wireless devices, where power and signals will be delivered across living tissues via some sort of telemetry system, this might not be the case for first-generation systems. The goal of many first-generation visual prostheses will be to demonstrate proof of concept, that the system can produce functional (but probably very limited) vision. These first-generation systems are likely to have lead wires that run from the neural interface that is implanted in the host tissues to some sort of a connector, mounted at some stable site on the body. Even if signals and power are delivered across the skin via some form of telemetry, the size of today's VLSI circuitry is such that it is still likely that some form of lead wire system will still be used between the implanted neural interface and the implanted telemetry and stimulating VLSI circuitry. Thus, it is quite likely that any first-generation visual prosthesis will have to deal with the problem produced by tethering from these lead wires. The problem of tethering can be best appreciated by considering a cortically based visual prosthesis, where an array of electrodes is implanted on or in the visual cortex, and a connector is attached behind the ear. Even if the mechanical problems outlined above are dealt with, any motion of the brain with respect to the skull will result in the generation of relative motion between the brain and neural interface, which could produce potential chronic inflammation. These problems can be partially mitigated by various strain-release lead wire architectures and by anchoring of the lead wires to the cortex near the implant site. However, these measures will only will reduce, not eliminate, the tethering problems.

11.3.5.2   Biological biocompatibility

While the goal of achieving a biological invisibility of the neural interface is daunting, the problem is mitigated to some degree by two factors: a judicious choice of the materials from which the interface is constructed and the well-documented immunologically privileged nature of the nervous system to foreign materials. The implantation of any neural interface, at any site in the nervous system, can be thought of as creating a wound at that site, no matter how benign. There is an extensive literature on the mechanisms and kinetics of wound healing that go well beyond the scope of this chapter, and the reader who is developing neural interfaces is encouraged to become familiar with this literature. To summarize for the purposes of this chapter, the nature of the wound healing response will depend upon the nature and extent of the wound and the materials from which the wounding device are manufactured. We will consider below the issue of materials.

Again, the quest for safe implant materials has been pursued for many decades and has been inspired by the development of implantable joints, pacemakers, catheters, cosmetic applications, and orthopedic devices, to name but a few more prominent devices. There is also an extensive literature on the biocompatibility of a wide spectrum of materials, but implant systems currently under development use a subset of these materials: silicone, silicon, platinum, iridium, gold, polyamide, parylene, Teflon, and PMMA (poly[methyl methacrylate]). These materials are used for various purposes in the construction of a neural interface: structural, charge carrying, electronic-to-ionic transduction, and insulation. These materials have been used because they provide the architectural and functional properties from which a neural interface can be constructed and because they are generally regarded as being quite biocompatible.

11.4 Subretinal neuroprosthetic approach

As some of the neural retina, with the exception of the photoreceptor layer, is preserved for both age-related macular degeneration[12] and retinitis pigmentosa,[10,11] a reasonable approach to the restoration of vision under these pathologies is the replication of the function of the photoreceptors. (Another group, however, has reported more severe retinal degeneration in animal models.[13]) From a simplistic engineering viewpoint, the photoreceptors of the outer nuclear layer appear as a tightly packed array of light-to-current transducers, a function readily replicated by arrays of semiconductor phototransducers. Two research groups have focused on replacing the function of photoreceptors with an engineered system: (1) the Optobionics group, led by Alan Y. Chow and Vincent Y. Chow; and (2) a consortium centered in Tubingen, Germany, led by E. Zrenner.[18] In the general sense, both groups have proposed a similar approach. An array of phototransducing elements is placed in the subretinal space, between the pigmented epithelium and the degenerating outer nuclear layer. Each element (or subunit) consists of a semiconductor-based photodiode and an electrode. Light striking the photodiode causes a photocurrent that flows between the electrode and a reference electrode, the entire back surface of the array. The result is a voltage gradient that is intended to stimulate the dendrites of the bipolar cells in the outer plexiform layer. Because the stimulating electrodes are placed in the subretinal space, this tactic has been termed the *subretinal approach*.

The scheme is tantalizingly simple, requiring no external power or control signals. Further, the fabrication of the photodiodes can be tailored to provide either positive or negative current in response to illumination, with the intention of mimicking the operation of ON and OFF bipolar cells. Both groups have suggested that, in theory, no signal processing would be necessary, as they claim that the entire neural network of the visual pathway would be stimulated in a physiologically normal manner. The success of such an approach depends upon three assumptions. First, the bipolar cells of the dysfunctional retina persist and function in a somewhat physiologically normal manner. Second, the photodiodes can produce sufficient current under normal illumination levels. Third, the electrodes can be placed in close enough proximity to the bipolar cells such that the induced photocurrent excites the bipolar cells. As will be discussed in this section, the validity of these assumptions, particularly the second, remains very controversial.

11.4.1 Optobionics group

The Optobionics Corporation, a privately held company located in Wheaton, IL, has developed the Artificial Silicon Retina (ASR), which entered into clinical trials of its safety in the summer of 2000. (Optobionics, Artificial Silicon Retina, and ASR are trademarks of the Optobionics Corporation.) This device represents the commercialization of the scientific research of Chow and Chow, two brothers who continue to provide the scientific leadership of the company. Much of their research has centered on the safety and efficacy of a subretinal neuroprosthetic device.

In their early research, Chow and Chow showed that a simple combination of an electrode placed in the subretinal space and a photodiode could become the basis of a visual prosthesis.[19] Using a single, very large gold electrode (36 mm² surface area or about twice the size of the macula in humans) acutely implanted into the subretinal space of an anesthetized rabbit, they observed that illumination of a photodiode connected to the electrode induced cortical activity. The cortical activity, measured by electrodes placed on the scalp of the rabbit, had amplitude and temporal characteristics similar to the visually evoked potential measured in the unoperated eye. Further, the magnitude of the cortical-evoked potential could be modulated by adjusting the intensity of the stimulus or the size of the photodiode. With a charge density of 100-nC/cm², resulting from 12-klux illumination (similar to a bright, sunny day) on a 29-mm² photodiode, the magnitude of evoked potential was similar to the visually evoked potential measured in the unoperated eye.

Having shown a proof of concept, their research moved into the design and test of an array of photodiodes and electrodes manufactured by standard semiconductor methods. In a series of articles, they have reported on the biological response to an implanted array and the ability to evoke neural activity with the array.[20–25] The test ASR has remained relatively unchanged except in its thickness throughout these studies. At the macroscopic scale, the device is a 1- to 3-mm-diameter disk (0.8- to 7-mm² area), initially being 250 μm thick and reduced to 50 μm thickness in the most recent report. At the microscopic scale, the test ASR is an array of photodiode and electrode elements, each element being approximately 20 × 20 μm by the thickness of the disk. Insulating moats (10-μm width) separate neighboring elements, resulting in a density of approximately 1100 elements/mm². Each element consists of a photodiode created by variously doped layers of silicon, sandwiched between two layers of gold. On the front side of the array (side closest to the light), the gold electrode is etched to the same dimension of the photodiode. Light reaching the photodiode passes through the gold layer, which is relatively transparent due to its thinness. The gold on the backside of the array forms a single sheet and acts as a return electrode. As gold does not readily adhere to silicon, a chromium adhesion layer is place between the gold and the silicon. The photodiode is responsive to light in the 500- to 1100-nm wavelength range, which covers most of the typical range in humans (360 to 830 nm) plus adding an infrared (IR) response.

In these reports, histological examinations and the electroretinogram (ERG) were used to assay the tissue response to the implant. In chronically implanted cats, they observed a significant loss of cells in the outer nuclear layer but insignificant changes in the cell density for both the inner nuclear layer and retinal ganglion layers.[24,25] Additionally, the appearance of giant cells indicated an inflammatory response to the implanted device. As AMD and RP are characterized by the loss of the outer nuclear layer, its loss in these animal experiments is considered inconsequential. Outside the site of implant, the retina had a normal laminar arrangement. The lack of a significant immune response to the implant is not surprising as the device is constructed from well-tolerated materials. Photic stimulation of the implanted eye resulted in an ERG similar to the unimplanted eye. This also is not an unexpected result, as the implant is relatively small and implant materials are well tolerated. If the device led to an out-of-control immune response, it likely would have been apparent in the histology long before affecting the ERG.

The ERG response for IR stimulation was used to verify the implant continued to be functional.[24] The ASR is much more sensitive to IR stimulation than the native photoreceptors. These data indicated that the test ASR continued to function as a photodiode array for up to 11 months, the longest time duration reported. However, the magnitude of the ERG took 1 to 2 months to reach its maximum and began to degrade after 4 months. From the data provided, the ERG response had a magnitude of up to 75 μV, and the sign of the response depended on the type of photodiode utilized (negative or positive current generating). The latter finding is a clear indication that this ERG response is simply a stimulus artifact. If the ERG represented neural activity and the two polarities were equally effective at eliciting activity, the two ERGs would be equal and similarly signed. If one polarity were more effective, the more likely case, the ERG would appear when the illumination is initiated for one polarity and when the illumination is terminated for the other polarity. At the end of the study, the arrays were explanted and examined. It was found that the gold electrode material was degraded or absent for active devices, suggesting that the gold electrodes were being dissolved into the tissue by electrical stimulation. Alternative electrode materials, platinum and iridium oxide, have been investigated but no specific recommendations are available at this time.[23]

In addition to animal experiments, Optobionics has initiated FDA-approved clinic trials to investigate the safety of the ASR in humans. The device currently in clinical trials is a 2-mm-diameter disk of 25 μm thickness. Each photodiode and electrode element has the same 20×20-μm dimensions with 10-μm moats separating adjacent elements. Given the approximately 3500 elements on the device, the total photodiode (and electrode) surface area is around 1.4 mm². In comparison, the devices used in their animal experiments, which had a complete lack of neural response for what should be effective stimulation, had a photodiode area of up to 7 mm². The only device where a neural response was assuredly observed had a surface area of around 30 mm² and then under aggressive stimulation.[19] The results of these clinical trials are only available in abstract form at this time.[26] The implants are reported to be well tolerated and have not resulted in infection, inflammation, or retinal detachment. Additionally, the implants are reported to have improved the recipient's visual function; however, these results are only available as recipient testimonials, not controlled scientific studies. Care must be taken to distinguish the improvement due to the activity of the device and the improvement that occurs simply by the introduction of the device. It has been established that a transient, functional rescue of a dysfunctional retina occurs in animal models as the result of surgical intervention into the subretinal space, even if the intervention does nothing more than expose the subretinal space. Nevertheless, the test of the biocompatibility of the materials and the development of the implant techniques make the ASR a groundbreaking device.

11.4.2   Southern German group

The southern German consortium, centered in Tubingen, has primarily investigated various engineering design issues for a subretinal implant. Their test device, called a *multiphotodiode array* (MPDA), is very similar to the one described in the previous section.[27] The individual photodiode and electrode elements are the same size, 20×20 μm, but the moats between elements are only 5 μm wide, leading to a density of 1600 elements/mm². Instead of covering the entire inner surface of the photodiode, the electrode of each element is an 8×8-μm square located at the center of the photodiode. The smaller electrode size allows more incident light to reach the photodiode and the electrode layer to be thicker. Gold, iridium, and titanium nitride have been investigated as electrode materials. The remaining inner surface of the element is insulated with silicon oxide, which is transparent to light. Details of the silicon doping used to create the photodiodes are not available.

This group took a very different approach, and arguably more useful approach, to the proof of concept. Using retinas isolated either from newly hatched chickens or Royal College of Surgeons (RCS) rats, they have shown that photic stimulation of the MPDA modulates the activity of retinal ganglion cells.[27,28] This method allows a much finer assay of the impact of electrical stimulation than the ERG and cortical potentials, which measure the mass action of large groups of neurons. In these experiments, a retina is removed from an animal and is sandwiched between the MPDA and a multisite recording electrode, with the ganglion cell layer closest to the recording electrode. To activate the MPDA, a spot of light of known size and illuminance is shined through the recording electrode and retina. Their recording electrode, consisting of an array of small metal electrodes sparsely distributed across a glass slide, is transparent to light. With this device, one can pick up the spiking activity of individual retinal ganglion cells which is easily distinguishable from the graded potentials of the photoreceptors and bipolar cells. To assure the ganglion cell activity is the result of the MPDA stimulation, the photoreceptor layer was damaged for these experiments. In the case of the chick retina, this was done by an unspecified method. In the case of the RCS rat, it is not necessary to damage the photoreceptor layer; this strain of rats is a well-known model for retinitis pigmentosa and, in mature individuals, has almost no residual photoreceptors. Tests with sham devices were used to verify the results were not due to residual photoreceptors.

In both of these species, the activity of individual ganglion cells could be modulated by light, presumably by the activity of the MPDA. However, very strong illuminance was necessary, between 10 and 100 klux (outside on a bright day). Further, as the diameter of the light was reduced, it became more difficult to evoke any ganglion cell activity. For example, with a 70-klux, 0.250-mm-diameter spot of light illuminating approximately 80 photodiodes, a barely perceptible change in the activity of one ganglion cell could be detected but only through repeated trials. When only a few photodiodes were illuminated, ganglion cell activity could not be induced regardless of the illuminance level. From these results, this group has concluded that a passive photodiode array cannot provide a useful sense of vision under normal lighting conditions.[29]

This group has extensively studied the biological tolerance to the materials being implanted and how coatings can enhance the tolerance.[30] Using dissociated retinal cells from a neonatal rat, the cell adhesion to various coated and uncoated implant materials were tested. To enhance adhesion, three different coating materials were examined, poly-L-lysine, poly-D-lysine, and laminin. When applied to an iridium surface, all three coating materials significantly improved the cell adhesion over an uncoated surface. Of these, the best adhesion was seen with poly-L-lysine, with approximately 75% of cells adhering. Therefore, all implant materials were coated with poly-L-lysine prior to testing. Of the implant materials tested (silicon, silicon oxide, silicon nitride, iridium, and titanium nitride), only titanium nitride exhibited a significantly poor cell adhesion. In comparison to the control (poly-L-lysine on glass), only 30% of the cells adhered after 28 days of culturing. This poor adhesion was not considered to be due to release of toxic materials and subsequent cell death, as cells readily adhered to a poly-L-lysine-coated glass coverslip introduced into the same Petri dish. As this group had proposed using a titanium nitride electrode, which they report has the greatest safe charge injection capacity, this was a discouraging finding. One of their electrode materials, gold, was not tested. The geometry of the MPDA does not appear to affect cell adhesion. With an iridium electrode, the adhesion was the same as the control. When using a titanium nitride electrode, the cell adhesion was improved but still was poor in comparison to the control. It is unclear whether the MPDA was active or inactive during these tests. An interesting but apparently untested comparison could be made between an active and inactive MPDA, thereby investigating whether the electrodes release toxic substances when active.

In addition to these *in vitro* tests, *in vivo* tests of the long-term function and biological tolerance of the MPDA have been performed in both rabbit and pigmented rat (Long Evans).[27,28] The ERG response to infrared stimulation (300 mW/cm^2 at 940 nm) was used to validate the survival of the implant. The details of the time course of the ERG magnitude are not available but are reported to be stable for up to 20 months, the longest time duration tested. Concurring with their excised retina experiments, there was no sign that IR stimulation resulted in activation of the inner retina. Without activation of the inner retina, the MPDA cannot provide functional restoration of sight. The white-light-induced ERG and postmortem histological examination assayed the biocompatibility of the implant but only in the rat model. A reduction of the ERG response was reported in three of five rats but the magnitude and time course were not detailed. Four months after implantation, the retina above the implant site exhibited a nearly complete loss of the outer nuclear layer as well as atrophy in the outer plexiform layer. As mentioned earlier, the loss of the photoreceptor layer is considered inconsequential. The thickness of the inner nuclear layer and inner plexiform layer were reported to be unchanged but it is unclear what it was compared against. The cell densities in the inner nuclear layer and the ganglion cell layer were similar to those observed away from the implant site. However, the stain utilized was insensitive to cell type and the authors clearly concede that its is possible, but unlikely, that all neurons have been replaced by other cell types. Additionally, the excised tissue was tested for glia fibrillary acidic protein (GFAP), a general marker expressed by glia in degenerating retina and Muller cells and astrocytes in the retina. In comparison to controls, GFAP expression was more prevalent in the implanted retinas. The reason for the increased expression and its impact are not clear at this time. From these results, this group has concluded that a subretinal neuroprosthesis will survive in the long term and is well tolerated. To validate the functional restoration of vision, they have proposed implanting the MPDA in RCS rats, which have a degenerated photoreceptor layer. However, they recognize the lack of a generally accepted and unambiguous method of verifying the restoration of function.

Due to the relatively strong illuminance necessary to induce neural activity by the passive MPDA, this group is also activity investigating an active subretinal device.[28,31] Such a device would be very similar to the epiretinal approach but would place the array of stimulating electrodes in the subretinal space. This approach is still in the proof of concept phase at this time. In their proof of concept tests, an isolated retina is placed on a multisite stimulating electrode, photoreceptor side closest to the electrodes, and the activity of a retinal ganglion cells is monitored with single microelectrodes. They have found that ganglion cell activity can be induced with a relatively low median charge of 0.4 nC. However, due to the small size of their electrodes (80 μm^2), the charge density is 500 $\mu C/cm^2$, well above the safe limit for most materials. At this point, an active subretinal device must be considered in the investigational phase.

11.4.3   Summary

Having a test device in clinical trials notwithstanding, the subretinal approach has many unresolved problems. Principal among these is whether a passive device can generate sufficient current to evoke ganglion cell activity under normal lighting conditions. In their most recent report, the southern Germany group concluded that it was not possible and has moved to an active device.[29] The Optobionics group believes it is possible and has moved to clinical trials of the device's safety.

In the reports where adequate details are given regarding the nature of the illumination, an effect on the subsequent neural pathway occurred only for relatively bright lights over relatively large regions of the test prosthesis. The Optobionics group reported that large, electrically evoked cortical potentials could be had with 12-klux illuminance applied to a photodiode having a 30-mm^2 surface area.[19] Using a finer assay of subsequent neural activity, retinal ganglion cell activity, the southern Germany group reported that a small but noticeable modulation could be had with a 70 klux, 250-μm-diameter spot.[28] Although the 10- to 100-klux range of illuminance is physiologically relevant, this level of illuminance is typically only seen outside in bright sunlight. More typical ranges of illuminance encountered are 1 to 1000 lux. The Illuminating Engineering Society of North America (IESNA) recommends illuminance levels of 260, 520, 700, and 2200 lux for a hotel room, library reading room, laboratory, and hospital operating room, respectively. Further, the lack of a noticeable neural response to IR stimulation casts further doubt on the applicability of a passive device. Specifically, because the published IR-induced ERG data do have a b-wave component, which indicates activity in the inner retina, neural activation is unlikely.[24,28]

Assuming one could develop sufficient current at the commonly encountered illuminance or if one assumes that the device is operational only in bright light, a passive subretinal device has a significant design flaw in that it cannot adapt to varying light levels. Assuming a linear response, such a device would be functional over 1 to 2 orders of magnitude of illumination. In contrast, our natural vision functions over 7 orders of magnitude. Nevertheless, it can be argued that a functional device that operates with some limitations would be an improvement over blindness.

Another area of concern is the destruction of the remnant photoreceptor layer and well as a significant impact on the outer plexiform layer resulting from the implant.[25,28] These cytological changes are presumed to be the result of the device blocking the free flow of nutrients and waste between the neural tissue and its blood supply. Both groups have proposed mesh-like subretinal implants, which would allow free flow, but the fabrication method for such devices is not clear. Unlike the outer retina, the inner retina appears unaffected by the implant. However, the studies to date simply examine the density of cells in each of the inner nuclear layer and retinal ganglion cell layer without examining the possibility of a functional consequence of the implant. Additionally, expression of GFAP by the glial cells is increased, but the consequence is unclear.

11.5 Epiretinal neuroprosthetic approach

As an alternative to stimulating the remnant bipolar cells in the dysfunctional retina, other research groups have proposed placing stimulating electrodes on the inner surface of the retina and stimulating the remnant retinal ganglion cells. Reports indicate that significant portions of the retinal ganglion cells survive even in the end stage of pathologies such as age-related macular degeneration[12] and retinitis pigmentosa,[10,11] giving credence to this approach (however, a significant disorganization of the retina has been reported in animal models[13]). The principal research groups investigating this approach are (1) a group led by Humayun and de Juan, which was located at Johns Hopkins University but recently has moved to the University of Southern California; (2) a collaboration between MIT and Harvard, led by Rizzo and Wyatt; and (3) a consortium centered in Bonn, Germany, initially led by Eckmiller and now led by Eysel.[32] In the general sense, all three groups have proposed a similar approach. An array of surface electrodes, attached to the inner surface of the retina between the vitreous humor and the inner limiting membrane, provides the interface to the neural retina. Driven by signal-processing electronics, patterns of these electrodes are electrically stimulated to produce a similar pattern of phosphenes, a consequence of the visuotopic organization of the ganglion cells. In a clinical device, the signal-processing electronics will be placed within the eye with both the power and the control of the signal-processing electronics provided by an optical or radio frequency link. As the stimulating electrodes are placed on the surface of the retina, this tactic has been termed the epiretinal approach.

This approach tries to make use of the advantages of a retinal prosthesis, chief of which is a simple, linear visuotopic organization, while recognizing what the other neuroprosthesis groups have long ago recognized, the need for more power and signal-processing capability than is available from simple phototransducers. The success of this approach depends upon four assumptions. First, the retinal ganglion cells of the dysfunctional retina persist and function in a somewhat physiologically normal manner. Second, the neural interface can be permanently attached to retinal without significant damage. Third, useful percepts can be had at safe stimulation levels. Fourth, patterned percepts can reliably be created.

11.5.1 Humayun and de Juan group

The Humayun and de Juan group initially began their work as part of collaboration between Johns Hopkins University and North Carolina State University. More recently, they have moved to the University of Southern California. Additionally, this group has a commercial relationship with Second Sight, a privately held company located in Valencia, CA, which is testing a retinal neuroprosthesis in clinical trials. This group has primarily investigated the biological aspects of an epiretinal implant, including a large number of experiments in human volunteers.

In their early research, they showed that electrical stimulation of the inner retina results in neural activity of the retina, or a general proof of concept.[33] These experiments were performed both with isolated bullfrog eyecup preparations and with *in vivo* rabbit retinas. Using platinum, bipolar stimulating electrodes, manually placed on the inner retinal surface, they excited the tissue with biphasic current pulses of between 50 and 300 µA amplitude, lasting 75 µs per half phase. Each electrode had around 0.13 mm^2 surface area and the tips of the bipolar pair were 200 µm apart. A second bipolar electrode, manually placed between the stimulating electrodes and the optic disk, sensed the neural response. In response to electrical stimulation, they observed a 2-msec long waveshape that had the appearance of an extracellular measured action potential. No response was observed if the stimulating electrodes were placed 1 mm off the retinal surface or if the recording electrodes and stimulating electrodes were placed on opposite sides of the optic disk. They reported that, for current levels as low as 50 µA in the bullfrog and 150 µA in the rabbit, a neural response could be observed. This is equivalent to charge densities of 3 and 9 µC/cm^2, respectively. This is well within the safe current density levels (<100 µC/cm^2) for platinum.[17] Their data indicate that the average magnitude of the neural response grew from 10 µV (peak to peak) to 20 µV as the current was increased from 150 µA to 300 µA in the rabbit preparation. This growth likely indicates that more ganglion cells were being excited at larger current levels but their report claimed the growth was not significant. In addition to using normal rabbits, they also tested a rabbit model of a degenerate photoreceptor layer, caused by intravenous injection of sodium iodate well before testing. With this preparation, they were able to observe a neural response but the requisite current level increased to 200 µA.

Having shown that electrical stimulation of the retina can produced localized neural activity, this group began investigating two key issues for a clinical device: (1) the possibility of selective stimulation of particular structures through changes to the stimulus paradigm, and (2) the biocompatibility of the implant materials, particularly the method of adhering the implant to the retina. The first area was explored by computation modeling studies of electrical stimulation of retina.[34] In particular, these studies were directed at distinguishing which neural structure is activated by electrical stimulation, the soma of retinal ganglion cells or passing axons from other retinal ganglion cells, the latter being between the stimulation source and the somas. To provide visuotopically organized stimulation, it is preferable to stimulate the somas and not the axons. The results of the study suggest that there is a penchant for stimulating the soma but the preference was slight. No specific recommendations for changes in the stimulus paradigm came from this work.

The biocompatibility issue was investigated with electrode arrays, both passive electrodes and actively stimulated electrodes, chronically implanted in animals. In both studies, the array consisted of large platinum disks placed in a silicone matrix. For the passive studies, the entire array was 3 × 5 mm by 1 mm thick. Specific details of the active electrodes are not available. The results from the passive implants suggest that the implant materials are well tolerated, indicated by both histological and electrophysiological examination.[35] However, the metrics employed solely indicate that the implanted retina continues to function and does not ensure that the retina under the array functions normally. No histological comparisons were made between the retina under the implant, approximately 15 mm² in area, and the remainder of the retina, over 1000 mm². In the second set of studies, the electrodes were stimulated with 0.05 or 0.1 µC/cm² charge densities, 10 to 12 hours a day for up to 60 days. The stimulation did not cause any gross changes in the function of the retina but the finer assay of tissue histology has yet to be done.[36] Interestingly, these researchers chose not to utilize a charge density where a pathological tissue response is expected, thereby providing a positive control. In both studies, one or two retinal tacks were used to hold the arrays in place. This group has additionally investigated the use of bioadhesives to attach the electrodes to the retina but did not make any specific recommendations.[37] As these studies are ongoing, the safety and efficacy of their epiretinal approach is yet to be established; however, it is clear that this group is addressing the key issues.

While performing this animal research, this group also studied the percepts resulting from electrical stimulation of the retina in blind human volunteers.[38–40] Initially, they investigated the most basic question: Does electrical stimulation by bipolar electrodes, manually placed on the retinal surface, evoke a percept in the blind? In a sense, this experiment repeated their earlier animal experiments but here they could ask the more germane question: what current level leads to a percept? They found that percepts could be evoked with charge densities of threshold between 160 to 3200 µC/cm² with platinum electrodes. These percepts were reported to range in size from a pinhead to the size of pea, at a distance of 30 cm. Further, the visuotopic organization of percepts, in a wide sense, was verified by stimulating widely separated regions of the retina. Interestingly, the size of the percept, and its brightness, did not appear to vary with retinal location. In a second set of experiments in blind volunteers, these researchers investigated the electrode spacing necessary to evoke distinct percepts. At the minimum spacing tested, 435 µm, the two phosphenes were distinct and separated by just over 1 degree. This spacing is consistent with the known magnification factor of the retina, approximately 250 µm per degree of visual space. In these experiments, the electrodes had diameters in the 50- to 100-µm range.

In a third set of experiments, using an array of electrodes, arranged either as a 5 × 5 grid or a 3 × 3 grid with the 400-µm-diameter platinum electrodes at 600 µm center-to-center separations, the percept of a fused line could be evoked by the stimulation of a line of electrodes on the array. Further, stimulating either a row or column of electrodes evoked distinct line percepts. When stimulating a U-shaped and box-shaped grouping of electrodes, the percepts were reported as an H shape and a box shape, respectively. The reliability of these shaped percepts was not reported. Due to incomplete or inconsistent information in their published reports, it is difficult to estimate the charge density necessary to evoke these percepts. It appears that threshold for the H-shaped percept required a charge density of about 300 µC/cm².[40] This charge density is the theoretic maximal safe charge density[41] but greatly exceeds the practical safe limit (25 to 75 μC/cm²) for long-term stimulation with platinum.[17] These investigators recognize that their charge densities would be unsafe in a long-term neuroprosthesis[32] and have proposed overcoming this problems by either using activated iridium electrodes, with a practical safe charge density of up to 3000 μC/cm², or using different electrode configurations.[40] With 600-μm center-to-center spacing of electrodes, the resulting phosphene centers would be separated by 2.4 degrees, a rather poor pixelization of the visual world. Yet, to increase the resolution, the electrode spacing would have to be reduced, resulting in reduced electrode size and increased current density.

11.5.2   MIT and Harvard group

The MIT and Harvard collaboration has primarily investigated the engineering aspects of an epiretinal implant but more recently they have begun to pursue human psychophysical experimentation as a proof of concept.[32] Their earlier work investigated the microfabrication and electronic design of the stimulating electrodes as well as the signal-processing electronics. This, of course, led to studies of the stimulation parameters necessary to evoke percepts and power requirements of the necessary devices.[42] Although their human psychophysical results have only been presented in abstract and preprint form, their results indicate that, while percepts can be induced by electrical stimulation of the retina, the ability to generate a patterned percept appears limited. The pattern of the percept did not follow the anticipated pattern in the majority of the cases and the reliability of the percept (similar description for the same pattern of stimulation) was only 66%.[18] Interestingly, a normally sighted volunteer had an 82% reliability.[44] It is unclear whether the normally sighted volunteer gave a more reliable report due to being more accustomed to visualizing objects or if this is an indication that the blind's dysfunctional retina is more disorganized than expected. Further, the charge densities necessary to produce a percept, up to 28 μC/cm,² approached the safe, long-term limit.[44]

   Their animal research has centered on both the long-term ability to evoke activity in the visual cortex, measured by evoked potentials, in rabbits and the question of what neural structures are excited by electrical stimulation of isolated rabbit retina.[45] This latter research has opened the possibility of investigating stimulation paradigms with arrays of electrodes that effectively produce patterned activity. In these experiments, a microfabricated electrode array was used for both stimulation and recording. The platinum black-coated, gold-stimulating electrodes were 10 μm in diameter and were placed on 25-μm centers. The recording electrodes had the same diameter and were placed on 70-μm centers. The retina was stimulated with anodic first, biphasic current pulses of between 0.01 and 20 μA amplitude, lasting 400 μsec per half phase. The threshold for observing an extracellular action potential ranged from 0.06 to 1.8 μA. However, the method appeared to very preferentially excite the passing ganglion cell axons and not the underlying retinal ganglion cells. Generally, the observed extracellular action potential was held to be due the ganglion cells soma becoming active due to antidromic propagation. Clearly, it is more desirable to provide somatic stimulation, allowing one to take advantage of the visuotopic organization of the ganglion cells bodies. However, it is not necessary that somatic stimulation be used.

11.5.3   Northern German group

The northern German consortium, centered in Bonn, has primarily investigated the signal-processing aspects of an epiretinal implant. To provide a more natural percept, this group started the development of specialized electronics, called a retinal encoder, that mimic the function of the retina and can automatically adjust its operation based upon conversational input from the user.[15] More recently, they have started an active program of primate experimentation but much of their results in this area are only available in abstract form. In these tests, they verified that their neural interface, called the Multi-Microcontact Array (MMA), could be secured to the retina by retinal tacks and that cortical activation could be had by stimulation.[46] However, these tests were of limited duration, lasting 6 to 8 hours, and their principal purpose appeared to be defining the precision of locating the MMA. The functionality of the array was tested only in a very gross sense and, then, no details the stimulus parameters were provided. Only limited histological examinations were performed.

These results have been expanded upon by optically imaging the activation of visual cortex in the cat while stimulating patterns of electrodes on the retina. The results indicate the existence of a relationship between the stimulation pattern and the pattern of activity in primary visual cortex.[47] However, the occurrence of a similar pattern of activity in visual cortex does not speak to the associated percept.

The northern German group has also performed computation analyses in an attempt to resolve the most likely neural structure activated by stimulation. In what is likely the most extensive study of all three groups, this group argues that one can selectively stimulate the desired target, the soma of retinal ganglion cells, by the simultaneous control of current at multiple electrode sites.[48]

11.5.4   Summary

Despite its relatively late entry into vision neuroprosthesis research, the epiretinal approach has made great strides but there are many yet to be resolved problems associated with this approach. The ability to access the neural retina has been established with novel surgical approaches. A general proof of concept, that electrical stimulation of the inner surface of the retina evokes action potentials, has been shown with isolated animal retinas.[33,45] The Humayun group has shown the more specific proof of concept that patterned visual percepts can be invoked by patterned electrical stimulation, a result found in blind human volunteers with either AMD or RP.[38–40] However, in more controlled studies, the MIT group has found only a weak relationship between a percept and the anticipated percept due to electrical stimulation. Blind volunteers reported a percept that had a "reasonable relationship" to the anticipated percept only around 40% of the time. Further the reliability of the percept was low, with the same precept being described only 66% of the time when presented with the same pattern of electrical stimulation.[18] Using an alternative approach to investigate patterned stimulation, the northern German group has shown that stimulation of the inner retina by an array of electrodes results in a visuotopically organized pattern of activity in primary visual cortex, as observed by intrinsic optical imaging.[47] Although an excellent method of investigating the relationship between electrical stimulation and cortical activation, this method says little about the relationship between patterned stimulation and perception. Collectively, these results argue that patterned perception may be possible but that it is not necessarily assured and further investigations are warranted. Well-defined and appropriately implemented psychophysical tests, particularly ones addressing the reliability of the percept, must be performed before this question is fully resolved.

An associated issue concerns which neural structures are being activated. Given the organization of the retina, it is possible that the passing retinal ganglion cell fibers, which lay between the stimulating electrodes and the soma of the retinal ganglion cell, are being activated. A number of computational studies that have addressed the issue[34,48] have concluded that the cell bodies can be preferentially activated, but their results have not been verified in a biological preparation. The one attempt to study preferential activation in an excised retina used a vastly different electrode configuration and did not provide any means of extrapolating to the configurations proposed by other groups.[45] However, if this research were expanded upon, one could readily investigate how to preferentially stimulate the cell bodies of the retinal ganglion cells.

To ensure long-term function and stability of perception, the array of stimulating electrodes must be securely attached to the retina. As the retina is a thin, delicate structure, just the fixation of a passive device to the retina is problematic. Further, as the electrode array and signal processing are active devices, there are additional concerns regarding the neural tissue response to both chronic electrical stimulation and heating, the later due to the power dissipations by the electronics. Preliminary results are available that indicate that both the electronics and electrodes can be secured to the retina via retinal tacks without serious complications.[35] These results are just now being expanded upon to include the effect of chronic electrical stimulation on tissue.[36]

Although the key assumptions introduced earlier have only partially been addressed, the three principal research groups are clearly investigating the crucial issues. It appears that retinal tacks can attach thin film electrode arrays to the retina without serious complications. To induce a neural response, the charge density often exceeded the safe, long-term limit.[17] In reaction, most epiretinal devices utilize relatively large electrodes but this leads to relatively widely spaced phosphenes. Finally, the reliability of a patterned percept is presently under study for the epiretinal approach.

11.6 Optic nerve neuroprosthetic approach

The optic nerves provide the sole conduits of visual information from the retina to the lateral geniculate nucleus in the thalamus. In certain visual pathologies, such as age related macular degeneration or retinitis pigmentosa, where a subpopulation of ganglion cells has been spared, but in which cells of the outer retina are completely degenerate, the optic nerve offers a possible site for intervention via a neural interface.

A group of researchers from the Catholic University of Louvain in Brussels has seriously entertained this as a candidate intervention site for a visual prosthesis.[49,50] They have tested the concept in a series of experiments conducted over the past 4 years in a single human volunteer with retinitis pigmentosa. The team has implanted a self-sizing spiral cuff electrode array around the right optic nerve of this volunteer. The electrode array consists of four surface electrodes on the inner surface of the cuff. When implanted around the optic nerve, currents can be passed in a bipolar fashion between groups of these surface electrodes in order to achieve some degree of stimulation selectivity in the population of nerve fibers. The use of cuff electrodes to stimulate peripheral nerves has a long history.[51] In many cases, such electrodes can provide long-term stimulation, with little biological complications.[52] Recent improvements in the implanted system involve wireless telemetry of stimulus signals to the implanted array.

The surgical access used in the implantation is complex. The lead wires in the implant system were fed under the skin, down the neck to a telemetry unit implanted near the collarbone. In this one subject, there has been no report of complications from the surgery or breakage of the lead wires.

The subject has been stimulated intermittently over the 4-year period of the implantation, and studies of phosphene thresholds and their spatial location have been monitored. Wide ranges of phosphenes have been able to be evoked with this system. They span a region extending 85 horizontal degrees and 60 vertical degrees in front of the observer (the observer points to the perceived phosphene location). The phosphenes range in size from 1 to 50 square degrees, and for a given bipolar stimulation regime the location of the phosphene, its intensity, and its size are a function of the stimulation current. As the stimulus current is increased by a factor of three, the location of the perceived phosphene migrates from the edge of the phosphene space to the center of the space. The absolute thresholds for evoking a just perceptible phosphene vary with stimulus pulse duration and whether the stimulation is delivered as a train or as a single biphasic pulse. For single biphasic pulses of 213 μsec durations, the average thresholds were about 350 μA, while for 17 pulse trains, delivered at 160 Hz, the thresholds for perceptions dropped to about 15 μA. When the stimuli were delivered at a low repetition rate, the phosphenes were observed to flicker, but they fused into a steady percept when the stimulus rate was between 8 and 10 Hz. The steady percept produced by these stimuli fades out after 1 to 3 seconds. The brightness of all evoked phosphenes varied with stimulus intensity, and ranged from "dim" to "average" as the current strength was increased by a factor of three.

The researchers claim that the phosphene space is sufficiently stable that they have been able to predict a phosphene's size, location, and intensity from the stimulation parameters used to evoke the phosphene. This is clearly a critical issue if optic nerve stimulation is expected to restore any form of useful vision.

A useful visual sense can only be built from multiple phosphenes, evoked at predictable locations. The Belgium team has used interleaved stimulation to evoke patterns of from 4 to 24 phosphenes, and the subject has been able to identify simple objects using these phosphene fields by scanning the objects using a head-mounted camera.

It is clear from these observations that this approach is unlikely to be able to recreate a high-resolution visual sense in those implanted with the four-electrode cuff. However, the researchers suggest that visually guided mobility and some forms of task performance do not necessarily require a high-resolution visual sense. They suggest that with sufficient training, a subject implanted with an optic nerve array could use this limited visual input to achieve simple visually guided mobility (at least in familiar environments). However, a number of issues require additional research before this approach could be considered tenable. Issues that relate to this potential intervention site are (1) how significant and how viable is the population of optic nerve fibers that are still functional; (2) does electrical stimulation of the optic nerve spare it from continued degeneration or does constant stimulation accelerate the process; (3) how many phosphenes, evoked by optic nerve stimulation, are required to produce given levels of visual task performance; and (4) would other electrode array designs that penetrate into the optic nerve produce more focal stimulation of optic nerve fibers and better control of phosphene location and intensity?

Researchers at the University of Utah have conducted experiments addressing this last issue. They have developed a unique electrode array architecture that was designed to provide highly selective electrical access to a number of the nerve fibers in the sciatic nerve.[53] This device, the Utah Slanted Electrode Array (USEA), is shown in Figure 11.7, and its access to the fibers in the nerve is depicted in Figure 11.8. It is built from silicon and contains 100 electrodes designed to penetrate the epineurium that surrounds the nerve and the perineuria that surround the individual fascicles within the nerve. The length of each electrode varies along the length of the array, with 0.5-mm-long electrodes on one side of the array, and 1.5-mm-long electrodes on the opposite side of the array. Each electrode is electrically isolated from its neighboring electrodes with a moat of glass at its base, and each electrode has a lead wire connected to a bond pad at its base that is

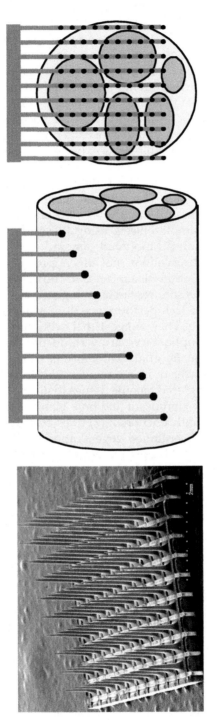

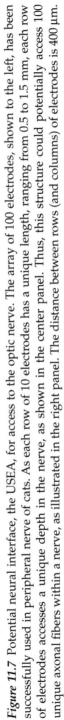

Figure 11.7 Potential neural interface, the USEA, for access to the optic nerve. The array of 100 electrodes, shown to the left, has been successfully used in peripheral nerve of cats. As each row of 10 electrodes has a unique length, ranging from 0.5 to 1.5 mm, each row of electrodes accesses a unique depth in the nerve, as shown in the center panel. Thus, this structure could potentially access 100 unique axonal fibers within a nerve, as illustrated in the right panel. The distance between rows (and columns) of electrodes is 400 μm.

Figure 11.8 Artist's concept of a cortically based visual neuroprosthesis. A small camera, located in the bridge of a pair of glasses, encodes the visual scene. This signal then drives a small array of electrodes implanted into primary visual cortex. The signal-processing electronics, not shown, convert the video signal into controlled current pulses on each of the electrodes, resulting in the percept of the same pattern seen by the camera.

brought out to a percutaneous connector. The tips of each electrode are metalized with platinum or iridium to facilitate the transduction of electrons in the wires into ions in the nerve. The entire structure, with the exception of the metalized tips, is insulated with silicon nitride or polymeric materials.

The Utah researchers have inserted the USEA into the sciatic nerve of the cat and measured the amount of electrical current required to evoke a twitch in the muscles innervated by the motor neurons of the sciatic nerve.[53] They found that single biphasic current pulses in the 5- to 20-μA range were generally sufficient to evoke muscle twitches. They also studied selectivity of stimulation by monitoring which muscles were excited when currents were passed through each of the implanted electrodes. They found that each of the muscles of the lower leg and ankle could be individually stimulated with currents passed through appropriate electrodes in the implanted array, and that individual muscles often could be activated by multiple electrodes in the array. Finally, these electrodes appeared to activate specific independent subsets of motoneurons. Such an electrode array architecture holds promise for an optic nerve interface.

11.7   Cortically based neuroprosthetic approach

As described earlier, the visuotopic organization of the visual pathways has motivated much of the effort at developing retinal, optic nerve, and cortically based vision neuroprostheses. An artist's conception of what a cortically based visual prosthesis might look like is shown in Figure 11.8. While a practical cortically based visual prosthesis has yet to be developed, the concept was first seriously entertained about 30 years ago. Before describing recent work in this field, we will explore the historical foundations of this approach to limited sight restoration.

The earliest work in this field was focused on implanting electrodes arrays on the surface of the primary visual cortex; this work was motivated by the pioneering findings of Hubel and Wiesel[54] and others that the visual cortex also had a visuotopic organization. Further, while access to the primary visual cortex requires specialized neurosurgical techniques, this region of cerebral cortex is surgically accessible without great difficulty, and it is protected by the skull from mechanical insult. Thus, the primary visual cortex seemed to be a good target for a first generation visual prosthesis.

11.7.1 Historical overview

Two efforts were launched in the 1960s and 1970s to explore this possibility: one by Brindley in England,[55] and one by Dobelle and colleagues at the University of Utah.[56] Both approaches proposed to stimulate visual phosphenes using arrays of platinum electrodes implanted under the dura, but on the surface of the visual cortex. The main difference between these two approaches was in the means by which stimulation signals were to be passed across the scalp. Brindley had developed an array of inductively coupled transmitting coils that were worn over the scalp and receiving coils attached to each electrode that were implanted under the scalp. Dobelle used a percutaneous connector mounted behind the ear, with each electrode connected to one of 64 pins in the connector.

Patients were implanted with these devices and extended periods of testing were conducted. The subjects were able to perceive points of light when currents were passed into the visual cortex, but two problems made the approach impractical. First, the currents required to evoke visual phosphenes were in the 1- to 10-mA range, current levels deemed unsafe for long-term chronic stimulation. Further, if future generation systems would require larger numbers of electrodes, summated currents could reach dangerous levels that could produce seizures. Second, the currents passed through electrodes evoked phosphenes that interacted in a nonlinear fashion. Specifically, the location of phosphenes that were evoked with current injections through individual electrodes was altered when currents were passed through groups of electrodes. This meant that the electrodes would have to be spaced at large intervals, making the concept of contiguous phosphene based percepts impossible. However, the experiments provided additional support for the visuotopic organization of human visual cortex and demonstrated that patterned electrical stimulation could evoked discriminatable patterned percepts: When stimulated with a subset of six electrodes, subjects could read Braille characters faster than with their sense of touch.[57]

The cortically based vision neuroprosthetic field then lay fallow for a number of years until a group of researchers in the Neuroprosthesis Program at the National Institute of Health conducted a series of acute experiments using electrodes that penetrated into visual cortex of human volunteers.[58,59]

The experiments were based upon the premise that, because the normal input to the visual cortex was layer 4C, electrodes that penetrated the visual cortex to this depth could produce much more focal stimulation of cortical neurons, with much lower current levels than were achieved with surface electrode arrays. This hypothesis was borne out in their experiments where it was shown that phosphenes could be evoked with electrical currents in the 1- to 10-μA region and that two distinct phosphenes could be evoked by current injections into electrodes as closely spaced as 500 μm. The electrodes used by the NIH team were of a map pin architecture. Each electrode was at the end of a 1.5-mm-long insulated needle, and each had a thin insulated wire bonded to the end of the needle. The interconnection of the wire and the electrode was insulated with a small bead (creating the map pin shape). The NIH team also used pairs and triplets of map pin electrodes to investigate the dependence of evoked phosphenes on electrode spacing. While these experiments demonstrated the benefits of penetrating electrodes over surface electrodes, the use of individually inserted map pin electrodes was an impractical approach to implantation of large numbers of electrodes.

11.7.2   Electrode arrays

The problem of creating high-count arrays of penetrating microelectrodes that could be easily implanted into cerebral cortical tissues was attacked independently by two research teams: one led by Wise at the University of Michigan,[60] and one led by Normann at the University of Utah.[61] Both groups wanted to capitalize upon newly developed silicon microfabrication technologies to build high-count arrays of penetrating microelectrodes because of the biocompatibility of silicon. Both teams are capable of making electrode arrays with hundreds of needles, and both teams have demonstrated that these arrays can be implanted chronically, often with little tissue insult. The difference between the arrays is in the way that they are fabricated. The Michigan array is built using conventional photolithographic techniques that result in very precise and reproducible array geometries. The Utah Electrode Array is built using micromachining techniques and requires more hands-on effort. Each needle in the Utah array has an approximately cylindrical cross section that rends the cortical tissues it is inserted into rather than cutting them. The Michigan arrays can have multiple electrode sites on each needle, and each needle has a rectangular cross section and may cut rather than rend the cortical tissues it is inserted into.

11.7.3   Animal experiments

These complex electrode array geometries were designed to be implanted into cerebral cortex to a depth of 1.5 to 2mm; however, the Utah team discovered that arrays containing very large numbers of individual electrodes cannot simply be pushed into the cortex of experimental animals. Rather, just as a staple can be inserted into a block of wood if it has sufficient momentum and velocity, these arrays require high velocity insertion to fully insert them. To achieve this end, the Utah team has built a pneumatically actuated insertion tool.[62] This insertion tool is capable of full insertion of a 100-electrode array in fewer than 200 μsec (an insertion velocity on the order of 7 m/sec).

Both the Utah and the Michigan electrode arrays have been used in scores of electrophysiological experiments in a variety of sites in the nervous system in rats, turtles, cats, monkeys, guinea pigs, and ferrets. They have been used in both stimulation and recording applications in both acute and chronic preparations. Histological studies indicate that there is usually a little localized bleeding associated with their implantation, but this usually resolves itself quickly, and single- and/or multi-unit recordings can be made within an hour or two of implantation. The fact that single-unit recordings can be made with these complex devices provides the best evidence that the neurons near the electrode tips have not been significantly damaged by the insertion process. The fact that recordings can be made on a chronic basis indicates that the materials used in their manufacture are biocompatible.

The efficacy of the Utah Electrode Array as a means to stimulate neural tissues has been studied in a series of chronic behavioral experiments conducted in cats.[63] Rather than training cats to respond to a visual task, Rousche decided to train cats to respond to auditory stimuli (cats seem to always attend to auditory stimuli while this is not the case for visual stimuli). Cats were trained to press a lever whenever they detected an auditory tone delivered through a loudspeaker that was initiated by a different lever press. When proficient at this task, the cats were implanted in the auditory cortex with a Utah array. After the implant, the cats were periodically electrically stimulated via the implanted array rather than aurally stimulated. If the electrical stimulation evoked a percept, the cats pressed the lever. These experiments demonstrated that the cats were able to detect presumed electrically evoked auditory percepts for current injections in the 1- to 10-μA range, a range well within the safe limits of chronic stimulation.

11.7.4 Human experimentation

The success of these animal experiments suggests that the stage may be set for short-term human experimentation with both the Michigan and the Utah electrode arrays. While such experimentation has yet to be conducted, issues to be resolved are:

> Can phosphenes be evoked in every electrode that is implanted in visual cortex?
> What is the variation in threshold currents required to evoke such phosphenes?
> What is the nature of the phosphenes from electrode to electrode?
> Does stimulation of each electrode always evoke spots of light or do some electrodes evoke spots of darkness?

What is the stability of the phosphene thresholds over time?

What is the visuotopic organization of the stimulated electrodes and the evoked phosphenes?

What is the minimum separation between stimulated electrodes that evokes two separate phosphenes?

How does current injection in one electrode affect phosphene percepts evoked in neighboring electrodes?

Do simple spatial patterns of electrical stimulation evoke patterned percepts that can be discriminated from other patterns of electrical stimulation?

The answers to these questions will provide the most complete proof of concept to date that electrical stimulation of visual cortex via an array of penetrating electrodes could eventually restore a useful visual sense in individuals with profound blindness. This will set the stage for a series of chronic human experiments where electrodes will be stimulated with more complex spatial patterns, and eventually with signals originating from a portable video camera.

11.8 Conclusions

This chapter has provided a brief overview of the physiological foundations upon which visual neuroprosthetic systems have been built. The chapter has described four intervention sites where a neural interface could be effectively used to stimulate neurons in the visual pathways: the subretinal sites, epiretinal sites, the optic nerve, and the primary visual cortex. Research in the field of visual neuroprostheses has been invigorated over this past decade due to recent advances in microfabrication of physical devices, VLSI electronic circuitry, and wireless telemetry. These technological innovations, coupled with improved understanding in the systems-level sensory neuroscience, are making the field of neuroprostheses into a reality. The authors of this chapter hope that the readers understand and appreciate four main points that were elaborated upon in the chapter:

- The use of small, passive photodiode arrays cannot produce sufficient currents to focally excite second-order retinal neurons with physiological levels of retinal illumination. Any such visual neuroprosthesis will require active electronics to boost the relatively small signals that are produced by small photodiodes.
- In order to optimally excite neurons, it will be important to position the stimulating electrodes very close to the neurons one is trying to excite (within a few microns). This means that the electrode must have dimensions that are similar to the size of the neurons one is trying to stimulate. Because of the location of the neurons within tissues, this constraint argues for a penetrating (or needle-shaped) electrode geometry and argues against the use of planar or surface electrodes.

- Stimulation of neurons is best accomplished with highly localized current injections. This again is best achieved with a penetrating electrode geometry.
- The recent work in visual neuroprosthetic systems has not yet resulted in the development of functional clinical systems. However, this work has resulted in the development of new classes of basic research instruments (electrode arrays) that will enable researchers to better understand the physiology of neuronal stimulation and sensory signal processing. These devices can be used in animals, and are now beginning to be used in human volunteers.

It is clear from what we have presented in this chapter that the field of visual neuroprostheses is rapidly emerging and that clinical systems are likely to become available over the next decade. What is also clear is that research should be continued in developing systems that can be applied at the retinal, optic nerve and cortical levels. Successful visual neuroprosthetic systems in all three interventional sites will give the ophthalmic surgeon or the neurosurgeon entirely new approaches that will provide limited but functional restoration of sight in those who have lost this highly prized sensory modality.

References

1. Nicholls, J.G., Martin, A.R. et al., *From Neuron to Brain: A Cellular and Molecular Approach to the Function of the Nervous System*, Sinauer Associates, Sunderland, MA, 1992.
2. Kandel, E.R., Schwartz, J.H. et al., *Principles of Neural Science*, McGraw-Hill, New York, 2000.
3. Balazsi, A.G., Rootman, J. et al., The effect of age on the nerve fiber population of the human optic nerve, *Am. J. Ophthalmol.*, 97(6), 760–766, 1984.
4. Naito, J., Retinogeniculate projection fibers in the monkey optic chiasm: a demonstration of the fiber arrangement by means of wheat germ agglutinin conjugated to horseradish peroxidase, *J. Comp. Neurol.*, 346(4), 559–571, 1994.
5. Fitzgibbon, T. and Taylor, S.F., Retinotopy of the human retinal nerve fibre layer and optic nerve head, *J. Comp. Neurol.*, 375(2), 238–251, 1996.
6. Hubener, M., Shoham, D. et al., Spatial relationships among three columnar systems in cat area 17, *J. Neurosci.*, 17(23), 9270–9284, 1997.
7. Van Essen, D.C., Anderson, C.H. et al., Information processing in the primate visual system, an integrated systems perspective, *Science*, 255(5043), 419–423, 1992.
8. Merigan, W.H. and Maunsell, J.H., How parallel are the primate visual pathways?, *Annu. Rev. Neurosci.*, 16, 369–402, 1993.
9. Kobatake, E. and Tanaka, K., Neuronal selectivities to complex object features in the ventral visual pathway of the macaque cerebral cortex, *J. Neurophysiol.*, 71(3), 856–867, 1994.
10. Stone, J.L., Barlow, W.E. et al., Morphometric analysis of macular photoreceptors and ganglion cells in retinas with retinitis pigmentosa, *Arch. Ophthalmol.*, 110(11), 1634–1639, 1992.

11. Santos, A., Humayun, M.S. et al., Preservation of the inner retina in retinitis pigmentosa: a morphometric analysis, *Arch. Ophthalmol.*, 115(4), 511–515, 1997.

12. Kim, S.Y., Sadda, S. et al., *Morphometric Analysis of the Macular Retina from Eyes with Discform Age Related Macular Degeneration*, ARVO, Fort Lauderdale, FL, 2001.

13. Jones, B.W., Chen, C.K. et al., *Severe Remodeling of the Mouse Neural Retina Triggered by Rod Degeneration*, ARVO, Fort Lauderdale, FL, 2002.

14. Warren, D.J., Fernandez, E. et al., High-resolution two-dimensional spatial mapping of cat striate cortex using a 100-microelectrode array, *Neuroscience*, 105(1), 19–31, 2001.

15. Eckmiller, R., Learning retina implants with epiretinal contacts, *Ophthalmic Res.*, 29(5), 281–289, 1997.

16. Jones, K.E. and Normann, R.A., An advanced demultiplexing system for physiological stimulation, *IEEE Trans. Biomed. Eng.*, 44(12), 1210–1220, 1997.

17. Robblee, L.S. and Rose, T.L., Electricochemical guidelines for selection of protocols and electrode materials for neural stimulation, in *Neural Prostheses: Fundimental Studies*, A.W.F. and McCreery, D.B., Eds., Prentice-Hall, Englewood Cliffs, NJ, 1990, pp. 25–66.

18. Rizzo, J.F., III, Wyatt, J. et al., Retinal prosthesis, an encouraging first decade with major challenges ahead, *Ophthalmology*, 108(1), 13–14, 2001.

19. Chow, A.Y. and Chow, V.Y., Subretinal electrical stimulation of the rabbit retina, *Neurosci. Lett.*, 225(1), 13–16, 1997.

20. Chow, A.Y. and Peachey, N.S., The subretinal microphotodiode array retinal prosthesis, *Ophthalmic Res.*, 30(3), 195–198, 1998.

21. Peyman, G., Chow, A.Y. et al., Subretinal semiconductor microphotodiode array, *Ophthalmic Surg. Lasers*, 29(3), 234–241, 1998.

22. Chow, A.Y. and Peachey, N., The subretinal microphotodiode array retinal prosthesis, II, *Ophthalmic Res.*, 31(3), 246, 1999.

23. Peachey, N.S. and Chow, A.Y., Subretinal implantation of semiconductor-based photodiodes, progress and challenges, *J. Rehabil. Res. Dev.*, 36(4), 371–376, 1999.

24. Chow, A.Y., Pardue, M.T. et al., Implantation of silicon chip microphotodiode arrays into the cat subretinal space, *IEEE Trans. Neural Syst. Rehabil. Eng.*, 9(1), 86–95, 2001.

25. Pardue, M.T., Stubbs, Jr., E.B. et al., Immunohistochemical studies of the retina following long-term implantation with subretinal microphotodiode arrays, *Exp. Eye Res.*, 73(3), 333–343, 2001.

26. Chow, A.Y., Peyman, G.A. et al., *Safety, Feasibility and Efficacy of Subretinal Artificial Silicon Retina™ Prosthesis for the Treatment of Patients with Retinitis Pigmentosa*, ARVO, Fort Lauderdale, FL, 2002.

27. Zrenner, E., Miliczek, K.D. et al., The development of subretinal microphoto-diodes for replacement of degenerated photoreceptors, *Ophthalmic Res.*, 29(5), 269–280, 1997.

28. Zrenner, E., Stett, A. et al., Can subretinal microphotodiodes successfully replace degenerated photoreceptors?, *Vision Res.*, 39(15), 2555–2567, 1999.

29. Zrenner, E.,   Will retinal implants restore vision?, *Science*, 295(5557), 1022–1025, 2002.

30. Guenther, E., Troger, B. et al., Long-term survival of retinal cell cultures on retinal implant materials, *Vision Res.*, 39(24), 3988–3994, 1999.

31. Stett, A., Barth, W. et al., Electrical multisite stimulation of the isolated chicken retina, *Vision Res.*, 40(13), 1785–1795, 2000.

32. Rizzo, J.F., Wyatt, J.L. et al., *Accuracy and Reproducibility of Percepts Elicited by Electrical Stimululation of the Retinas of Blind and Normal Subjects*, ARVO, Fort Lauderdale, FL, 2001.

33. Humayun, M., Propst, R. et al., Bipolar surface electrical stimulation of the vertebrate retina, *Arch. Ophthalmol.*, 112(1), 110–116, 1994.

34. Greenberg, R.J., Velte, T.J. et al., A computational model of electrical stimulation of the retinal ganglion cell, *IEEE Trans. Biomed. Eng.*, 46(5), 505–514, 1999.

35. Majji, A.B., Humayun, M.S. et al., Long-term histological and electrophysiological results of an inactive epiretinal electrode array implantation in dogs, *Invest. Ophthalmol. Vis. Sci.*, 40(9), 2073–2081, 1999.

36. Weiland, D., Fujii, G.Y. et al., *Chronic Electrical Stimulation of the Canine Retina*, ARVO, Fort Lauderdale, FL, 2002.

37. Margalit, E., Fujii, G.Y. et al., Bioadhesives for intraocular use, *Retina*, 20(5), 469–77, 2000.

38. Humayun, M.S., de Juan, Jr., E. et al., Visual perception elicited by electrical stimulation of retina in blind humans, *Arch. Ophthalmol.*, 114(1), 40–46, 1996.

39. Humayun, M.S. and de Juan, Jr., E., Artificial vision, *Eye*, 12(pt. 3b), 605–607, 1998.

40. Humayun, M.S., de Juan, Jr., E. et al., Pattern electrical stimulation of the human retina, *Vision Res.*, 39(15), 2569–2576, 1999.

41. Brummer, S.B., Robblee, L.S. et al., Criteria for selecting electrodes for electrical stimulation, theoretical and practical considerations, *Ann. N.Y. Acad. Sci.*, 405, 159–71, 1983.

42. Wyatt, J.L., Jr. and Rizzo, III, J.F., Ocular implants for the blind, *IEEE Spectum*, 33, 47–53, 1996.

44. Rizzo, J.F., Personal communication, 2002.

45. Grumet, A.E., Wyatt, Jr., J.L. et al., Multi-electrode stimulation and recording in the isolated retina, *J. Neurosci. Meth.*, 101(1), 31–42, 2000.

46. Gerding, H., Hornig, R. et al., *Implantation, Mechanical Fixation, and Functional Testing of Epiretinal Multi-Microcontact Arrays, MMA in Primates*, ARVO, Fort Lauderdale, FL, 2001.

47. Eysel, U.T., Walter, P. et al., *Optical Imaging Reveals Two-Dimensional Patterns of Cortical Activation after Local Retinal Stimulation with Sub- and Epiretinal Visual Prostheses*, ARVO, Fort Lauderdale, FL, 2002.

48. Hornig, R. and Eckmiller, R., *Movement of Stimulation Focus by Multi-Electrode Clusterstimulation for Retina Implants: Computational Results*, ARVO, Fort Lauderdale, FL, 2001.

49. Veraart, C., Raftopoulos, C. et al., Visual sensations produced by optic nerve stimulation using an implanted self-sizing spiral cuff electrode, *Brain Res.*, 813(1), 181–186, 1998.

50. Veraart, C., Raftopoulos, D. et al., *Optic Nerve Electrical Stimulation in a Retinitis Pigmentosa Blind Volunteer*, Society for Neuroscience, Los Angeles, CA, 1998.

51. Mortimer, J.T., Agnew, W.F. et al., Perspectives on new electrode technology for stimulating peripheral nerves with implantable motor prostheses, *IEEE Trans. Neural Syst. Rehabil. Eng.*, 39(15), 145–153, 1995.

52. Romero, E., Denef, J.F. et al., Neural morphological effects of long-term implantation of the self-sizing spiral cuff nerve electrode, *Med. Biol. Eng. Comput.*, 39(1), 90–100, 2001.

53. Branner, A., Stein, R.B. et al., Selective stimulation of cat sciatic nerve using an array of varying-length microelectrodes, *J. Neurophysiol.*, 85(4), 1585–1594, 2001.

54. Hubel, D.H. and Wiesel, T.N., Receptive fields, binocular interaction and functional architecture in the cat's visual cortex, *J. Physiol.*, 160, 106–154, 1962.

55. Brindley, G.S. and Lewin, W.S., The visual sensations produced by electrical stimulation of the medial occipital cortex, *J. Physiol. (London)*, 194(2), 54–59, 1968.

56. Dobelle, W. and Mladejovsky, M., Phosphenes produced by electrical stimulation of human occipital cortex, and their application to the development of a prosthesis for the blind, *J. Physiol. (London)*, 243(2), 553–576, 1974.

57. Dobelle, W.H., Artificial vision for the blind: the summit may be closer than you think, *ASAIO J.*, 40(4), 919–922, 1994.

58. Bak, M., Girvin, J.P. et al., Visual sensations produced by intracortical microstimulation of the human occipital cortex, *Med. Biol. Eng. Comput.*, 28(3), 257–259, 1990.

59. Schmidt, E.M., Bak, M.J. et al., Feasibility of a visual prosthesis for the blind based on intracortical microstimulation of the visual cortex, *Brain*, 119, 507–522, 1996.

60. Hoogerwerf, A.C. and Wise, K.D., A three-dimensional microelectrode array for chronic neural recording, *IEEE Trans. Biomed. Eng.*, 41(12), 1136–1146, 1994.

61. Jones, K.E., Campbell, P.K. et al., A glass/silicon composite intracortical electrode array, *Ann. Biomed. Eng.*, 20(4), 423–437, 1992.

62. Rousche, P.J. and Normann, R.A., A method for pneumatically inserting an array of penetrating electrodes into cortical tissue, *Ann. Biomed. Eng.*, 20(4), 413–22, 1992.

63. Rousche, P.J. andNormann, R.A., Chronic intracortical microstimulation (ICMS) of cat sensory cortex using the Utah Intracortical Electrode Array, *IEEE Trans. Rehabil. Eng.*, 7(1), 56–68, 1999.

chapter twelve

Motor prostheses

Richard T. Lauer and P. Hunter Peckham

Contents

12.1 Introduction

Damage or disease of the central nervous system results in a variety of deficits including sensory loss, tonic contraction of muscle (spasticity), cognitive impairment, impairment of biological functions, and the loss of volitional control over the extremities. Current treatment methodologies for these deficits include the use of pharmacological agents, physical therapy and rehabilitation, and surgical intervention. Also, other treatment methodologies could be available at some time in the future, such as neural regeneration. One method that is currently available and often overlooked for the treatment of deficits resulting from central nervous system trauma, however, is the use of neural prostheses that effect motor function.

A neural prosthesis is a device that uses electrical stimulation to interface directly to the nervous system to restore function. A motor prosthesis is defined as a device that electrically stimulates a nerve (or nerves) innervating a muscle (or series of muscles) for restoring functional movement or biological function. The purpose of this chapter, therefore, will be on how the principles of electrical stimulation, as realized in a motor prosthesis, can be used to overcome motor and functional losses. To this end, this chapter will be structured to provide the reader with the following information:

- Clinical applications of the motor prosthesis
- Characteristics/design of the motor prosthesis
- Commercial and research motor prostheses
- Future avenues of development for motor prostheses

12.2 Clinical applications of motor prostheses

The motor prosthesis, as stated in the introduction, can be used to restore loss of movement or biological function after damage to the central nervous system. Damage to the central nervous system can take several forms, the result of either trauma or disease, and the effects of the damage can vary greatly. In fact, even with the same trauma or disease the effects on the central nervous system can vary greatly from individual to individual. This presents a unique problem for the use of motor prostheses as a rehabilitation tool. Therefore, before discussing motor prostheses and ongoing research in this area, it is first necessary to examine some of the diseases and traumas that can affect the central nervous system and define what the general needs of the individual will be given the pathology.

12.2.1 Injury to the spinal cord

The first area of central nervous system trauma to be discussed will be injuries sustained to the spinal cord. The reasons for presenting this topic first are that the greatest levels of success with applications of motor prostheses have been achieved with spinal cord injuries. Injuries of the spinal

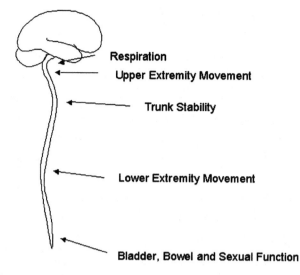

Figure 12.1 Overview of the human spinal cord and the motor/biological function associated with each level.

cord result in different types of disorders depending upon the level of the injury and the spinal cord tracts involved (Figure 12.1). Trauma to the spinal cord is usually the result of an automobile or diving accident, or a missile wound,[1] often secondary to a fracture or dislocation of the vertebral column. This results in severance or compression of the cord by the fractured bone and edema (tissue swelling). As would be expected, the most mobile areas of the vertebral column are the most susceptible to injury. An injury to the lumbar or thoracic area results in paraplegia (paralysis of the lower extremities), while an injury to the cervical area results in tetraplegia (paralysis of both the upper and lower extremities). Secondary to the loss of motor function below the level of the injury is the subsequent death of the spinal nerves from those segments at the level of injury (denervation), muscle spasticity, and muscle atrophy. Muscle atrophy is the degeneration of the muscle tissue with a resultant loss in muscle strength due to the loss of neuronal input. Because of this, the muscles that are to be stimulated by the motor prosthesis will undergo a strengthening regimen to build up mass and strength before the introduction of the system.[2]

The loss of voluntary movement and the resultant effects upon motor tasks are only one effect of spinal cord injury; there is also a subsequent loss in the autonomic functions of the nervous system. In the cases of high cervical level injuries, sustained at the second cervical level and higher, loss of the ability of the brain stem to control the diaphragm results in the inability to breathe voluntarily. In the case of both cervical and lumbar level lesions, bladder, bowel, and sexual functions are lost. The loss of these functions can lead to secondary complications, including urinary tract infections, impacted bowel and constipation, and difficulties with procreation.[3]

12.2.2   Injury to the brain

Injury to the brain can result in many different types of disorders depending upon the areas involved (Figure 12.2). Most often, in addition to motor deficits, there are also deficits in cognitive and sensory functions. The cognitive and sensory effects of brain damage are outside of the scope of discussion for the applications of a motor prosthesis. However, it is important to note that the use of motor prostheses as a rehabilitation tool does depend upon the cognitive capabilities of the individual. If cognitive function is impaired greatly by the extent of the injury to the brain, the effects of motor prostheses for the restoration of function can be hindered. However, motor prostheses can still be effective as tools for motor relearning. This will be discussed later in this chapter.

One common cause of damage to the brain is that resulting from a cerebrovascular accident (CVA) or stroke. A stroke occurs when a blood vessel within the brain is either blocked or ruptured, resulting in a loss of blood flow to the area of the brain supplied by the vessel.[1] This loss of blood flow leads to neuronal death and subsequent losses in motor, sensory, and cognitive function depending on the size and location of the lesion. These losses, however, may not be permanent and it is possible for individuals to learn to compensate and correct for these deficits with rehabilitation therapy.[4] The most common form of stroke involves the motor area of the brain in one hemisphere.[5] This results in hemiplegia (paralysis of the side of the body opposite to the hemisphere involved). Secondary to the loss of motor function are cognitive changes, depending on which hemisphere is involved; hyperreflexia (exaggerated stretch reflexes); muscle spasticity (tonic levels of muscle contraction independent of voluntary control); and muscle atrophy.

Another common form of brain injury is cerebral palsy. Cerebral palsy is the term used to define a wide range of movement disorders that occur during or shortly after birth. These disorders are caused by an accident (falls or other sudden trauma to the brain), brain infection (bacterial meningitis or viral encephalitis), or medical incidents during childbirth (such as birth asphyxia trauma, when oxygen flow to the brain has been cut off temporarily).[1,6] All of these conditions lead to neuronal death to parts of the brain similar to that experienced with stroke. Unlike stroke, however, the motor effects are quite different. Most often (80% of the time), these children experience what is known as spastic cerebral palsy.[7] In spastic cerebral palsy, the muscles are stiffly and permanently contracted, leading to difficulties in performing motor tasks. Other types of cerebral palsy include dyskinetic cerebral palsy, characterized by uncontrolled, slow, writhing movements; ataxic cerebral palsy, which effects balance and depth perception; or a mixed form that combines the other three types to some degree.[7]

Brain injuries are also sustained by a quick and sudden trauma to the head as can occur during an automobile or sporting accident. In this case, the trauma to the head results in the rupture of blood vessels on the surface of the brain and tissue swelling, causing an increase in intercranial pressure.

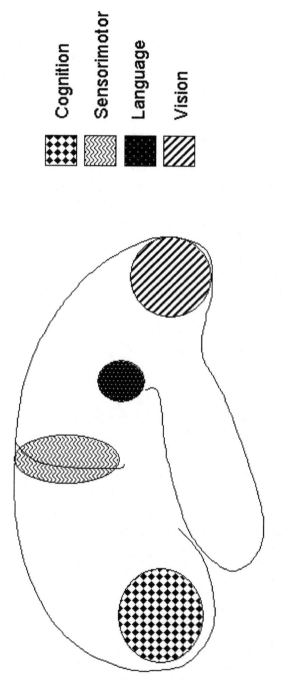

Figure 12.2 Schematic of cerebellum with approximate locations of important motor, sensory, and cognitive areas.

If not relieved, this increase in pressure results in neuronal death, and subsequent loss and impairment of motor function if the motor areas of the brain are involved.

12.2.3 Diseases that affect neural function

Numerous diseases of the nervous system can result in a loss of neuromuscular function. However, these can be roughly divided into those that lead to neuronal death, those that result in the loss of the myelin sheath around the neuron preventing the conduction of action potentials, and those that affect the generation or release of neurotransmitters. Of the three categories, for the application of motor prostheses only those diseases that cause neuronal death can be treated. More specifically, only those diseases that do not affect the nerve going to the muscle have benefited from the use of these systems. Current motor prosthesis technology is dependent upon the artificial generation of action potentials in the nerve going to the muscle to cause muscle contraction. Anything that affects action potential conduction in the nerve and the subsequent excitation of the muscle via the neuromuscular junction will prevent effective use of the motor prosthesis. Therefore, the applications of motor prostheses to the restoration of function in such diseases as multiple sclerosis and neuropathy are limited at this time.

Table 12.1 Summary of Neurological Conditions and the Associated Deficits

	Motor Deficits	Spasticity	Nerve Degeneration	Muscle Atrophy	Cognitive Deficits	Sensory Deficits	Respiratory Deficits	Bladder/Bowel Deficits
Stroke	X	O		X				
Cerebral Palsy	O	O		O				
Tetraplegia	X	O	O	X			X	X
Paraplegia	X	O	O	X				X
Multiple Sclerosis	O		O	O				
Neuropathy	O		O	O		O	O	O
Amyotrophic Lateral Sclerosis	O		O	O		O	O	O
Parkinson's Disease	O	O	O					

X Motor Prostheses Developed

O Possible Future Research

Note: Highlighted in the table are specific applications to which motor prostheses have been developed and employed. Also indicated are what are believed to be possible future avenues of motor prosthesis investigation.

Table 12.1 provides a summary of the expected deficits associated with brain injury, spinal cord injury, and certain diseases of the nervous system. Not all the deficits listed will occur with each injury or disease, and each individual will present a unique case. The highlighted areas in the table are those areas to which the motor prosthesis can and has been applied for use as a rehabilitation tool. Also given in the table are some suggestions of what are believed to be possible future applications for motor prostheses. As can be seen in this table, the motor prosthesis can be a valuable clinical tool for the restoration of motor function, the restoration of biological functions, and for motor relearning and training.

12.3   Motor prosthesis design

The applications of motor prostheses for the restoration of function have produced a variety of system designs, each directed toward a particular clinical use. However, all of the systems developed have certain character-istics in common and adhere to certain design criteria that are aimed at providing the greatest functional benefit and user acceptance. The major components of the motor prosthesis system are:

- The stimulus delivery system, consisting of the electrode and lead wires, that provide stimulation of the nerve
- The control unit, responsible for interpreting the operational com-mand generated by the user and converting that information into muscle stimulation
- The command interface, which records signals generated by the user and converts that information into operational commands for the motor prosthesis (Figure 12.3)

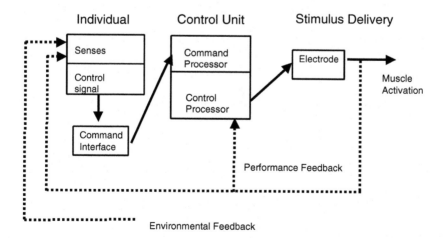

Figure 12.3 Block diagram showing the fundamental components of the motor pros-thesis and the interaction patterns with the intended user.

12.3.1 Stimulus delivery system

Currently, all motor prostheses operate by the electrical activation of nerves leading to muscles to elicit a muscle contraction. However, the methods by which the nerve can be stimulated are varied. One method by which a muscle can be electrically excited is with surface electrodes (Figure 12.4). These electrodes are placed on the surface of the skin directly over the point where the nerve and muscle are joined (motor point). These electrodes require the use of a conductive media and adhesives to ensure contact with the skin or, more commonly, are imbedded in an adhesive pad to ensure quick and clean placement.[8,9] The advantages to the use of the surface electrode are the relatively low cost of the electrode and the noninvasive method of placement. The disadvantages to the use of these electrodes are that only those muscles for which the motor point can be accessed from the skin can be electrically stimulated, and the placement of these electrodes requires an individual with knowledge and experience on the location of the motor point. In addition, the power requirements for such a system are usually higher given the large voltages (approximately 80 V) required in order to drive the current across the skin impedance.[10]

The second method by which the muscle can be electrically stimulated is to use percutaneous electrodes (Figure 12.4). The percutaneous electrode is inserted through the skin with a needle near the motor point of the muscle. The electrode can consist of a single strand of stainless steel wire or multiple strands.[11,12] These electrodes are barbed at the end, usually by bending the end of the wire, to ensure that the electrode is anchored after insertion and is not removed with the needle. By crossing the skin barrier, the use of the percutaneous electrode reduces the amount of power

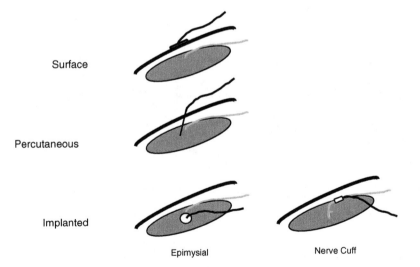

Figure 12.4 Schematic drawing representing the different types of electrode designs. Not represented in this drawing is the intramuscular (implanted) electrode type.

required to electrically stimulate the muscle and allows easier access to muscles that lie deeper under the skin. The drawback to this method, however, is that the skin has been compromised, which can lead to skin irritation and possible infection. This requires greater diligence on the part of the motor prosthesis user, or, more likely, the attendant or caregiver, to keep the entrance sites for the electrode clean. Percutaneous electrodes are also susceptible to breakage due to stress at the skin–muscle interface, requiring the reinsertion of new electrodes.

A final method by which the muscle can be electrically stimulated is with electrodes that are implanted completely within the body and do not cross the skin barrier (Figure 12.4). These are referred to as *implanted electrodes*, of which there are three major categories. The first type is the epimysial electrode, which consists of a platinum–iridium disk imbedded in a reinforced silicon pad and sutured near the motor point of the muscle.[13] The second type of implanted electrode is the intramuscular electrode, which is similar to the percutaneous electrode in that it is inserted with a needle directly into the motor point of the muscle. The electrode is constructed of stainless steel wire, but it has an umbrella-type of anchor at the end constructed of polypropylene to maintain its position in the muscle.[14] The final type of electrode is the nerve cuff electrode, which consists of one to three metallic (generally platinum–iridium) bands imbedded in a silicon sheet that encircle the nerve going to the muscle.[15–17]

The advantages to the use of the implanted electrodes are that in most cases they are selective in the muscle that they stimulate, being located near the motor point of the muscle. The exception to this is the nerve cuff electrode. The use of this electrode can ensure selectivity only if the nerve contains motor nerves that innervate a single muscle. A nerve containing fibers for multiple muscles would have all of the muscles excited by the use of this electrode, unless other techniques were employed to ensure some degree of selectivity.[18] Other advantages to the use of the implanted electrodes are that they require less power to elicit a contraction within the muscle, which is of prime consideration in the development of implanted motor prostheses. However, the primary drawback to their use is the level of invasiveness, as surgery and multiple incision sites are required for their placement and replacement is difficult.

12.3.2 Control unit

The control unit of the motor prosthesis is the key component to the system, and consists of two parts (Figure 12.3). The first part is the command processor. The command processor is used to interpret user-generated signals, as recorded from a variety of human–machine interfaces, as command signals to operate the various functions of the motor prosthesis. These functions can include control over the system state (on/off/idle/lock) and selection of activation patterns to be delivered to the muscles via the stimulus delivery system. The output of the command processor is relayed to the control processor of the unit. The control processor converts the signals generated by the command processor into actual function. The processor determines, based upon the activation pattern identified by the control processor, the specific muscle stimulation levels necessary to achieve the pattern. It is this information which is then delivered by the lead wires and electrodes to the muscle to generate purposeful contraction.

12.3.3   Command interface

The command interface, or human–machine interface, is the sensing system that records user-generated signals to operate the motor prosthesis (Figure 12.3). Numerous types of user-generated signals have been used in the past for control of motor prostheses. These include the use of muscle activity (electromyogram, or EMG) from a muscle under voluntary control,[19–22] joint angle transducers,[23,24] switches,[25,26] respiration,[27] voice activation,[28–30] and even a cortical signal.[31] Independent of the type of control signal used, the interface design has followed a set of criteria aimed at achieving optimal performance and user acceptance. Several researchers have outlined these criteria as they pertain to both prosthetic and motor prosthetic control.[23,32,33] What follows is a summary of these criteria, with subdivisions between the engineering aspects of the device and the user concerns.

The signal criteria set goals and expectations for the user-generated signal, allowing optimal performance of a motor prosthesis. The first criterion in this category for the command interface is to use the least number of input channels possible to control all functions of the system. This is because studies have shown that as the number of inputs increase, control and accuracy can be negatively affected.[34] The second criterion is to provide the greatest amount of information possible per input channel per unit of time. This is referred to as information content and is expressed mathematically as:[35,36]

$$I = T/\tau \log_2 n \tag{12.1}$$

where I is information content in bits, T is transmission time in seconds, τ is the minimum time required for a level change in seconds, and n is the number of discrete levels.

Some studies have also defined the information outflow rate (IOR), related to the information content.[37] The IOR considers the accuracy in achieving a discrete level and the transition rate. The equation is as follows:

$$IOR = (accuracy/transition\ rate) \times information\ content \tag{12.2}$$

Information content is the same as calculated in Eq. (12.1).

Another criterion of the signal is to reduce the amount of noise in the signal, as measured by the signal-to-noise ratio (SNR). A value of 20 decibels has been given as a possible acceptable value for motor prosthetic control.[38]

The final signal criterion refers to how quickly the input signal can change states. This is defined as the transition rate for the signal.

The next category of criteria for an acceptable command interface includes those that define the performance goals and the structure for the control interface. The first is accessibility, which is defined as having direct access to as many functions as possible. This implies the ability to multitask (parallel processing) instead of using menus and submenus to control functions (serial processing). The next criterion is durability. The interface should be able to withstand normal usage, and its lifetime should approach that of the user. Interference is the concept that the interface should not interfere or hinder the performance of normal activities or movement, nor should these activities interfere with the recording and interpretation of the signal. Processing is the measure of the interval between the initialization of a command and the appropriate response of the system, involving the transition rate and signal processing involved. A maximum value for this is approximately 200 msec, at which point the individual will begin to perceive the delay between thought and action.[39] The design of the interface should also be robust. This implies that the interface will provide the same level of performance without compensation by the user, regardless of changes in placement or in the quality of the signal. In addition, the interface should be repeatable, responding with the same output for a given input each time, with little drift or change. Finally, the interface should have good resolution, allowing the user to generate both large and small changes in position and/or force.

The final category of criteria are those that reflect the wants and needs of the user as determined by socioeconomic factors. Cosmesis is important in that the controller should not draw additional attention to the disability of the individual. The interface should also be as inexpensive as possible and should be easy to put on and take off. Ideally, the user should be able to accomplish this alone. Anything that is too complex or requires a lengthy donning period could lead to rejection. Related to this is the concept of ease of use, which implies that use in the controller requires a minimum training period and little concentration for use. One method of achieving this is to use the concept of extended physiological proprioception (EPP).[40,41] EPP is a prosthesis control technique that allows the user to accurately perceive the static and dynamic characteristics of the device through natural proprioceptive sensations. Finally, the interface should be free from all electrical hazards and the materials used should be biocompatible.

The objective with the design of the command interface to the motor prosthesis is to adhere to these criteria as closely as possible. Not all applications of the motor prosthesis will require the use of all of these criteria, nor has there been an interface developed which has met all of these criteria. However, how close an interface comes to meeting these criteria ensures how well the interface, and the motor prosthesis, will be accepted and used by the potential recipient.

12.4   The first motor prostheses

The previous sections of this chapter have provided the reader with an introduction to the motor prosthesis by defining clinical applications of the device and outlining motor prosthesis design. At this point, it would now be worthwhile to review the predecessors of the modern-day motor prostheses which were essential in defining the design criteria and provided insight into how electrical stimulus could be used in the treatment of various clinical pathologies.

12.4.1   The cardiac pacemaker

The cardiac pacemaker, although not considered by many to be a true motor prosthesis, is included here because the lessons learned and the technologies developed are the basis for most motor prosthetic developments. Damage or disease of the cardiac tissue can result in a number of conditions where the electrical impulses to contract the heart are not generated or the impulses have failed to be conducted throughout the cardiac tissue.[42] The cardiac pacemaker is an implanted device that delivers artificial electrical impulses to the cardiac tissue to enhance or control the contraction of the heart to correct for these conditions.

The design of the cardiac pacemaker is similar to that described for the motor prosthesis. The stimulus delivery system is made up of lead wires and electrodes, which can be located on the surface of the heart (epicardial electrodes), inside the muscle of the heart (intramyocardial electrodes), or within the cavity of the heart pressed against the lining (endocardial or intraluminal electrodes).[43,44] The lead wires travel to the control unit, which is an implanted device located some distance from the heart in the thoracic area. This device controls the generation of the electrical pulses delivered to the heart based upon a sensing system that takes the place of the command interface of a motor prosthesis.

The lead wire design of the cardiac pacemaker is similar to the lead wire design in the motor prosthesis. The lead wire consists of multiple stands of wire, helically wound, in an encapsulating sheath.[43,45] The multistrand approach to the lead wire ensures that electrical impulses can still be delivered even if one of the strands breaks due to stress on the wire. Stress on the wire, however, is reduced by the helical coil structure. This structure acts like a spring, allowing for stretching and bending, reducing breakage. The encapsulating sheath acts to further reduce stress on the wire by allowing the wire to move freely in the body by preventing tissue adhesions. The sheath also provides insulation of the wire from the body, ensuring proper electrical conduction and preventing unwanted biological reactions.

The electrode in the cardiac pacemaker has to withstand repeated movement of the tissue and the associated stresses resulting from that movement. The electrode–lead wire interface must also be able to withstand these stresses, especially given the fact that the stress will be concentrated at this point. Finally, the material from which the electrode is constructed has to be biologically compatible. The design that was developed for the electrode was a platinum or platinum alloy disk or probe imbedded in a support structure constructed from the same material as the encapsulating sheath.[43] The materials used ensured good biocompatibility, while the use of the platinum stimulating electrode provided the ability to continuously excite muscle tissue without electrode corrosion occurring. The support is continuous with that of the lead wire sheath, reducing stresses at the electrode wire interface and reducing the chance of breakage.

The control unit of the cardiac pacemaker is called the implanted pulse generator (IPG). It is constructed of a titanium or stainless steel package.[43] The metallic package of the implant allows it to act, in some cases, as the return anode for the delivery of electrical pulses. This principle is used in most implanted motor prostheses. Communication with the implant, for programming of the system or adjustment of parameters for electrical stimulation, is provided by a radiofrequency (RF) link. A receiver coil is located in the implant package, and, with an external coil placed over the implant, information can be sent to the system without the need for surgery to adjust the system.

The cardiac pacemaker, described here, did not begin to take that form until the 1950s and 1960s, although the first indications of using electrical stimulation to control the heart muscle can be found as early as 1930.[46] The reason for this rapid change in the design of the system and its widespread clinical use can be found in the introduction of the transistor in the 1940s. The transistor allowed smaller devices to be developed and eventually to be implanted within the body. At the same time, with the advances being made in the pacemaker design, individuals began to apply the concepts of electrical stimulation to the control of the extremities.

12.4.2 Hand and foot stimulators

The first motor prosthesis for the restoration of function is the system developed by Liberson and colleagues in 1961.[47] The system that was developed was used to correct for footdrop in individuals with hemiplegia. Footdrop, as is common after stroke, is the inability to lift the foot correctly during the swing phase of gait. This results in the toes not clearing the ground, causing the individual to trip and fall. Because of this, most individuals with hemiplegia will adopt a shuffling gait so as not to lift the foot off the ground. The first motor prosthetic system developed by Liberson corrected for footdrop by stimulating the peroneal nerve in the leg using surface-mounted electrodes. Stimulation of this nerve caused the contraction of the tibialis anterior muscle, resulting in dorsiflexion of the foot and allowing the toes to clear the floor. A switch was located in the sole of the shoe, which closed the circuit when the foot was lifted off the ground and stimulated the nerve. Approximately 100 individuals were treated with this system, achieving some level of gait improvement.

Long and Masciarelli accomplished the application of motor prostheses to the upper extremity shortly after the work of Liberson.[48] The system that was developed was used to control hand function in individuals with a cervical level injury at the fourth and fifth levels. This was accomplished by using electrical stimulation in combination with a hand and wrist splint. A surface electrode was placed over the motor point of the extensor digitorum communis to provide for finger extension. Finger flexion was achieved by using the splint to hold the middle and index fingers together and force them into opposition to the thumb using a spring. The splint was also used to stabilize the wrist. A potentiometer was mounted on the opposite arm to control the level of stimulation delivered to the extensor muscle. As stimulation was increased, the activation of the extensor muscle overcame the stiffness of the spring allowing the hand to open. When the stimulus level was decreased, the stiffness of the spring dominated the muscle, and the hand closed. This device was tested in one individual for a period of 16 months, allowing for an increased level of independence; however, the device did not see widespread clinical acceptance or use.

These first motor prosthetic systems established the fact that electrical stimulation could be effectively used to restore function to individuals with central nervous system injuries or diseases. However, these first systems were quite simple in that activation was only over a single muscle, the human–machine interfaces relied upon simple switches or potentiometers, and surface-stimulating electrodes were cumbersome and did not provide specific activation. The challenges facing development of the motor prosthesis included activating multiple muscles, developing better user interfaces, advancing the technology to remove as much external hardware as possible, and broadening the user population for these devices.

12.5   Limitations of a motor prosthesis

Before continuing on to a review of current commercial and research motor prostheses, it would be beneficial at this point to review the limitations of the motor prosthesis. These limitations are areas of future research, and effort is ongoing to overcome these limitations to increase the clinical applications of these devices and to provide greater function with the existing systems.

The first limitation of the motor prosthesis, introduced briefly in the discussion of spinal cord injury, is the problem of denervation. The motor prosthesis operates by electrically activating the nerve to the muscle. The reason for this is that by stimulating the nerve it is possible to activate parts of the muscle not accessible by the surface. This allows for the recruitment of more motor units and the generation of greater muscle force. In addition, the stimulation of the nerve requires a smaller charge, thus less energy, than that needed to recruit the muscle fiber directly.[49] Figure 12.5 illustrates the differences in charge required to activate a nerve vs. a muscle fiber, as indicated by the strength–duration curve. Charge in this

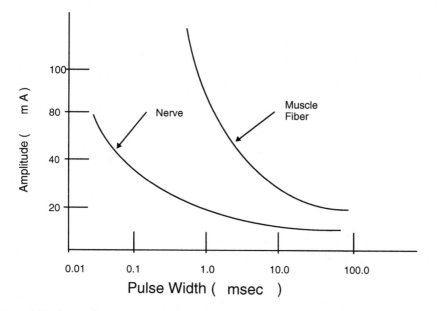

Figure 12.5 Strength–duration curve comparison of the charge required to excite a nerve compared to that of a muscle fiber. The differences in charge required to excite muscle fiber alone preclude this method as a means of restoring function with a motor prosthesis.

case is the relationship between pulse width and stimulus amplitude, represented mathematically as:

$$Q = PW \times A \qquad\qquad (12.3)$$

where Q is the charge, measured in columbs, PW is the pulse width of the delivered charge (measured in seconds), and A is the amplitude of the stimulus pulse, measured in amperes.

As can be seen in the graph, it requires a substantially greater charge to activate a muscle fiber than the nerve going to the muscle. These higher charge levels are also at the point where tissue damage can occur. Tissue damage in this case is the result of direct heating of the tissue or by the creation of toxic chemicals due to the electrochemical reactions being driven at the electrode-tissue interface.[50] Therefore, for these reasons motor prostheses are not used for those cases where the nerve to the muscle has atrophied or the nerve can no longer be stimulated due to the loss of the myelin sheath.

The second limitation of the motor prosthesis is the problem of muscle spasticity. Muscle spasticity is the tonic level of muscle activation that can occur with some central nervous system injuries or diseases. In this case, action potentials are spontaneously active in the nerve, which in turn generates muscle contraction, which prevents the use of the motor prosthesis

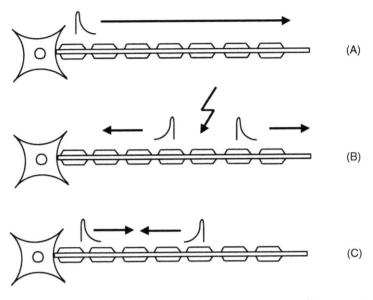

Figure 12.6 The principle of collision blocking. (A) The natural progression of an action potential from the neuron down the axon. (B) With the application of an electrical pulse, action potentials can be generated in both directions. (c) With a nerve cuff electrode, pulses can be generated so that they travel up the neuron toward the direction of the cell body so that they can block any impulse traveling down the nerve.

for the restoration of function because the current designs rely upon generating action potentials in the nerve to control contraction. However, given the ability of the motor prosthesis to generate an action potential, it can be seen how this can be applied to the problem of spasticity. The action potential generated on a nerve by the motor prosthesis travels in both directions from the stimulation site, both down to the muscle and back toward the spinal cord. However, with the nerve cuff electrode, it is possible to generate an action potential in only one direction.[51] If an action potential is traveling from the central nervous system to the muscle at the same time the artificially generated action potential is traveling up the nerve, both potentials will meet and cancel each other out. This principle, referred to as collision blocking (Figure 12.6), can be used to control spasticity and could be applied to some of the injuries and diseases discussed earlier.

The final limitation of the motor prosthesis is the limited number of feedback signals available. In Figure 12.3, the system design of the motor prosthesis indicated performance and environmental feedback being sent to the user. However, in most motor prosthesis applications, this feedback information is limited to visual and auditory feedback, with little or no information from sensory receptors (e.g., touch, temperature, pain) or muscle proprioceptors (e.g., joint position, force).[52,53] The user of the motor prosthesis must be continually monitoring the output of the system visually, and rely upon whatever machine information is being sent back through auditory and visual cues. This places a high cognitive demand upon the user of such a system, which makes the motor prosthesis, at times, difficult to use and implement. Several methods have been explored to provide sensory feedback though electrical stimulation of areas where sensory function is intact[54,55] or by recording from the nerves going to cutaneous receptors and then passing this information onto the motor prosthesis.[56,57] However, the issue of feedback to the user is one that has yet to be addressed in a satisfactory manner.

12.6 Commercially available motor prostheses

The results achieved with the first motor prostheses, and the clinical success of the cardiac pacemaker, have led to the development and marketing of several motor prosthetic systems for clinical use. Figure 12.7 shows the clinical deployment of two of the commercially available motor prosthetic systems in recent years. Although the field of motor prosthetics is relatively new, it can be seen that after their introduction there has been an increasing application of motor prosthetics in the rehabilitation field. If this curve is projected out at its current rate, there will be more than 100,000 motor prostheses in use by the year 2010. Although only two motor prosthetic systems are shown in Figure 12.7, several are commercially available for use. The goal of this section will be to review these systems. In order to facilitate this discussion, the commercial systems will be divided into three groups

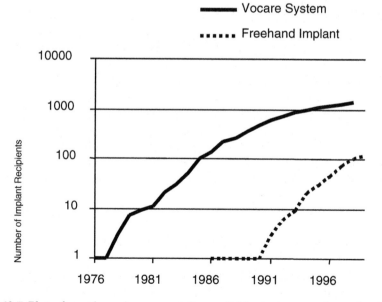

Figure 12.7 Plot of number of commercially available motor prostheses implanted since 1976. Only the information about two of the systems, one for the upper extremity and one for bladder and bowel restoration, are shown.

based upon their function: those for upper extremity movement, those for lower extremity movement, and those used for organ systems.

12.6.1 Upper extremity motor prostheses

Four motor prostheses are commercially available for the restoration of function in the upper extremity. Most of these systems are directed towards the restoration of hand function after a spinal cord injury; however, these systems have also found applications for motor relearning after stroke, and one is specifically marketed for this purpose.

The first system to be reviewed is the *Bionic Glove*, also known as the *Tetron Glove*, originally marketed by NeuroMotion, Inc.[58,59] This device was designed for individuals with a cervical level injury at the sixth and seventh level to augment the natural tendonesis grasp to provide active hand function. Individuals with this level of spinal cord injury will have active control over shoulder and elbow movement and active wrist flexion. However, there is no remaining control over the muscles for the hand to provide for grasp. Therefore, they rely upon the use of the tendonesis grasp to acquire objects. Tendonesis grasp is a naturally occurring phenomenon due to the length of the finger flexor tendons. The primary finger flexors are in the forearm, and the tendon for these muscles crosses both the wrist and the finger joints. As the wrist is extended (pulled up against gravity), tension on the finger flexor tendons increases, causing the tendon to pull the fingers into flexion. This grasp, while allowing the hand to close, is often insufficient in force to allow an individual to grasp an object.

The Bionic Glove augments this grasp by stimulation of the finger and thumb flexors and the finger extensors using surface-mounted electrodes. The system consists of three electrodes that are placed over the motor points of the muscles to be stimulated, with the return anode located proximal to the wrist. Metal studs are located on the back of each electrode, which can connect to a stainless steel mesh located inside a neoprene glove. When the glove is worn, the studs interface with the mesh, completing the stimulus delivery system. The control unit for the system is located on the forearm portion of the glove and provides for the electrical activation of the muscles. A position transducer attached to the wrist provides control over the unit. As the individual flexes the wrist, the finger extensor muscles are stimulated to provide hand opening. When the wrist is extended, stimulation to the extensors is decreased and the stimulation of the finger flexors is increased.

The Bionic Glove achieved limited clinical success, being implemented in 37 individuals by 1996.[59] The device was found to improve function during activity of daily living (ADL) assessments, providing the ability to grasp and manipulate larger and heavier objects could be accomplished without the system. The exercise regimen used to strengthen the atrophied muscles also demonstrated therapeutic benefits in that individuals, after several months of using the system, no longer required it for use in ADL tasks.[58]

The second commercially available system also uses a surface electrode delivery system for the restoration of hand grasp. This device is the *Hand-master system,* marketed by Neuromuscular Stimulation Systems, Ltd.[30,60–62] This device is designed for individuals with a cervical-level spinal cord injury sustained at the fifth level and for individuals with hemiplegia. The device consists of three to five electrodes used to stimulate the finger flexors and extensors and the thumb flexor muscles. However, unlike the Bionic Glove, these electrodes are permanently mounted in an orthotic brace. This makes the system easier to use than the Bionic Glove because individual electrode placement is not required each time the system is used. The orthotic brace also prevents movement at the wrist, but this stability is essential in individuals with a C5-level injury who have completely paralyzed wrist flexors and extensors.

The control unit of the system is external and independent from the brace, with a belt clip to allow the unit to be worn about the waist. Control over the application of the electrical stimulation is accomplished by a series of switches and a potentiometer. The switches are used to electrically stimulate either the finger flexors or extensors to control the degree of hand opening and closing, with the potentiometer used to control the degree of thumb flexion. Two additional push-button switches allow the user to increase the level of grasp force.

The location of the switches and the potentiometer are on the control unit and require the use of the opposite hand for control. While this arrangement may not be ideal for individuals with a spinal cord injury, it is applicable for individuals with hemiplegia. The Handmaster system, in fact, has been proven to be a valuable clinical tool for these individuals. Results in subjects tested with this system have indicated a reduction in spasticity as measured by improved ranges of motion about the hand and wrist joints, and better hand function.[60] This device is currently available in Europe, with approval by the Food and Drug Administration (FDA) to begin marketing and sales in the United States.

A third commercially available system is the *AutoMove AM800*, marketed by Stroke Recovery Systems, Inc.[63] This device, like the others, uses surface electrode technology to activate the muscles. However, unlike the other systems, this device is marketed solely as a motor recovery device for individuals who have sustained hemiplegia. The device uses three to five electrodes to activate the muscles of the wrist and hand. Control over the application of the stimulation is accomplished by the recording of muscle activity from the same electrode set. When the attempt to active a muscle is recorded (i.e., a twitch in the EMG signal), the command is sent to the control unit to active the muscle electrically. In this manner, the motor command from the individual is augmented by electrical stimulation. This makes it possible for the brain to relearn how to activate the muscle, reducing the effects of hemiplegia. The electrical activation of the muscles also reduces spasticity in these individuals. This device is currently available in the United States and is FDA approved.

A final motor prosthetic system that is commercially available is the *Freehand System*, marketed by NeuroControl Corp.[64-68] This device, unlike the other devices discussed, is an implanted system aimed at restoring hand function to individuals with a cervical level spinal cord injury at the fifth or sixth level. The device consists of an implanted stimulator, located in the chest, with lead wires for eight stimulating electrodes. These lead wires are tunneled under the skin and are connected to implanted electrodes located in or on the motor points of the muscles of the forearm and hand. The stimulation parameters, and the power to the implanted device, are provided through a RF coil placed on the skin over the implant. The external control unit, which contains the information on the muscle activation patterns and the batteries, communicates with the implant via the coil.

The Freehand system provides the individual with two grasp patterns, a palmar grasp and a lateral grasp. In the palmar grasp, the thumb is brought into opposition to the fingers and the fingers are flexed against the thumb. This grasp is useful in acquiring and holding objects such as drinking glasses and books. In the lateral grasp, the fingers are flexed and the thumb is flexed against the side of the index finger. This grasp is useful is acquiring and holding objects such as pencils, forks, and keys.

The user of the Freehand system has complete control over the degree of hand opening and closing, as well as the selection of grasp, through the use of a transducer mounted on the shoulder opposite to the arm implemented with the Freehand system. The shoulder transducer consists of two parts, a push button switch and a joystick. The switch is used to operate the binary functions of the system, such as grasp selection and turning the system on and off. The joystick, which spans the glenohumeral joint of the shoulder, provides the input signal to control hand opening and closing. The shoulder range of motion is mapped directly to the degree of hand opening and closing. By moving the shoulder through the range of motion, the user can position the hand at any point between full hand opening and full hand closure.

The Freehand system has achieved a modest level of success, being implemented in over 200 individuals worldwide. The device has been demonstrated to improve grasp force in individuals with sustained tetraplegia, to improve performance on ADL assessments, and to improve hand posture and joint range of motion.[67,68]

12.6.2   *Lower extremity motor prostheses*

Currently, there are five commercially available motor prosthetic systems for the restoration of function in the lower extremity. However, given the complications in the biomechanics of the lower extremity, the focus has primarily been on providing systems that correct for footdrop in individuals with sustained hemiplegia or for providing standing in individuals with sustained paraplegia. Walking has been achieved to a limited extent; however, these systems are dependent upon the use of a walker or crutches in conjunction with the system to provide upright stability.

Five systems are currently available to correct for footdrop: *WalkAid*, originally marketed by NeuroMotion, Inc.;[69,70] *Odstock Dropped Foot Stimulator* (ODFS), made available for use by the Department of Medical Physics and Biomedical Engineering, Salisbury District Hospital, United Kingdom;[71,72] *MicroFES*, made available for use by the Institute for Rehabilitation in Ljubljana, Slovenia;[73,74] *Footlifter*, marketed by Elmetec A/S;[75] and *PNS 2000*, marketed by BarMed Pty., Ltd. All of these systems provide for the electrical activation of the muscle with surface-mounted electrodes. These systems consist of one or two electrodes, placed over the peroneal nerve to cause foot dorsiflexion and knee flexion (withdrawal reflex). An external control unit is mounted on the calf or above the knee using Velcro straps. Control over the system is provided by either a foot-mounted switch to record when the foot is lifted off of the ground or a push-button switch to control the application of stimulation. These systems can also be placed into an exercise mode to improve muscle strength and reduce muscle spasticity. These devices have experienced a modest level of clinical acceptance, with 7500 to 10,000 devices currently in use.[75]

The *Parastep system*,[26,76,77] marketed by Sigmedics, is the only commercially available system for the restoration of standing and walking in individuals with sustained paraplegia. This device has been implemented in over 600 individuals to date.[78] The device consists of an external control unit, surface electrodes, and push-button switches that are mounted on a walker. This system is to be used in conjunction with the walker to provide upright stability and to move the individual forward. Six electrodes are used in this system, four to provide upright standing and two to elicit rudimentary stepping. Standing in the system is provided by the electrical activation of the quadriceps muscles to fix the knee in extension and activation of the paraspinals and the gluteal muscles to provide lower back stability and hip extension. Stepping is achieved by the stimulation of the peroneal nerve, as in the footdrop systems, to activate the withdrawal reflex. Control over stepping is achieved by the push-button switches, one for each leg. When the button is pressed, the leg is stimulated to achieve knee flexion and ankle dorsiflexion, while the quadriceps of the other leg are stimulated to accept the weight of the individual. The control unit for this system is worn around the waist of the individual and contains the power source and the electrical activation patterns for each individual.

12.6.3   Organ system prostheses

The commercial applications of motor prostheses for the restoration of biological function are focused on the restoration of bladder and bowel function and in the restoration of respiratory function in those individuals with a high cervical level spinal cord injury. Four systems have been identified for use in bladder control and/or bladder continence: *Brindley–Finetech system*, marketed by NeuroControl Corp.; *LEVATOR Turbo CS200 Continence Stimulator*, marketed by Ferraris Medical, Ltd.; *Interstim*, marketed by Medtronic,

Inc.; and a stimulator marketed by Dobelle Institute Avery Laboratories. All of these stimulators are available in the United States and generally operate along the same principles, stimulation of the sacral nerves to control bladder and urethral contraction.

The Brindley–Finetech Bladder Controller, also known as the VOCARE system in the United States,[79,80] consists of two electrode pairs and a stimulator implanted within the body and an external control unit worn around the waist. The electrodes are implanted near the second through fourth sacral nerves bilaterally in the sacral canal after they have been exposed by a laminectomy. The stimulating electrodes provide for the contraction of the bladder wall to induce urine flow. The location of the electrodes on the second through fourth sacral nerves also provides for contraction of the lower bowel and relaxation of the anal sphincter in approximately 50% of the recipients of the system, promoting regularity and greatly reducing the time of the bowel program.[3] Control over the system is maintained by the user, with a switch selector on the external control unit to allow the individual to control micturition and defecation.

The system has achieved a great deal of clinical success, being implemented in over a thousand individuals worldwide.[3] Implementation of this system is also frequently accompanied by a dorsal rhizotomy of the second through fifth sacral nerves. This improves continence and prevents involuntary urethral sphincter contraction with the electrical contraction of the bladder wall. The problem with this, however, is a loss of reflex erection, ejaculation, perineal sensation, and an alteration in reflexive defecation. These drawbacks have somewhat limited the acceptance of this system, primarily by males.

Four systems are commercially available for use in the restoration of respiration or to provide for coughing in individuals with high cervical level spinal cord injuiries: *Quik-Coff* system, marketed by B&B Medical Technologies outside of the United States; *Diaphragm Pacing* system, marketed by Dobelle Institute Avery Laboratories; *Atrostim* system, marketed by Atrotech Oy, Finland; and *T154*, marketed by MedImplant, Austria. These devices, except for the Quik-Coff system, are implanted devices that stimulate the phrenic nerve to cause contraction of the diaphragm to restore respiration without ventilatory support. These implanted devices, except for the T154, are available in the United States for use in high tetraplegia.

The Quik-Coff system is a surface-electrode system developed to assist individuals with cervical level spinal cord injuries with coughing.[81] Surface electrodes are placed on the abdominal wall, in conjunction with an abdominal binder, to activate the external and internal oblique muscles. An individual who wishes to initiate or needs assistance with a cough uses a push-button switch to activate the muscles. Coughing can be accomplished on a single cough attempt or multiple attempts (up to four) separated by 1-, 2-, 3-, or 4-second intervals. The control unit of the system is a small, battery-powered device the size of a pager that can be worn around the waist with a belt clip. Results with this system are encouraging, indicating that expiratory pressure can be increased up to two and a half times of what can be achieved with no assistance.[81] This can greatly reduce pulmonary complications associated with spinal cord injury.

12.7 Current research motor prosthetic systems

Ongoing studies in the area of motor prostheses are focused at expanding the current applications of the systems. This is achieved by providing for more channels of activation to recruit more muscles, improving the electrical stimulation technology, and expanding the clinical indications for motor prostheses. This section will attempt to provide an overview of current research in this area, again dividing the systems into those for use in the upper extremity, the lower extremity, and for the restoration of organ system function.

12.7.1 Upper extremity motor prostheses

Several investigators are working in the area of upper extremity motor prostheses with the goal of either providing for a greater restoration of function or improving motor prostheses technology. One series of systems aimed at providing both greater function with the motor prostheses and advancing motor prostheses technology are the implanted stimulator telemeters (ISTs) being developed at Case Western Reserve University. These newer systems are an expansion of the Freehand system mentioned earlier and are the first implanted, sensor-driven motor prostheses ever to be developed. The IST systems provide for a greater restoration of function by allowing more muscles to be incorporated into the electrically stimulated hand grasp, up to a maximum of 16 muscles. Examples of improved function include the fact that individuals using these systems are provided with electrical stimulation of the triceps muscle to allow for active elbow flexion and extension, which can greatly increase their usable workspace.[82] In addition, these individuals are provided with electrical activation of the finger intrinsic muscles,[83] which has improved grasp strength and helped maintain a more natural hand posture.

The technology of the IST systems is also an improvement upon motor prosthesis technology in that the RF link with the implanted system provides for two-way communication. This has allowed for the implantation of the command interface for the users of this device. The implanted joint angle transducer (IJAT) available with this system is a Hall effect sensor that is implanted into the bones of the wrist.[24,55] As the individual moves their wrist in flexion and extension, the IJAT system provides an output of the joint angle that is used to control the degree of hand opening and closing in the same hand. The IST system also allows for electrodes to be placed on muscles under voluntary control and uses the EMG signal to provide control over the hand grasp[55] in those cases where the individual does not have active wrist extension.

Another system designed at providing greater function with motor prostheses and advancing motor prosthesis technology is the system developed at Tohoku University in Japan.[27-29,84-88] This system is designed as a multipurpose system that can be used to restore complete upper extremity function regardless of the level of injury, to provide for standing and walking, and for therapeutic use. The system provides for up to 64 channels of stimulation delivered through the use of percutaneous electrodes. Control over this system is provided either with a respiration (sip-puff) controller or through voice recognition software. Stimulation patterns to be delivered to the muscles are based upon EMG activation patterns recorded during the accomplishment of different tasks in able-bodied individuals.[84,86] This device has been demonstrated to be useful in the restoration of function in individuals with a cervical level injury at the fourth level where shoulder and elbow control are also required and for walking in individuals with thoracic level injuries.

The disadvantages of the research systems described above are that they all involve placement of materials inside the body that may be considered unacceptable by some of the end users of these devices. Therefore, some investigators are focusing on the development of surface stimulation systems for use in the upper extremity. The *ETHZ-ParaCare system*, now being marketed in Europe as the *Compex Motion system*, is a new system that provides for four channels of stimulation using surface-mounted electrodes.[78,89] The control unit device is a small, pager-sized system that can be clipped onto a belt or placed inside a pocket, with overall control of the system by the user being provided by electromyographic signals under voluntary control. The control unit can be programmed using a graphical interface, and the interface and the memory cards inside the control unit are interchangeable. This allows the system to be individually tailored to each individual and applied quickly and easily. The four channels of stimulation in the system can be expanded in multiples of four, although how this is accomplished or the maximum number of available channels has not been reported. Currently, ten individuals have used this system for walking, hand grasp, and treatment of shoulder subluxation.[90]

A final series of investigations into the uses of motor prostheses for the upper extremity are those that have examined the effects of electrical stimulation on motor relearning.[5,60,91-94] In addition to shoulder subluxation, the effects of hemiplegia on motor control can include the inability to activate natural movement patterns in the upper extremity to interact with the environment, as well as paralysis of certain muscles. Investigations at several universities and institutions.[91-94] are directed toward systems that can record trace activity of muscles in the forearm and hand and then apply electrical stimulation to the same muscle to achieve full muscle recruitment. This is the same concept as discussed in the section on the AutoMove AM800. By amplification of muscle activity, the goal is to retrain motor cortical areas using positive feedback mechanisms. This should provide for a more efficient recruitment of the muscle to achieve grasping and movement.

Other investigators are also addressing paralysis and spasticity in the hemiparetic hand.[5,60,89,91] A common occurrence with hemiplegia resulting from stroke is the inability to active the finger flexors to achieve hand closure, in conjunction with a tonic level of activity in the extensor muscle groups of the forearm in hand. These systems are designed to address these two needs separately to provide for efficient hand grasp. The principle of collision blocking, as discussed earlier, is used in some systems to prevent the tonic level of contraction for the finger extensors by preventing the conduction of the action potential along the motor neuron to the extensor muscle. When finger extension is required, these systems turn off the blocking pulses, allowing finger extension to occur. When finger flexion is required, these systems then block the action potentials to the finger extensors while at the same time stimulating the finger flexors to close the hand. Other systems address this problem by increasing the stimulation to the finger flexors without blocking the extensors, with the idea that the flexors are stronger than the extensors and will overpower them to allow for hand closing. The difficulty with many of these systems, however, is that stimulation is achieved through the use of surface or percutaneous electrodes, both of which have been unacceptable to the end user of the device because of the unpleasant sensation elicited by the electrical stimulation.

12.7.2   Lower extremity motor prostheses

Current research trends for the lower extremity motor prosthesis, such as in the upper extremity, are directed toward increasing the functional capabilities of the motor prosthesis and advancing the technology. To this end, several different investigations have been identified that are aimed at improving walking and standing using motor prostheses. The research being conducted at the Salisbury District Hospital[95] is directed toward providing standing in individuals with sustained paraplegia using sacral root stimulation. The system that has been developed, the lumbosacral anterior root stimulator implant (LARSI), uses 12 intradural electrodes placed on the second lumbar through the second sacral anterior roots in the cauda equina. Postsurgical stimulation of each of the roots individually is performed to identify joint movement generating capabilities and nerve stimulation combinations to achieve upright standing. Currently, two individuals have received the system and, even though standing is possible, they have encountered difficulties in attaining a good posture due to excessive hip flexion occurring with the sacral root stimulation. Current research by this group is underway to reduce hip flexion to achieve upright, hands-free standing.

The *Praxis-24 system* being developed by Neopraxis Pty., Ltd.[96,97] is an implanted system that provides for up to 22 channels of electrical stimulation. This system is designed to provide for upright standing and walking for individuals with sustained paraplegia, in addition to providing for bladder control. Electrodes placed adjacent to the motor nerves in the quadriceps, hamstrings, ankle, and gluteal muscles provide standing and walking functions. Additional electrodes are placed near the motor nerves for the psoas muscle to provide for hip flexion. Control over the system is accomplished using gyroscope and accelerometers to record limb position, with overall control being maintained by the user through a touch-sensitive LCD screen on the control unit.

Bladder function in this system is achieved by the placement of three electrodes near the first through third sacral spinal nerves, exposed by a laminectomy. Another electrode is placed over the conus medullaris at the twelfth thoracic/first lumbar level to remove the need for the dorsal rhizotomy. This system has currently been implanted in two individuals with a modest degree of success. These individuals have been able to achieve standing while being able to manipulate objects with one hand. In addition, the system has been able to provide bladder emptying in one subject with pressures of 50 to 70 centimeters of water.[97]

Another system that is directed toward walking and standing in individuals with sustained paraplegia is the 16-channel neural/epimysial stimulator that is being developed under the European "Stand-Up-And-Walk" (SUAW) project.[98] This work is early in the research phase and limited information is available; however, the results have been encouraging and warrant mention here. The device is an implanted system using a combination of muscle- and nerve-stimulating electrodes, with overall control over walking being maintained by the individual with push-button switches. This device has been implemented in two individuals, with one individual being able to achieve standing and short distance walking, while the other individual is still in the training period for the device.

A final system under development for standing and transfers in individuals with sustained paraplegia is the implanted system under development at Case Western Reserve University.[99,100] The implantable receiver used in this system is identical to the early upper extremity system that led to the development of the Freehand system. The system provides for eight channels of stimulation, delivered by the use of epimysial and intramuscular electrodes. The electrodes are placed bilaterally in the vastus lateralis, gluteus maximus, and semimembranosus muscles to provide for knee and hip extension. The other two electrodes are placed near the twelfth thoracic and first lumbar nerve roots to provide for stimulation of the erector spinae to provide trunk extension. Control over the system is maintained by the user with a push-button switch, independent of the control unit for the device, which is a portable unit worn around the waist.

This system has been implemented in 13 subjects to date,[100] all of whom have been able to achieve upright standing with the use of a walker. In most cases, the legs have supported approximately 80 to 95% of the body weight, allowing the users to perform limited, one-handed manual tasks.[101] Ongoing research is being directed toward achieving longer standing times in this system before muscle fatigue requires the individual to turn the system off. Work is also directed toward trying to achieve the distribution of all of the body weight to the legs to allow for hands-free standing and limited mobility with the use of a walker.

The clinical research on the standing and walking motor prostheses and the advances being made there are mirrored by the ongoing research aimed at correction of gait in individuals with hemiplegia. At least three series of investigations are ongoing to correct for footdrop in hemiplegia. Two of these research systems, like the commercially available systems, rely upon stimulation of the peroneal nerve to elicit ankle dorsiflexion. However, the stimulation and control methods are different. The system developed at the University of Aalborg[102] uses nerve cuff electrodes to record activity from the sural nerve which innervates the heel region. By monitoring activity in this nerve, it has been possible to record heel strike and the toe-off phase of gait. This signal has then been used in place of the external switches used in the other systems to provide stimulation to the peroneal nerve when required. The systems being developed in Ljubljana[103] have focused more on the implantation of the stimulating electrodes, with placement of the electrode on the peroneal nerve to avoid issues with electrode placement and unwanted recruitment of sensory nerves in the skin. This system has also been expanded to stimulate other muscles in the leg to provide for standing and walking for individuals with sustained paraplegia.[104,105]

The third clinical system under development is the footdrop system under investigation at the Cleveland FES Center.[106,107] This system differs from many of the others discussed in that stimulation of the muscles to correct for foot drop is not achieved by stimulation of the peroneal nerve. Instead, percutaneous electrodes are used to individually excite muscles of the lower leg, as well as those around the knee and hip, to improve gait. In addition, studies with this system have examined the effect of electrical stimulation on motor relearning in the lower extremity. The results of this system are better than what has been achieved with the other systems in that stimulating the individual muscles instead of the peroneal nerve to elicit the withdrawal reflex eliminates concerns about accommodation to the stimulation over time.[108]

12.7.3 Organ system prostheses

Although the commercial applications of motor prostheses for the restoration of organ system functions have been limited at this time, there are numerous investigations ongoing into using motor prostheses for the restoration of these functions that should result in a variety of commercial systems in the future. One series of investigations is directed toward using motor prostheses for the treatment of upper airway disorders. Research at the National Institutes of Health[109–111] is directed towards the development of motor prostheses that can be used to stimulate the genioglossus nerve to open the hypopharynx to correct for sleep apnea. Sleep apnea is a condition in which the tongue muscles relax during sleep, resulting in restriction and blockage of the hypopharynx. Studies are also being conducted on stimulation of the thyroarytenoid muscles to prevent aspiration in dysphagia and on stimulation of the thyroarytenoid muscle to correct for spasmodic dystonia, a voice disorder that prevents effective communication.

Another area of investigation for the use of motor prostheses for the restoration of organ system function is in the area of cardiac assist devices. The cardiac assist device differs from the cardiac pacemaker in that skeletal muscle is used to repair or reinforce cardiac muscle. The skeletal muscle is conditioned to convert the fibers to a fatigue-resistant variety and is electrically stimulated to contract when the heart contracts to increase blood pressure and blood flow. The most common muscle used is the latissimus dorsi, which is removed from its origin on the back and hip, inserted into the chest, and wrapped around the heart. Investigations at the Milwaukee Heart Institute[112] have been directed toward the development of a motor prosthesis that can sense the contraction of the heart and then provide the stimulation of the latissimus dorsi to assist in the contraction. This device, called the *LD Pace II*, has undergone successful animal testing and has been implanted in one human subject to date.[112]

The use of motor prostheses for the restoration of respiration after spinal cord injury is another area of investigation. Numerous studies have investigated the possibility of electrically stimulating the phrenic nerve to elicit contraction of the diaphragm and restore breathing.[113–116] The difficulty in using these systems, however, are that individuals are reluctant to undergo surgery for a system that may damage the nerve. Research being conducted at Case Western Reserve University has addressed this concern by developing two different motor prosthetic systems.[117,118] The first system electrically stimulates the intercostal muscles using intramuscular electrodes to provide for respiration in the cases where the phrenic nerve is lost or damaged or where surgery is undesired by the individual. The second system also employs the use of intramuscular electrodes, but these are placed on the motor point of the diaphragm. By using these approaches, these research systems have been able to successfully address the problems of phrenic stimulators developed in the past. The diaphragm stimulator has already been successfully implanted in one human subject.[118]

A final area of investigation for the use of motor prostheses for restoration of biological functions is in the area of the restoration of bladder, bowel, and sexual functions. The objective of these studies has been the development of systems that restore all of these functions and accomplish this without the need for a dorsal rhizotomy. This involves primarily stimulation of the pudendal afferent pathways to prevent incontinence and to increase bladder volume. Studies at the University College London[119,120] and Case Western Reserve University[121] show promise in the use of this method to avoid the dorsal rhizotomy, with four individuals being implanted with the sacral posterior and anterior root stimulator implant (SPARSI) at the University College London. Recipients of the SPARSI system have been shown to have a significant reduction in reflex incontinence with stimulation of the second through fourth sacral roots[120] with accompanying increases in bladder volume.[120]

12.8 Future avenues of investigation

The previous sections of this chapter have attempted to present the clinical applications of motor prosthetics and the systems that have been or are being developed to address these needs. As can be seen from these previous sections, a great deal of work has been accomplished in the area of motor prosthetics; however, it is far from being complete. The clinical need to increase the user base of these systems not only warrants investigations into how to apply these systems to address central nervous system pathologies, but also investigations in how to improve the technology and functionality of these systems. This section will attempt to introduce where work is just starting or needs to be accomplished in order to address the clinical needs for these systems.

12.8.1 New clinical applications

The previous sections on the commercial and research motor prostheses may leave the reader with the false impression that motor prostheses are only effective for use in individuals with a spinal cord injury, with some minor applications to individuals recovering from stroke. The reason for this one-sided application of the motor prosthesis, however, lies in one of in the fundamental limitations of these systems. Current technologies for the system rely upon an intact motor neuron to the muscle in order to generate a muscle contraction. If the nerve is damaged or lost, then the usefulness of the current motor prosthesis technology for the treatment of central nervous system deficits is lost.

The limitation that an intact motor nerve must be present in order to generate a muscle contraction has limited the application of motor prosthesis technology in the areas of neuropathy, amyotrophic lateral sclerosis, and other diseases where the myelin sheath of the nerve has degenerated and thus the nerve is no longer capable of conducting an action potential or where the nerve fibers themselves are attacked by the pathogen and thus lost. However, in these conditions, there is a definite need for a device that can restore functional movement or biological function. Solutions to this problem may lie in the areas of nerve and cellular regeneration. Even if full nerve regeneration cannot be accomplished so that the connection between the brain and the muscle is restored, the ability to regenerate nerves or reinnervate muscle can be extremely beneficial. If a nerve can be restored to a muscle, even if no longer connected to higher nervous system functions, the motor prosthesis can be used to excite this nerve to cause the muscle to contract.

Motor prostheses can also be used, although not in a traditional sense, to aid in the area of neural regeneration in the spinal cord. It has been suggested[122–124] that the motor prosthesis, or more importantly the electrical currents generated by the motor prosthesis, can be used to guide axonal growth for regenerated nerves so that they make functional connections between the central and peripheral nervous systems. Motor prostheses could also perform an important role in maintaining muscle mass and strength while nerve regeneration was occurring so that, once the connection was completed, the individual would not have lost the ability to control the extremities in a functional manner, requiring months of therapy to rebuild muscle mass and tone.

Another possible role for motor prostheses, which has been suggested several times in this chapter, is in a role of reducing muscle spasticity that can occur in cases of sustained hemiplegia or cerebral palsy. The method of collision blocking, introduced earlier in this chapter, has been demonstrated to be theoretically possible to prevent action potential conduction. The goal with a motor prosthetic system, therefore, would be to make this theory a reality in a working clinical system. Such a system not only would have to prevent the conduction of unwanted action potentials but would also have to allow muscle contraction to occur when the user desired to generate a movement.

12.8.2 *Technology development*

The hypothetical clinical applications of the motor prosthesis, and even current ongoing research, will require further advances in the technology of the motor prosthesis in order to accomplish these goals. One area that will require attention will be on the development of implanted components. Although research is being directed at this time on non-implanted, non-invasive systems, future directions in motor prosthetic development will require the use of implanted technologies. This will be necessary in order to address the cosmesis and ease of use requirements for the system. A series of external devices that may draw unwanted attention to the user or that will require a long period to don and doff will most likely lead to rejection of the device. However, the increasing amount of technology that will be required to be implanted into the body will require a great deal of surgical time and increase the risk to the recipient of the system. Technologies will be required that allow for placement of the devices into the body quickly and easily.

The *BION system*, developed at the AE Mann Institute for Biomedical Engineering, is aimed at advancing the technology of the motor prosthesis by addressing these needs.[125-127] The key component of the system is the BION, which is a small, injectable, single-channel stimulator. This stimulator is encased in a ceramic package, with the power and the simulation parameters being provided to the BION by a RF link. The objective of this system is to remove much of the implanted technology and lead wires associated with the other systems and instead have all communication be conducted by the RF link. This system has been reported to be capable of providing up to 255 independent channels of stimulation and has already been applied for uses in shoulder subluxation.[127] Shoulder subluxation is a condition common in hemiplegia where the paralysis of the rotator cuff muscles results in partial dislocation of the humeral head, resulting in pain and an inability to use the shoulder effectively. This system has also been applied to strengthening of the quadriceps muscle in individuals with osteoarthritis, although tests are still in the early clinical stage[127] and with stimulating the pudendal nerve to treat overactive bladder.[127]

The demands on the development of implanted technologies for motor prosthesis user will also be complicated by the fact that these systems will be required to restore greater and greater function. This places demands on the amount of hardware that is implanted and raises issues on how these different components of the system will communicate. For example, if the technology used by the Freehand or Praxis system is developed further to allow for up to 32 or 64 channels of stimulation, this will result in numerous lead wires that will need to be tunneled under the skin, with all of them originating from a central processor, an arrangement that will be highly undesirable to the physician who has to implant such a system. Likewise, external systems will have multiple lead wires running from various areas on the individual to a central processor, which will most like result in a lack of cosmetic appeal. These issues will most likely be addressed through the development of modular, implanted systems that will be able to communicate with a limited need for wires or antennas.

The increasing complexity of these newer systems, with more functions and more components, will place greater demands upon the power supply to these systems. Although battery technology is developing due to advances in telecommunications and computer science, these batteries are not designed with biological issues in mind. Most are not biocompatable, nor do they meet the voltage and current requirements to allow for muscle stimulation for periods of months to years. Battery technology will need to be directed toward the specific goals of biocompatibility and the ability to replace the battery easily in the implanted system or to allow for quick and efficient recharging. One possibility for the development of a power source for use in the human body are the developments in the area of biofuel cells.[128] These cells will make use of the body's own chemistry and structure, with the introduction of some components, to provide the power to drive these systems. Research in this area is evolving and realization of these systems *in vivo* has yet to be accomplished.

The increasing complexity of newer motor prosthetic systems will also require increasing complexity in the command interface. Information requirements for the system will increase, with subsequent decreases in the number of input channels available. Generating these high information content signals will also have to be accomplished while at the same time ensuring that the interface is easy and understandable to use. The most likely avenue of investigation for command interfaces that will be able to meet the needs of the complex system will be to derive command signals from the cortex. Numerous investigators are examining the feasibility of extracting control information from the motor and pre-motor areas of the brain.[129–131] These studies have resulted in the control over a robotic arm in primate studies. The animal in this case moves their arm and the robot mirrors this movement. These studies are instrumental in demonstrating that by recording from just a few neurons in the motor and pre-motor areas, it is possible to reconstruct arm movement. Areas of future investigation with these studies will need to address the ability to think of movement and have a system respond instead of having the system mirror movement. Issues of biocompatibility and human implantation need to be explored, as do ways of allowing cortical signals to control not only functional movement but system state (e.g., turning the system on/off, selection of exercise modes), as well.

12.9   Summary and conclusion

The objective of this chapter was to provide the reader with a sufficient background in the area of motor prosthetics in order to allow them to begin to conduct their own research in this area. To this end, a review of past motor prostheses was presented, as well as an overview of the technology that has been employed to date. As can be seen from this review, motor prostheses have achieved a great deal of clinical success; however, they are still underutilized by rehabilitation practitioners. In order for motor prosthetics to continue to develop as a field, it is necessary for research to continue into expanding the clinical applications of these devices, to continue to develop the technology so that it becomes an integral part of the individual for which the system is designed, and to educate individuals as to what has been developed and what is currently in the research stages so that these devices can see greater clinical use.

References

1. Heimer, L., *The Human Brain and Spinal Cord*, 2nd ed., Springer-Verlag, New York, 1995, chap. 24.
2. Peckham, P.H., Mortimer, J.T., and Marsolais, E.B., Alteration in the force and fatigability of skeletal muscle in quadriplegic humans following exercise induced by chronic electrical stimulation, *Clin. Orthop.*, 114, 326, 1976.
3. Creasey, G., Restoration of bladder, bowel, and sexual function, *Top. Spinal Cord Inj. Rehabil.*, 5, 21, 1999.
4. Kalra, L., The influence of stroke unit rehabilitation on functional recovery from stroke, *Stroke*, 25, 821, 1994.
5. Chae, J. et al., Neuromuscular stimulation for upper extremity motor and functional recovery in acute hemiplegia, *Stroke*, 29, 975, 1998.
6. McColl, M. and Bickenbach, J., *Introduction to Disability*, W.B. Saunders, London, 1998.
7. Dzienkowski, R. et al., Cerebral palsy: a compressive review, *Nurse Practitioner*, 45, 1996.
8. Triolo, R., Kobetic, R., and Betz, R., Standing and walking with FNS: technical and clinical challenges, in *Human Motion Analysis: Current Applications and Future Directions*, Harris, G. and Smith P., Eds., IEEE Press, New York, 1996, p. 318.

9. Benton, L.A. et al., *Functional Electrical Stimulation: A Practical Guide,* 2nd ed., Rancho Los Amigos Medical Center, California, 1981.

10. Edelberg, R., Electrical properties of the skin, in *Biophysical Properties of the Skin,* Elden, H.R., Ed., Wiley, New York, 1971, p. 513.

11. Memberg, W.D. et al., An analysis of the reliability of percutaneous intramuscular electrodes in upper extremity FNS applications, *IEEE Trans. Rehabil. Eng.,* 1, 126, 1993.

12. Scheiner, A., Polando, G., and Marsolais, E., Design and clinical application of a double helix electrode for functional electrical stimulation, *IEEE Trans. Biomed. Eng.,* 41, 425, 1984.

13. Gradjean, P.A. and Mortimer, J.T., Recruitment properties of monopolar and bipolar epimysial electrodes, *Ann. Biomed. Eng.,* 14, 53, 1986.

14. Memberg, W., Peckham, P., and Keith, M., A surgically implanted intramuscular electrode for an implantable neuromuscular stimulation system, *IEEE Trans. Rehabil. Eng.,* 2, 80, 1994.

15. Naples, G. et al. A spiral nerve cuff electrode for peripheral nerve stimulation, *IEEE Trans. Biomed. Eng.,* 35, 905, 1988.

16. McNeal, D.R. and Bowman, B.R., Selective activation of muscles using peripheral nerve electrodes, *Med. Biol. Eng. Comput.,* 23, 249, 1985.

17. Barone, F.C. et al., A bipolar electrode for peripheral nerve stimulation, *Brain Res. Bull.,* 4, 421, 1979.

18. Veraart, C., Grill, W.M., and Mortimer, J.T., Selective control of muscle activation with a multipolar nerve cuff electrode, *IEEE Trans. Biomed. Eng.,* 40, 640, 1993.

19. Scott, T., Peckham, P.H., and Kilgore, K., Tri-state myoelectric control of bilateral upper extremity neuroprostheses for tetraplegic individuals, *IEEE Trans. Rehabil. Eng.,* 4, 251, 1996.

20. Peckham, P., Mortimer, J., and Marsolais, E., Controlled prehension and release in the C5 quadriplegic elicited by functional electrical stimulation of the paralyzed forearm musculature, *Ann. Biomed. Eng.,* 8, 369, 1980.

21. Saxena, A., Nikolic, S., and Popovic, D., An EMG-controlled grasping system for tetraplegics, *J. Rehabil. Res. Dev.,* 32, 17, 1995.

22. Hart, R., Kilgore, K., Peckham, P.H., A comparison between control methods for implanted FES hand-grasp systems, *IEEE Trans. Rehabil. Eng.,* 6, 1, 1998.

23. Johnson, M. and Peckham, P.H., Evaluation of shoulder movement as a command control source, *IEEE Trans. Biomed. Eng.,* 37, 876, 1990.

24. Johnson, M. et al., Implantable transducer for two-degree-of-freedom joint angle sensing, *IEEE Trans. Rehabil. Eng.,* 7, 349, 1999.

25. Nathan, R., Control strategies in FNS systems for the upper extremities, *Crit. Rev. Biomed. Eng.,* 21, 485, 1993.

26. Graupe, D. and Kohn, K., Transcutaneous functional electrical stimulation of certain traumatic complete thoracic paraplegics for independent short-distance ambulation, *Neurol. Res.,* 19, 323, 1997.

27. Hoshimiya, N. et al., A multichannel FES system for the restoration of motor functions in high spinal cord injury patients: a respiration-controlled system for multijoint upper extremity, *IEEE Trans. Biomed. Eng.,* 36, 754, 1989.

28. Handa, Y. et al., A voice controlled functional electrical stimulation system for the paralyzed hand, *Jpn. J. Med. Electron. Biol. Eng.,* 23, 292, 1985.

29. Handa, Y. and Hoshimiya, N., Functional electrical stimulation for the control of the upper extremities, *Med. Prog. Technol,* 12, 51, 1987.

30. Nathan, R. and Ohry, A., Upper limb functions regained in quadriplegia: a hybrid computerized FNS system, *Arch. Phys. Med. Rehabil.*, 71, 415, 1990.

31. Lauer, R., Peckham, P.H., and Kilgore, K., EEG-based control of a hand grasp neuroprosthesis, *NeuroReport*, 8, 1767, 1999.

32. Scott, T., Peckham, P.H., and Keith, M., Upper extremity neuroprostheses using functional electrical stimulation, in *Clinical Neurology: International Practice and Research*, Brindley, G. and Rushton, D., Eds., Balliere Tindall, London, 1995.

33. Lauer, R.T. et al., Applications of cortical signals to neuroprosthetic control: a critical review, *IEEE Trans. Rehabil. Eng.*, 8, 205, 2000.

34. Mortimer, J. et al., Shoulder position transduction for proportional two axis control of orthotic/prosthetic system, in *The Control of Upper-Extremity Prostheses and Orthoses*, Herberts, P. et al., Eds., Charles C Thomas, Springfield, IL, 1974.

35. Shannon, C. and Weaver, W., *The Mathematical Theory of Communication*, The University of Illinois Press, Urbana, 1949.

36. Schwartz, M., *Information Transmission, Modulation and Noise*, McGraw-Hill, New York, 1980.

37. Schmidt, E. et al., Fine control of operantly conditioned firing patterns of cortical neurons, *Exper. Neurol.*, 61, 349, 1978.

38. Meek, S. and Fetherston, S., Comparison of signal-to-noise ratio of myoelectric filters for prosthesis control, *J. Rehabil. Res. Dev.*, 29, 9, 1992.

39. Welford, A., *Fundamentals of Skill*, Butler & Tanner, London, 1971.

40. Doubler, J. and Childress, D., An analysis of extended physiological proprioception as a prosthesis-control technique, *J. Rehabil. Res. Dev.*, 21, 5, 1984.

41. Doubler, J. and Childress, D., Design and evaluation of a prosthetic control system based on the concept of extended physiological proprioception, *J. Rehabil. Res. Dev.*, 21, 19, 1984.

42. Guyton, A., *Textbook of Medical Physiology*, 8th ed., W.B. Saunders, Philadelphia, 1991, chap. 22.

43. Neuman, M.R., Therapeutic and prosthetic devices, in *Medical Instrumentation: Application and Design*, 2nd ed., Webster, J., Ed., Houghton-Mifflin, Boston, 1992, chap. 13.

44. Roy, O.Z., The current status of cardiac pacing, *CRC Crit. Rev. Bioeng.*, 2, 259, 1975.

45. Harthorne, J.W., Pacemaker leads, *Int. J. Cardiol.*, 6, 423, 1984.

46. Furman, S. and Escher, D., *Principles and Techniques of Cardiac Pacing*, Harper & Row, New York, 1970.

47. Liberson, W.T. et al., Functional electrotherapy: stimulation of the peroneal nerve synchronized with the swing phase of the gait of hemiplegic patients, *Arch. Phys. Med. Rehabil.*, 42, 101, 1961.

48. Long, C. and Masciarelli, V., An electrophysiological splint for the hand, *Arch. Phys. Med. Rehabil.*, 44, 449, 1963.

49. Mortimer, J., Motor prostheses, in *Handbook of Physiology: The Nervous System II*, Brookhart, J.M. and Mountcastle, V.B., Eds., American Physiological Society, Bethesda, MD, 1981, p. 155.

50. Akers, J.M. et al., Tissue response to chronically stimulated implanted epimysial and intramuscular electrodes, *IEEE Trans. Rehabil. Eng.*, 5, 207, 1997.

51. van den Honert, C. and Mortimer, J.T., Generation of unidirectionally prop-agated action potentials in a peripheral nerve by brief stimuli, *Science*, 206, 1311, 1979.

52. Chae, J. et al., Functional neuromuscular stimulation, in *Rehabilitation Medicine: Principles and Practice*, 3rd ed., DeLisa, J.A. and Gans, B.M., Eds., Lippincott-Raven, Philadelpiha, 1998.

53. Zafar, M. and Van Doren, C., Effectiveness of supplemental grasp-force feedback in the presence of vision, *Med. Biol Eng. Comput.*, 38, 267, 2000.

54. Johnson, K.O. et al., Perspectives on the role of afferent signals in control of motor neuroprostheses, *Med. Eng. Phys.*, 17, 481, 1995.

55. Bhadra, N., Kilgore, K.L., and Peckham, P.H., Implanted stimulators for restoration of function in spinal cord injury, *Med. Eng. Phys.*, 23, 19, 2001.

56. Inmann, A. et al., Signals from skin mechanoreceptors used in control of a hand grasp neuroprosthesis, *NeuroReport*, 12, 2817, 2001.

57. Haugland, M. and Sinkjaer, T., Interfacing the body's own sensing receptors into neural prosthesis devices, *Technol. Health Care*, 7, 393, 1999.

58. Cameron, T. et al., The effect of wrist angle on electrically evoked hand opening in patients with spastic hemiplegia, *IEEE Trans. Rehabil. Eng.*, 7, 109, 1999.

59. Prochazka, A. et al., The bionic glove: an electrical stimulator garment that provides controlled grasp and hand opening in quadriplegia, *Arch. Phys. Med. Rehabil.*, 78, 608, 1997.

60. Weingarden, H.P. et al., Hybrid functional electrical stimulation orthosis system for the upper limb: effects on spasticity in chronic stable hemiplegia, *Am. J. Phys. Med. Rehabil.*, 77, 276, 1998.

61. Weingarden, H.P. et al., Upper limb functional electrical stimulation for walker ambulation in hemiplegia: a case report, *Am. J. Phys. Med. Rehabil.*, 76, 63, 1997.

62. Triolo, R. et al., Challenges to clinical deployment of upper limb neuroprostheses, *J. Rehabil. Res. Dev.*, 33, 111, 1996.

63. Kraft, G.H., Fitts, S.S., and Hammond, M.C., Techniques to improve function of the arm and hand in chronic hemiplegia, *Arch. Phys. Med. Rehabil.*, 73, 220, 1992.

64. Peckham, P.H. et al., Efficacy of an implanted neuroprosthesis for restoring hand grasp in tetraplegia: a multicenter study, *Arch. Phys. Med. Rehabil.*, 82, 1380, 2001.

65. Stroh-Wuolle, K. et al., Satisfaction with and usage of a hand neuroprosthesis, *Arch. Phys. Med. Rehabil.*, 80, 206, 1999.

66. Smith, B. et al., An externally powered, multichannel, implantable stimulator-telemeter for control of paralyzed muscle, *IEEE Trans. Biomed. Eng.*, 45, 463, 1998.

67. Kilgore, K.L. et al., An implanted upper-extremity neuroprosthesis. Follow-up of five patients, *J. Bone Joint Surg. Am.*, 79, 533, 1997.

68. Wijman, C.A. et al., Functional evaluation of quadriplegic patients using a hand neuroprosthesis, *Arch. Phys. Med. Rehabil.*, 71, 1053, 1990.

69. Stein, R.B., Functional electrical stimulation after spinal cord injury, *J. Neurotrauma*, 16, 713, 1999.

70. Wieler, M. et al., Multicenter evaluation of electrical stimulation systems for walking, *Arch. Phys. Med. Rehabil.*, 80, 495, 1999.

71. Taylor, P. N et al., Clinical use of the Odstock dropped foot stimulator: its effect on the speed and effort of walking, *Phys. Med. Rehabil.*, 80, 1577, 1999.

72. Taylor, P.N. et al., Patients' perceptions of the Odstock dropped foot stimulator (ODFS), *Clin. Rehabil.*, 13, 439, 1999.

73. Kralj, A., Acimovic, R., and Stanic, U., Enhancement of hemiplegic patient rehabilitation by means of functional electrical stimulation, *Prosthet. Orthot. Int.*, 17, 107, 1993.

74. Kralj, A. and Bajd, T., *Functional Electrical Stimulation: Standing and Walking after Spinal Cord Injury*, CRC Press, Boca Raton, FL, 1989.

75. Grill, W.M. and Kirsch, R., Neuroprosthetic Applications of electrical stimulation, *Asst. Technol.*, 12, 6, 2000.

76. Graupe, D. et al., Ambulation by traumatic T4–12 paraplegics using functional neuromuscular stimulation, *CRC Crit. Rev. Neurosurg.*, 8, 221, 1998.

77. Graupe, D. and Kohn, K.H., Functional neuromuscular stimulator for short-distance ambulation by certain thoracic-level spinal-cord-injured paraplegics, *Surg. Neurol.*, 50, 202, 1998.

78. Popovic, M.R. et al., Functional electrical stimulation for grasping and walking: indications and limitations, *Spinal Cord*, 39, 403, 2001.

79. Creasey, G.H. et al., An implantable neuroprosthesis for restoring bladder and bowel control to patients with spinal cord injuries: a multicenter trial, *Arch. Phys. Med. Rehabil.*, 82, 1512, 2001.

80. Brindley, G.S., The first 500 patients with sacral anterior root stimulator implants: general description, *Paraplegia*, 32, 795, 1994.

81. Linder, S.H., Functional electrical stimulation to enhance cough in quadriplegia, *Chest*, 103, 166, 1993.

82. Bryden, A.M., Memberg, W.D., and Crago, P.E., Electrically stimulated elbow extension in persons with C5/C6 tetraplegia: a functional and evaluation, *Phys. Med. Rehabil.*, 81, 80, 2000.

83. Lauer, R.T. et al., The function of the finger intrinsic muscles in response to electrical stimulation, *IEEE Trans. Rehabil. Eng.*, 7, 19, 1999.

84. Kameyama, J. et al., Electromyographic study relating to shoulder motion: control of shoulder joint by functional electrical stimulation, *Tohoku J. Exp. Med.*, 187, 339, 1999.

85. Kameyama, J. et al., Restoration of shoulder movement in quadriplegic and hemiplegic patients by functional electrical stimulation using percutaneous multiple electrodes, *Tohoku J. Exp. Med.*, 187, 329, 1999.

86. Handa, Y., Yagi, R., and Hoshimiya, N., Application of functional electrical stimulation to the paralyzed extremities, *Neurol. Med. Chir. (Tokyo)*, 38, 784, 1998.

87. Ichie, M. et al., Control of thumb movements: EMG analysis of the thumb and its application to functional electrical stimulation for a paralyzed hand, *Frontiers Med. Biol. Eng.*, 6, 291, 1995.

88. Handa, Y. et al., Functional electrical stimulation (FES) systems for restoration of motor function of paralyzed muscles — versatile systems and a portable system, *Frontiers Med. Biol Eng.*, 4, 241, 1992.

89. Popovic, M.R. et al., Surface-stimulation technology for grasping and walking neuroprosthesis, *IEEE Eng. Med. Biol. Mag.*, 20, 82, 2001.

90. Mackenzie-Knapp, M., Electrical stimulation in early stroke rehabilitation of the upper limb with inattention, *Aust. J. Physiother.*, 45, 223, 1999.

91. Chae, J. and Yu, D., A critical review of neuromuscular electrical stimulation for treatment of motor dysfunction in hemiplegia, *Assist. Technol.*, 12, 33, 2000.

92. Dimitrijevic, M. and Soroker, N., Mesh-glove. 2: Modulation of residual upper limb control after stroke with whole hand electrical stimulation, *Scand. J. Rehabil. Med.*, 26, 187, 1994.

93. Francisco, G. et al., Electromyogram-triggered neuromuscular stimulation for improving the arm function of acute stroke survivors: a randomized pilot study, *Arch. Phys. Med. Rehabil.*, 79, 549, 1996.

94. Carmick, J., Use of neuromuscular electrical stimulation and dorsal wrist splint to improve the hand function of a child with spastic hemiparesis, *Phys. Ther.*, 77, 661, 1997.

95. Wood, D.E. et al., Is paraplegic standing by root stimulation a practical option? Conclusions from the LARSI project, in *Proc. 6th Annual Conference of the IFESS*, Cleveland, OH, 2001, p. 13.

96. Davis, R., MacFarland, W., and Emmons, S., Initial results of the Nucleus FES-22 implanted stimulator for limb movement in paraplegia, *Stereotact. Funct. Neurosurg.*, 63, 192, 1994.

97. Davis, R., Patrick, J., and Barriskill, A., Development of functional electrical stimulators utilizing cochlear implant technology, *Med. Eng. Phys.*, 23, 61, 2001.

98. Guiraud, D. et al., One year implanted patients follow up: SUAW project first results, in *Proc. 6th Annual Conference of the IFESS*, Cleveland, OH, 2001, p. 55.

99. Davis, J.A. et al., Preliminary performance of a surgically implanted neuro-prosthesis for standing and transfers — where do we stand?, *J. Rehabil. Res. Dev.*, 38, 609, 2001.

100. Davis, J.A. et al., Surgical technique for installing an eight-channel neuropros-thesis for standing, *Clin. Orthop.*, 385, 237, 2001.

101. Uhlir, J.P., Triolo, R.J., and Kobetic, R., The use of selective electrical stimula-tion of the quadriceps to improve standing function in paraplegia, *IEEE Trans. Rehabil. Eng.*, 8, 514, 2000.

102. Haugland, M. and Sinkjaer, T., Interfacing the body's own sensing receptors into neural prosthesis devices, *Technol. Health Care*, 7, 393, 1999.

103. Bosnjak, R., Dolenc, V., and Kralj, A., Biomechanical response in the ankle to stimulation of lumbosacral nerve roots with spiral cuff multielectrode — preliminary study, *Neurol. Med. Chir. (Tokyo)*, 39, 659, 1999.

104. Bajd, T. et al., Use of functional electrical stimulation in the lower extremities of incomplete spinal cord injured patients, *Artif. Organs*, 23, 403, 1999.

105. Kralj, A. et al., FES gait restoration and balance control in spinal cord-injured patients, *Prog. Brain Res.*, 97, 387, 1993.

106. Daly, J.J. et al., Feasibility of gait training for acute stroke patients using FNS with implanted electrodes, *J. Neurol. Sci.*, 179, 103, 2000.

107. Daly, J.J. and Ruff, R.L., Electrically induced recovery of gait components for older patients with chronic stroke, *Am. J. Phys. Med. Rehabil.*, 79, 349, 2000.

108. Pearson, J.A. and Wenkstern, B., Habituation and sensitization of the flexor withdrawal reflex, *Brain Res.*, 43, 107, 1972.

109. Grill, W.M. et al., Emerging clinical applications of electrical stimulation: opportunities for restoration of function, *J. Rehabil. Res. Dev.*, 38, 641, 2001.

110. Bidus, K.A., Thomas, G.R., and Ludlow, C.L., Effects of adductor muscle stimulation on speech in abductor spasmodic dysphonia, *Laryngoscope*, 110, 1943, 2000.

111. Barkmeier, J.M. et al., Modulation of laryngeal responses to superior laryngeal nerve stimulation by volitional swallowing on awake humans, *J. Neurophysiol.*, 83, 1264, 2000.

112. Curbelo, R. et al., LD Pace II, an easily programmable device for cardiomyoplasty, *Med. Eng. Phys.*, 23, 45, 2001.

113. Hogan, J.F., Koda, H., and Glenn, W.W., Electrical techniques for stimulation of the phrenic nerve to pace the diaphragm: inductive coupling and battery powered total implant in asynchronous and demand modes, *Pacing Clin. Electrophysiol.*, 12, 847, 1989.

114. Glenn, W.W. et al., Fundamental considerations in pacing of the diaphragm for chronic ventilatory insufficiency: a multi-center study, *Pacing Clin. Electrophysiol.*, 11, 2121, 1988.

115. Sauermann, S. et al., Computer aided adjustment of the phrenic pacemaker: automatic functions, documentation, and quality control, *Artif. Organs*, 21, 216, 1997.

116. Girsch, W. et al., Vienna phrenic pacemaker — experience with diaphragm pacing in children, *Eur. J. Pediatr. Surg.*, 6, 140, 1996.

117. DiMarco, A.F. et al., Evaluation of intercostal pacing to provide artificial ventilation in quadriplegics, *Am. J. Respir. Crit. Care Med.*, 163, A151, 2001.

118. DiMarco, A.F., Neural prostheses in the respiratory system, *J. Rehabil. Res. Dev.*, 38, 601, 2001.

119. Kirkham, A.P. et al., The acute effects of continuous and conditional neuromodulation on the bladder in spinal cord injury, *Spinal Cord*, 39, 420, 2001.

120. Craggs, M.D. et al., SPARSI: an implant to empty the bladder and control incontinence without posterior rhizotomy in spinal cord injury, *Brit. J. Urol. Int.*, 85, 2, 2000.

121. Grill, W.M., Bhadra, N., and Wang, B., Bladder and urethral pressures evoked by microstimulation of the sacral spinal cord in cats, *Brain Res.*, 836, 19, 1999.

122. Grill, W.M. et al., At the interface: convergence of neural regeneration and neural prostheses for restoration of function, *J. Rehabil. Res. Dev.*, 38, 633, 2001.

123. Borgens, R.B., Electrically mediated regeneration and guidance of adult mammalian spinal axons into polymeric channels, *Neuroscience*, 91, 251, 1999.

124. Borgens, R.B., Blight, A.R., and McGinnis, M.E., Functional recovery after spinal cord hemisection in guinea pigs: the effects of applied electrical fields, *J. Comp. Neurol.*, 296, 634, 1990.

125. Loeb, G.E. et al., BION system for distributed neural prosthetic interfaces, *Med. Eng. Phys.*, 23, 9, 2001.

126. Cameron, T. et al., Micromodular implants to provide electrical stimulation of paralyzed muscles and limbs, *IEEE Trans. Biomed. Eng.*, 44, 781, 1997.

127. Dupont, A. et al., Clinical trials of BION injectible neuromuscular stimulators, in *Proc. 6th Annual Conference of the IFESS*, Cleveland, OH, 2001, p. 7.

128. Chen, T. et al., A miniature biofuel cell, *Am. Chem. Soc.*, 123, 8630, 2001.

129. Chapin, J. et al., Real-time control of a robot arm using simultaneously recorded neurons in the motor cortex, *Nat. Neurosci.*, 2, 664, 1999.

130. Maynard, E. et al., Neuronal interactions improve cortical population coding of movement direction, *J. Neurosci.*, 19, 8083, 1999.

131. Schwartz, A., Motor cortical activity during drawing movements: population representation during sinusoid tracing, *J. Neurophysiol.*, 70, 28, 1992.

section six

Emerging technologies

chapter thirteen

Neurotechnology: microelectronics

Danny Banks

Contents

0-8493-1100-4/03/$0.00+$1.50
© 2003 by CRC Press LLC

13.1 Introduction

In recent decades, the development of neural prostheses has benefited
greatly from technological advances, particularly in the area of greater
miniaturization and integration of microelectronics. The development of
glass microelectrodes early in the 20th century and the later development
of metal wire microelectrodes in the 1950s[1] gave the neurophysiologist new
insights into the operation of the nervous system. As early as the 1960s, it
was being proposed that integrated circuit (IC) fabrication techniques
could be employed to develop improved devices. By the early 1970s,
devices had been fabricated and a design had been published for an
implantable monolithic wafer electrode.[2,3] In 1975, Wise and Angell published an improved version[4] of their 1970 design, which included transistor
buffer amplifiers.

Since this time, microelectromechanical systems (MEMS) techniques
have improved markedly and have been used to develop a variety of different microelectrode structures. These techniques are reviewed in other
chapters of this volume. The problems relating to the integration of microelectronic circuitry onto these MEMS structures for neuroprosthetic applications are often regarded as problems that will be overcome with the continual
reduction of feature size (driven by demand from the microelectronics industry) and complementary metal-oxide semiconductor (CMOS)-compatible
MEMS processes (driven by demand from the MEMS industry). Nonetheless,
the integration of MEMS technologies with microelectronics for neuroprosthetic applications presents particular problems. It is the intention of this
chapter to present the reader with an overview of the available microelectronics technologies and approaches that have been used to integrate these
with MEMS-based microelectrode structures. This is done within the context
of some generic "MEMS neuroprosthetic systems," outlined below. First,
however, it is necessary to make a case for increasing integration in neuroprosthetic applications.

13.1.1 The need for integration

The use of MEMS-based microelectrode structures offers clear advantages. They allow more precise stimulation with lower stimulation current requirements than would otherwise be obtainable with larger devices; this was recognized at an early stage in attempts to develop a prosthesis that stimulated the visual cortex.[5] Large numbers of electrode sites provide for redundancy and the possibility to compensate for drift of individual neurons in and out of range of the recording or stimulation electrode.[6,7] Miniaturization also facilitates the placement of large numbers of electrode sites in small volumes of relatively inaccessible tissue, which provides advantages over other approaches.[8]

In and of itself, miniaturization of the electrode structure alone is not an ideal solution. Neuroprostheses have suffered, and may still suffer, from reliability problems when a large amount of wiring is used to interconnect the many different components involved: cuff electrodes, controller, etc. By introducing an even larger number of electrodes, through miniaturization, these problems are compounded. Additionally, the MEMS structure itself is at a disadvantage when connected by a large number of cables. MEMS-based microelectrodes have often been designed to be floating structures, so that they are able to move with the nervous system. Attaching wires to the device has a tethering effect. Not only are the connections liable to break through mechanical stress, but the device is now more likely to move relative to the neurons that it is stimulating/recording from, causing unwanted tissue damage. This is of great concern in retinal prostheses where the mechanical properties of the retina have been likened to those of wet tissue paper.

Furthermore, the use of MEMS-based devices for recording in neuropysiological investigations where implantation and mobility of the specimen have not been of such concern[9] has revealed an additional problem — that of dealing with the sheer volume of data generated by systems incorporating large numbers of microelectrode recording sites capable of recording details of individual action potential spikes. This is of particular concern in implanted neuroprosthetic systems where real-time control is imperative yet computing power, as well as electrical power to supply the computing element, is limited.

Thus, the need for intelligent MEMS-based microelectrode devices has been established that offer reduction or elimination of interconnecting leads and distribution of computing power. It then becomes necessary to consider the approach to be taken to achieve this desirable goal. The details of this are considered below, but the researcher/designer is faced with one initial decision to make: whether or not to select a *monolithic* approach, in which the microelectronics is integrated into the same substrate from which the microelectrodes are machined, or a *hybrid* approach, where the electrodes and microelectronic circuitry are fabricated using different and often incompatible technologies, and then later assembled into one unit.

The hybrid approach enables the designer to select the most appropriate technology for each component. The Fraunhöfer Institute has, for example, achieved some remarkable results with regeneration electrode devices fabricated on polymer substrates.[10,11] However, technology for fabricating electronic components using polymers is in its infancy. Therefore, signal processing circuitry is fabricated on a silicon chip that is bonded to the polymeric electrode structure.[12]

There are, however, several disadvantages to the hybrid approach. The first is that of cost. Once the batch production regime of microfabrication has been abandoned and components have to be handled individually, costs increase dramatically. For research purposes, however, this is not of such a great concern as the economies of scale are not there, and one will often wish to examine different combinations of components as potential solutions to a problem. Of greater concern is that of reliability. Each additional component that must be assembled into a system increases the chance of system failure, and each additional bonded joint is a potential point of failure.

The disadvantage of the monolithic approach is that a compromise technology has to be developed, and this may not necessarily lead to the best possible system performance. Monolithic solutions will generally take longer to develop than hybrid solutions, and the initial expenditure on research will be greater; they will generally lag hybrid approaches in performance but should eventually yield less expensive and more reliable solutions.

It is now possible to summarize the reasons for, goals, and requirements of greater integration of microelectronics with MEMS-based electrode structures:

- Reduce or eliminate interconnecting leads.
- Distribute intelligence.
- Minimize power consumption.
- Minimize device area (volume).
- Improve reliability and longevity of implants.

Device area is usually referred to rather than volume because microelectronic circuits are usually fabricated on one side of a silicon wafer and have negligible height in comparison to the other two dimensions. Power consumption must be minimized not only from the point of view of the difficulty of transmitting electrical power across the skin, but also from the point of view of power dissipated from the implanted device into the surrounding tissue. Najafi and Wise[13,14] provide an example of this consideration.

13.1.2 *Noise sources in recording*

The neuroprosthesis will consist of either recording electrodes or stimulation electrodes, or both. When considering stimulation, the circuit design problems that arise are concerned with selecting the electrode combination to be used, efficiently delivering a stimulating current, controlling the shape and charge balance of the stimulating pulse, and ensuring that the device has sufficient power to perform the required stimulation. These circuit design considerations are dealt with in further detail below. Recording differs from stimulation in that, while the latter involves the delivery of relatively large signals into a resistive/reactive load, the former involves trying to detect relatively small voltage signals through a large source impedance. Before discussing the circuitry that has been proposed to solve the problems that this gives rise to, it is useful to lay out the problems to be solved by the recording circuitry in more detail.

The noise sources encountered when recording can be classified as those intrinsic to the electrode design, over which the designer of the microelectronic circuitry has no control, and those over which the designer of the microelectronic circuitry has control. In the first camp, one has the source impedance of the electrode-tissue interface and the spreading resistance of the microelectrode recording site. In his analysis of noise sources related to electrode design, Edell[15] concluded that the spreading resistance of the electrode site was what would provide a physical limit to recording site size and hence selectivity. One other noise source over which the designer of the microelectronics has no control is the slow base-line drift, or even the half-cell potential, of the recording-site tissue interface.[13] Those aspects over which the designer of the microelectronic circuitry has considerable control are those relating to noise in the signal-processing circuitry itself, electromagnetic pick-up (or interference [EMI]), and stimulation artifacts. One must also consider the problem of bias currents passing across the microelectrode–tissue interface.

Consideration of signal processing, or amplifier noise, and EMI provides additional support for integrating amplifiers as close as possible to the microelectrode recording sites. EM pick-up is caused by induction of currents from an external source of EMI (e.g., radiofrequency sources, although mains electricity supply lines are also problematic). The normal solution to this is to employ a differential amplifier with a high common-mode rejection ratio (CMRR); this is a measure of how well the amplifier responds when the same signal appears on both inputs (for a differential amplifier, the output should be zero) and is defined as the ratio of the gain in difference mode to the gain in common mode (the same signal applied to both inputs; see Bogart[24]). This, of course, requires that the signals induced in the two leads of the amplifier by the same EMI source be as similar as possible. EMI is partially related to the area of any circuit loops into which currents are induced, so normally an effort is made to minimize the area of any loops in the circuit; however, this is usually more applicable to printed circuit board (PCB) design than integrated circuit (IC) design.

Capacitive coupling between the tracks that connect the recording sites to the inputs of the amplifiers is also a potential source of interference (cross-talk), where a large signal on one recording site appears as an artifact in the record of an adjacent site. Najafi et al.[16] have analyzed how this affects electrode design and scaling for silicon-based devices, concluding that cross-talk should not be expected to be a limiting factor for purely recording devices, and data are presented to assist in the design of mixed recording/stimulating devices.

The importance of having the highest possible gain on the first amplification stage, situated as electrically close as possible to the recording site, should also be mentioned when discussing noise sources. Consider two amplification stages, the first with gain A_1 and the second with gain A_2. The total voltage noise due to the first amplifier stage (referred to input) will be V_{N1}, and the noise referred to input of the second stage will be V_{N2}. Ideally, V_{N1} will be insignificant compared to the voltage noise level present in the signal, but this may not always be achievable. We consider these signals to be random processes and thus sum their squares. Thus, the total noise contribution arising from the amplifiers will be $V_N = \sqrt{A_1^2 A_2^2 V_{N1}^2 + A_2^2 V_{N2}^2}$.

From this it is possible to see that any gain above unity provided by A_1 will tend to make the V_{N1} term dominant, thus the noise arising from subsequent circuitry may be neglected, although the designer should not take this as an allowance for bad design of subsequent analog stages. Note, for instance, that noise appearing on the power rails, or through other mechanisms, due to poorly designed analog or digital stages can feed back and degrade the performance of the initial amplification stage.

The question of interference from unwanted bioelectric sources, such as EMG, straddles the divide between microelectronics design and microelectrode design. The use of insulating nerve cuffs is generally effective in excluding or significantly reducing interference from extraneural sources. However, their effect of amplifying signals within the nerve trunk being recorded from may well counteract a primary reason for using MEMS-based microelectrode devices: increased recording selectivity.

Edell's analysis has elegantly demonstrated that by recording differentially between two adjacent sites on the device, utilizing the common-mode rejection characteristics of the amplifier, the magnitude of the signal drops off with the square of distance when the distance to the source becomes much greater than the distance separating the two recording sites.[15] When the source is closer to the pair of recording sites, the magnitude of the signal would be inversely proportional to distance (approximately). This suggests that differential recording between adjacent sites could be effectively employed to reduce noise and increase specificity of recording. Unfortunately, a simple differential configuration also has a "dead spot" directly between the two recording sites. On a typical MEMS-based microelectrode structure, these recording sites will either be on the same bar or shank, or on adjacent bars or shanks. In the first case, the recording sites will be most sensitive to the sources along the axis of the shank — those most likely to have been damaged by insertion or where connective tissue would have grown around a chronic implant. In the second case, the healthiest tissue, from which one would wish to record, lies precisely in the dead spot between the two recording sites. Clearly, then, a simple differential configuration cannot be readily adopted. Alternative recording site/reference electrode layouts could be investigated for differential recording. This would, however, have an impact on the micro-electronics design. Generally, it would be desirable to situate the differential stage as electrically close as possible to the electrode sites involved to benefit from the CMRR of the amplifiers. With good analog circuit design, however, the designer may prefer to use computing power to combine the signals appearing at different electrode sites once they have been digitized. One reason that this may be beneficial is that it would be a much more flexible approach, perhaps enabling the processor to track individual signal sources to some extent. Information lost due to the implementation of differential recording in hardware cannot be recovered again.

13.1.3 MEMS neuroprosthetic system outlines

From the literature, it is possible to generalize three different situations for MEMS-based neuroprosthetic systems and to identify similarities and differences between them that affect the circuit design. These are systems located in the central nervous system (CNS), systems located in the peripheral nervous system (PNS), and those located in the eye (retinal prostheses). With this in mind, it is possible to outline generic systems, and use these as a basis for discussion of the individual circuit elements involved. Many of these circuit elements will overlap from one system to another, but there are some unique problems to be solved for the different applications; particularly when one considers the retinal prosthesis.

13.1.3.1 CNS applications

In this generalization, prostheses applied to the brain are considered. The spinal cord represents something of a transitional area, exhibiting some of the characteristics one may find relating to prosthetics applied to the brain and some of those relating to the PNS. Specific examples of such prostheses include visual prosthetics applied to the visual cortex,[5,17,18] prosthetics applied to the auditory brain stem,[19,20] and those applied to the motor cortex.[21,22] It is useful to note that the latter represent, to some extent, an attempt to link CNS and PNS prosthetics in the case of spinal cord injury. It has been proposed that lost sensation and control in paralyzed limbs can be restored by recording from the motor cortex and interpreting and transmitting these commands to a PNS element of the prosthesis. This would, in turn, electrically stimulate the PNS/muscles to achieve the desired effect. Proprioceptive signals recorded from the PNS would be used to control this stimulation and those returned to the CNS element of the prosthesis would provide conscious feedback of the process by stimulating the sensory cortex of the patient.[23] This is, perhaps, one of the most challenging systems so far proposed for MEMS-based neuroprosthetics. The principle characteristics of the environment encountered by prosthetics applied to the brain, from an electronics point of view, are the relative mechanical stability, the potential of mounting connectors on the skull, the relative lack of EMG and other interference, and complexity.

We can, therefore, identify a generic CNS neuroprosthetic system consisting of a fairly substantial two- or three-dimensional electrode array, penetrating up to a few millimeters into the brain. Although the complexity of brain means that the signal processing and identification overhead involved can be relatively large, the location of the device in the relatively stable and shielded environs of the skull means that a relatively large area of silicon real estate can be located relatively near to the electrode structure itself. Additionally, the skull provides a good anchor point for percutaneous connectors, which can be employed (e.g., work of Dobelle et al.[18]), although some may still prefer the radiofrequency (RF) induction coil or other transcutaneous approaches. The length of wiring is of less concern when all elements of a system can be mounted in close proximity, but the number of connections required, particularly to percutaneous connectors, is still a matter of interest. This is because of the pulsation of the brain with blood flow and the potential tethering effect of the connector or other elements of the system. The sort of generic system that we consider can be illustrated by the works of Dobelle[18] and Kennedy et al.[22] (see Figure 13.1a).

13.1.3.2 PNS applications

The PNS differs from the CNS in that the complexity of the system decreases as one considers more distal locations; there is even some evidence for functional organization within the nerve trunk itself. Nerve cuff solutions have been applied to the PNS with considerable success in the past, but more complex systems for standing and walking have suffered from problems of cable failure.[25] Where such systems have employed more proximally located cuff electrodes (at the ventral roots, for example), the success of the system has been limited by the selectivity of the electrodes employed.[26]

It is noted that while the PNS has evolved to withstand many decades of repeated mechanical stress from joint flexion and muscle contraction, the same is not true of metal wires. Locating prosthetic systems beyond the relatively protected confines of the spinal column can be problematic; even here, space is at a premium, implying a requirement for a miniaturized system.

The PNS prosthesis will typically employ two-dimensional regeneration electrodes or probe-type devices,[16,27,28] although at least one group has proposed that a three-dimensional array would be required to interface in a highly selective manner to individual nodes of Ranvier.[29] One of the applications typically proposed for such a system is standing and walking in paraplegics. Here, it is proposed, the highly selective nature of the MEMS-based microelectrode structures can be drawn upon. The system could be located in or near the spinal column with electrodes situated in the ventral roots for stimulation and devices situated in the dorsal roots providing sensory feedback for closed-loop control.[6,30] Another proposed application is to provide control and sensation for prosthetic limbs.[10,27,28]

The generic PNS prosthesis therefore consists of a number of microelectrode structures at several discrete locations within the PNS. These will typically perform minimal signal processing, due to the difficulty of supplying

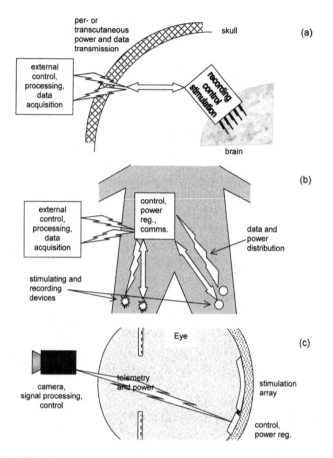

Figure 13.1 (a) CNS (brain) prosthesis, characterized by co-location of recording and stimulating sites with control and signal processing circuitry and a relatively stable situation for per- or transcutaneous signal transmission. (b) PNS prosthesis, characterized by distributed microelectrode devices with the requirement for more distributed intelligence and localized power regulation; per- or transcutaneous power and data transmission is still required. (c) The retina prosthesis. For the purposes of this work, the spinal cord lies somewhere between the CNS and PNS outline.

such locations with cabling for power and signal transmission. Ideally, such devices would not project from the nerve trunks into which they had been implanted, due to mechanical considerations, and wireless communication and power would be employed. Control would be centralized, typically in a unit implanted beneath the skin with a coil for RF coupling; this latter item is not dealt with in this chapter (see Figure 13.1b). Injectable stimulators[31,32] provide a well-developed example of this approach.

13.1.3.3   The eye

The retinal prosthesis has received much attention recently,[33–37] and a typical system is outlined in Figure 13.1c. In such a system, an electrode array designed to stimulate the retinal ganglion cells is implanted either on top of or beneath the retina. Power and communications are provided by an optical link either to the array itself or to a control circuit located within the eye away from the array but connected by cables. The retinal prosthesis is dealt with in some detail in Chapter 11 in this volume, and it possesses some unique characteristics when it comes to the integration of microelectronic circuitry.

Notably, a wireless approach appears to be the most practical when transmitting power and data to the array. This implies that any circuitry should consume as little power as possible, and the additional complication of localized heating caused by power dissipated by the electronics also must be looked at anew. Additionally, the delicate nature of the retina means that any implant must be minimally invasive and mechanically compatible (i.e., soft, flexible, and shaped to the curvature of the eye) to minimize the damage caused. This tends to point to the mutually incompatible goals of a polymeric electrode structure and one with integrated electronics.

13.1.4 *Power dissipation in integrated circuits*

The power available for an implanted system will invariably be limited unless a percutaneous connector is used, in which case the system will be relatively generously supplied. Even so, power conservation is a question that needs to be addressed. Power may be dissipated in one of several ways:[38,39]

- Static power dissipation
- Dynamic (switching) power dissipation
- Short circuit power dissipation
- Leakage power

Power is dissipated whenever current flows through a device that has a voltage drop across it, given by the equation:

$$\text{Power} = I\,V$$

This static power dissipation is most prominent in analog circuits, where resistors and bias chains are part of the circuit design. Analog circuits normally require that transistors be biased into a partially "on" state (conducting current, with a voltage drop across them) in order to fulfill their function. This is best combated by care in circuit design, ensuring that voltages and quiescent currents are as low as possible.

Dynamic, or switching, power dissipation is the principle mode of current loss in CMOS digital logic circuits. An ideal switch would go instantaneously from an open circuit condition (maximum voltage dropped across it, but no current flowing) to short circuit (no voltage dropped across it, but maximum current flowing). In such a case, no power would be dissipated, as either V or I would be zero. However, switching involves the charging and discharging of capacitances within the circuit. This means that there is a period in which the voltage across a switch is falling (or rising) while the current is rising (or falling), leading to power dissipation in the device. The power dissipated by a circuit switching at a frequency f with a capacitance c is given by:

$$\text{Power} = c\,f\,v^2$$

Thus, losses can be reduced by reducing parasitic capacitances and by reducing the clock frequencies used, but more significantly they can be reduced by reducing the supply voltage at which ICs operate.

In many circuit configurations, particularly digital logic, complementary transistors appear across the power rails (see Figure 13.6, for example). Different threshold voltages and operating parameters mean it is possible when switching that both transistors could be on at the same time. This momentary short circuit condition dissipates power (short circuit dissipation) and is related to the switching frequency as well as the technology employed.

Leakage current power dissipation occurs when subthreshold leakage currents prevent transistors from turning completely off. Until recently, this has not been a significant problem in ICs, but with ever-reducing feature sizes it may become an additional consideration in circuit design.

13.1.5   MOS or bipolar?

The designer is faced with a number of possibilities when considering the design of integrated circuitry for MEMS-based neuroprostheses: a bipolar approach, MOSFET, JFET, or a mixed approach (usually bi-CMOS). It is now becoming possible to integrate all these technologies with MEMS, although CMOS is currently favored. Within this chapter, the focus is almost entirely on CMOS circuitry, however it is necessary at this point to introduce and dismiss other approaches. It is also intended to give the reader a very brief introduction to transistors.

13.1.5.1   Semiconductor basics

The reader is referred to introductory books on microelectronic circuit design and fabrication for a more complete treatment of this topic.[24,39–42] Semiconductor devices are commonly fabricated from silicon into which different impurities have been introduced to alter its electrical characteristics. For the purposes of this section, we consider *n-type* and *p-type* silicon. A diode is formed by the interface between a region of n-type and p-type silicon and allows electrical current to flow in only one direction (from p to n).

When integrated circuits are fabricated, the impurities that will form the devices are introduced into different areas of the silicon wafer. Silicon with introduced impurities is termed *doped*. Conducting layers, separated by insulating layers, are then built up on top of the devices thus formed, to interconnect them in a meaningful manner. Typically these will be one or two

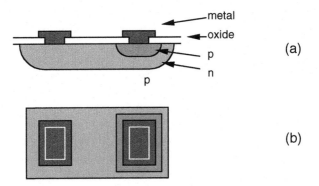

Figure 13.2 (a) Diode in cross section. (b) Plan view (layout). Not to scale.

layers of polysilicon (this is silicon, but with a different structure than that of the silicon wafer), and two or more (up to six or more) layers of metal, usually aluminum. The circuitry is usually no more than a few microns thick and is located on the surface of a wafer typically half a millimeter thick. Thus, it may be considered to be more or less two dimensional. Devices are usually depicted in cross section when describing their function but are typically depicted in plan view when laying out a design. Figure 13.2 illustrates a diode to familiarize the reader with this type of depiction.

13.1.5.2   Bipolar

The bipolar junction transistor (BJT) consists of a thin *base* region, sandwiched between the *emitter* and *collector* regions. These may be doped either n–p–n or p–n–p (collector–base–emitter), leading to a device with two diode junctions. During operation, a current applied to the base is used to control the larger current that flows between collector and emitter. When constructing analog circuits (e.g., amplifiers and filters), bipolar technology possesses the advantages that it can achieve higher gains than similar field effect transistor (FET) circuits and is not as noisy as metal-oxide semiconductor field effect transistor (MOSFET) approaches. One of the principle disadvantages when considering neuroprosthetic applications, particularly microelectrode recording, is the base current. This is very large in comparison to MOSFETs and causes polarization and electrode drift. Additionally, BJT circuits typically consume more power than FET circuits. The BJT has generally been superseded by MOSFETs in digital logic circuitry, due to far lower power requirements of the latter.

13.1.5.3   The MOSFET

A field effect transistor consists of a channel between *source* and *drain*, controlled by a *gate*. In the *depletion-type* FET, a normally conducting channel is restricted by applying a voltage to the gate, and in the *enhancement-type* FET a normally nonconducting channel is made to conduct by application of a voltage to the gate. In the metal-oxide semiconductor field effect transistor (MOSFET), the

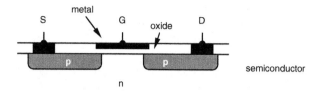

Figure 13.3 Structure of a p-channel enhancement-type MOSFET. The channel lies beneath the gate electrode (G). Applying a negative voltage to the gate induces positive carriers (holes) in the channel region, connecting the two p-doped regions. Not to scale.

gate electrode is separated from the channel by a thin insulating layer of silicon dioxide (Figure 13.3). The properties of the device, for a particular fabrication process, are primarily determined by the width and length of the channel (the ratio W:L being known as the *aspect ratio*). These are usually specified in λ (lambda) units; one λ unit being half the minimum feature size of the process. This is to facilitate scaling of the design as fabrication technologies improve.

Circuits can be constructed from n-channel MOSFETS (n-MOS), p-channel MOSFETS (p-MOS), or both types (complementary MOS, or CMOS). CMOS is implemented using only enhancement-type transistors. Broadly speaking, MOS approaches exhibit much lower power consumption that BJTs, but are generally slower (p-MOS is the slowest and CMOS the fastest of the three). That said, CMOS devices now effectively challenge BJTs in all but the most demanding of applications. In analog terms, MOS solutions suffer from *flicker noise*, which is inversely proportional to frequency and is significant in the frequency range exhibited by neural signals (below 10 kHz); p-MOSFETs are slightly less noisy than n-MOSFETs. However, they also exhibit very high input impedances and exceedingly low (gate) bias currents in comparison to BJTs. An additional characteristic of MOSFETs is the *threshold voltage*. This is the voltage that must be applied between gate and source (V_{GS}) in order for the transistor to operate ("turn on," in CMOS terms). IC fabrication processes that mix CMOS and BJTs are now increasingly available. Known as bi-CMOS, this technology allows the designer to select the most appropriate solution to their problem by mixing MOSFETs and BJTs as required.

13.1.5.4 The JFET

The junction field effect transistor (JFET) is available only as a depletion-type FET. It operates on the basis of a reverse-biased diode junction, which closes off the conducting channel as the bias is increased. The principle application of the JFET is in the input stages of amplifiers; they exhibit better input characteristics than BJTs and better noise performance than MOSFETs.

13.2 Elements of a system

Diverse designs are employed in neurotechnology. This section introduces the reader to the circuit elements that have been employed and also to one or two likely candidates. The intention is to enable the identification of components that may be implemented in a system and the deciphering of system block diagrams or circuit diagrams. The treatment focuses first on the special case of the neuron–transistor junction, before discussing sequentially: CMOS circuits in general, general system building blocks (such as oscillators), circuits specific to recording (e.g., preamplifiers and filters), circuits specific to stimulation (e.g., current output DACs), and concluding with brief discussion of elements more specific to CNS, PNS, and retinal prostheses. Due to overlap, the reader will find that some topics discussed in earlier sections are also relevant to later sections. Najafi[23] has provided a very readable introduction to neuroprosthetic systems incorporating MEMS devices, which is illustrated by specific examples. See also Chapter 6.

13.2.1 The neuron–transistor junction

Simple FET follower circuits have been used as headstage buffer amplifiers for conventional microelectrodes in the past; their purpose being to provide an interface (impedance match) between the very high impedance recording site and relatively long cables with significant parasitic capacitance that would otherwise attenuate the signals of interest. One logical approach for integrated microelectrode arrays would, therefore, be to integrate FET followers directly beneath metal recording sites, such that the gate conductor of the transistor formed the microelectrode itself. Such devices have been reported,[43] although design is complicated by a number of problems, including the need to provide noble metal gate metallization and ionic contamination of the gate.[43] Additionally, follower circuits typically have a gain of slightly less than unity so the implementation of a follower circuit as the first amplification stage introduces additional electrical noise while slightly attenuating the signal of interest. Given these problems it may be preferable to position recording sites slightly away from the amplification circuitry, especially as modeling studies have suggested that, on the length scales typically found in probes for CNS applications, parasitic capacitance and cross-talk can be kept at a manageable level even when multiple interconnection layers are implemented.[16]

Research led principally by Fromherz and Offenhäusser[44-48] has, however, focused on this type of interface. In this particular approach, there is no metallization on the gate of the FET used in recording. Stimulating electrodes are fabricated by thinning the insulation film above doped conducting tracks in the substrate. This approach is illustrated in Figure 13.4. The main advantage of this approach is that the electrode sites on the array are chemically virtually indistinguishable from the rest of the surface. This has been put to good use in terms of surface modification for patterning and studying cells in culture and is potentially interesting for biocompatibility purposes. Other advantages are that there are no problems with exposed fabrication layers at electrode sites (e.g., adhesion layers beneath noble metal films or internal laminations of insulating/passivating films)

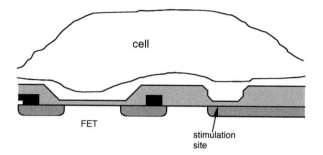

Figure 13.4 Neuron–transistor junction and capacitive stimulation electrode.[44-48] Not to scale.

and no concerns about long-term stability of metal electrodes or electro-chemical activity at low frequencies, such as baseline drift, thus potentially simplifying fabrication. At present, this approach appears only to be useful for *in vitro* applications; however, studies into their electrical properties have been undertaken,[48-50] and the advantages noted make a brief summary of these relevant to the present topic.

Flicker noise in MOSFET circuits is a significant design consideration for neural signal recording. When a metal recording site is used, especially at a distance from the MOSFET, then the designer can implement relatively large transistors to reduce flicker noise problems. Najafi,[14] for instance, used 145-μm-wide, 10.4-μm-long transistors; Ji,[51] 120-μm-wide, 15-μm-long transistors. (In both cases, input p-MOS transistors were used as they are less noisy than n-type devices.) With a direct neuron–transistor junction, the gate region (hence, the size of the transistor) is limited to the desired area of the recording site. Offenhäusser et al.[46] reported gate dimensions of 28 × 12 μm down to 10 × 4 μm. In this case, recording is limited to neural signals of greater than about 100 μV amplitude. In comparison, Bai et al.[52] have reported overall noise levels as low as 20 μV for very complex three-dimensional recording arrays, and it is anticipated that noise levels can be reduced to below those of the metal recording sites themselves. In the *in vitro* situation, however, neurons placed or grown directly over FETs can couple tightly to the substrate. In this case, recorded signals can rise from several hundred microvolts to millivolt amplitudes. Such sealing to metal microelectrodes appears to be difficult to achieve. Unfortunately, when considering chronic neural implants, the microelectrode structure is typically surrounded by connective tissue that will prevent this type of seal from being achieved.[53-55]

The electrical mechanism by which the nonmetallized stimulation sites of these devices couple to cells placed or grown over them has been studied.[48-50,56] As may be expected, this is primarily a capacitive mechanism, complicated by the placement and coupling of the cell body to the substrate. This contrasts with metal microelectrode sites, however, in that when stimulation is applied using the latter, they have to be operated in a charge-balanced pseudo-capacitive regime in order to prevent degradation of the site through electrochemical activity. While the designer suffers from this disadvantage with metal microelectrode stimulating sites, this is compensated for by a number of advantages. The capacitance of these sites can generally be made to be greater than with insulated sites. The effective separation of the plates is typically a few angstroms[3] due to the electrochemical nature of the interface, surface roughening techniques can be used to increase the effective plate area of the capacitor, and multivalent oxides (e.g., of iridium or tantalum) can be used as the interface layer. All these considerations allow high charge densities to be achieved, resulting in more flexible stimulation options.

13.2.2   General

A number of general considerations must be kept in mind when designing CMOS circuitry for implantable MEMS for neuroprosthetics. First, capacitors implemented on-chip tend to take up relatively large areas of silicon, which is particularly valuable in PNS applications. Neural signals are of relatively low frequency compared to those usually encountered in modern electronics, and relatively large capacitors are generally required to create appropriate filter circuits. Furthermore, while it is often possible to fabricate adjacent capacitors that match relatively closely on an IC, absolute tolerances tend to be poor. Where precision is required, devices may be laser trimmed, but this adds considerably to the cost of the process. As a consequence, capacitors are best avoided when possible.

Similarly, resistors are difficult to fabricate. Some processes provide solutions to the creation of high value resistors on chip. Where this option is not available, the designer has to resort to the unsatisfactory solution of a very long polysilicon track, and this is only an option for low value resistors. To overcome this problem in the design of integrated circuits, the active load is used. This is a transistor biased on by connecting the gate to the drain pin (Figure13.5c). This acts like a resistor, the exact performance being controllable by the choice of aspect ratio.

Transistor-level circuit diagrams can be quite daunting at first sight; however, they generally consist of a few common motifs. These are reproduced in Figure 13.5, using shorthand notation for enhancement-type n-MOS and p-MOS FETs. Note that source and drain are not distinguished; this is because the transistors will normally be symmetrical and the substrate bias is taken care of in very-large-scale integration (VLSI) design (for discrete devices, the substrate may be biased to source or drain).

Following from the active load configuration, the single transistor source follower and the common source amplifier are shown in Figures 13.5d and e. In the case of the source follower, the output will approximate the input, the gain being close to but less than unity provided that the load resistance is much greater than 1 divided by the transconductance of the transistor (*transconductance*, g_m, expressed as the ratio of output current to the input voltage). In the common source amplifier, the current flowing

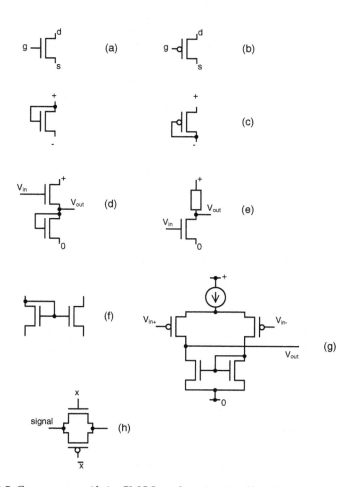

Figure 13.5 Common motifs in CMOS analog circuits. Shorthand notation for (a) n-MOSFET; (b) p-MOSFET; (c) active loads; (d) source follower with active load; (e) common source amplifier; (f) current mirror; (g) long-tailed pair (p-MOS) with current minor active load; and (h) transmission gate.

through the drain resistor is controlled by the gate voltage ($i_d = g_m V_{GS}$). Typically, the gain of such a configuration will be much lower than that achievable using a common emitter BJT configuration. Note that in each case, an active load can be used, and closely matched transistor characteristics provide the designer with more options than would be available when using discrete devices.

A further use of the availability of matched transistors is the current mirror circuit (Figure 13.5f). Because the characteristics of these transistors are almost identical and they are both biased by the same voltage, then the current flowing through transistor A must be the same as that flowing through B. Notice that by changing the aspect ratio of one of the transistors it is possible to have a multiple of the current flowing through the other.

This is often made use of in analog circuits; a single biased transistor is used to generate a reference current and a series of mirroring transistors act as current sources for different parts of the circuit. Thus, variations in supply to one part of the circuit are mirrored throughout and correct operation is maintained.

The most common input arrangement for an IC amplifier is the differential pair configuration (Figure 13.5g); this is also known as a "long-tailed pair." In this case, p-type MOSFETs have been used as the input transistors. The current source in the tail would typically be implemented as part of a current mirror, or as a biased transistor. A current-mirror load is shown in Figure 13.5g. The effect of this is that both sides of the amplifier contribute to the output, despite its being taken from only one arm. It is at this point that additional transistors may be added to adjust the gain of the first stage and introduce feedback within the amplifier design itself.

One final analog building block of interest is the CMOS transmission gate (Figure 13.5h). In this case, the MOSFETs are acting as analog switches. They will either be high impedance (off) when x is the most negative voltage in the circuit or low impedance (on) when x is the most positive. The controlling signal would typically be a logic signal, so the analog signal being switched would have to lie within the voltage range used for logic. In the case of an implantable system, one would endeavor to minimize the number of power rails and run analog and digital parts of the circuit from the same supply, although this is not without its drawbacks. Transmission gates can be implemented in CMOS, n-MOS, or p-MOS technologies, although there is no bipolar equivalent. If CMOS is not available, the properties of the gate will be degraded by the threshold V_{GS} required for the transistor to turn on. The transmission gate itself can form the basis of the analog multiplexer and may also be used to discharge charge build up or apply test signals to the inputs of amplifiers. One must be aware, however, of capacitive feedthrough, both from input to output and from gate to channel.

Compared to analog circuitry, CMOS logic is much simpler to understand. CMOS gates are shown in Figure 13.6. More complex gates are created by adding combinations of transistors in series and in parallel. The function of the circuit can be determined by treating each transistor as a switch, with p-type MOSFETs being "on" when a logic 0 is applied to the gate, and n-type being on when a logic 1 is applied. Complex logic functions are thus built up by implementing series and parallel combinations, for each series combination in the p arm, there must be an equivalent parallel combination in the n arm, and vice-versa. Feedback from output to input would usually indicate a latch function of some type.

13.2.2.1 *Clocks and oscillators*

Digital logic will normally require a clock signal to control it; this can be a multiphase clock, with several different signals being distributed across the circuit to effect appropriate control, and it may also be necessary to ensure that clock edges do not overlap so it can be guaranteed that certain transistors

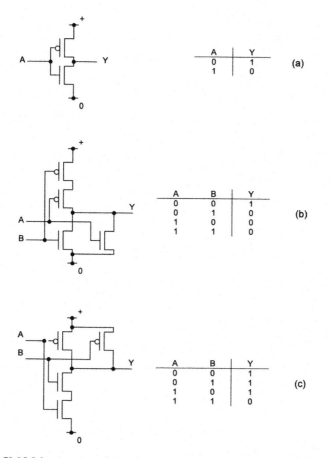

A	Y
0	1
1	0

(a)

A	B	Y
0	0	1
0	1	0
1	0	0
1	1	0

(b)

A	B	Y
0	0	1
0	1	1
1	0	1
1	1	0

(c)

Figure 13.6 CMOS logic gates: (a) inverter, (b) NOR gate, (c) NAND gate.

turn off before others turn on. The simplest form of oscillator makes use of the inherent delay present in logic circuits. A string of inverters, with the output fed back to the input, will oscillate at a frequency related to the total delay through the chain. More common, however, is the relaxation oscillator, which comes in various forms based on the same principle. A capacitor is slowly charged up and then rapidly discharged, creating a sawtooth waveform which is shaped to a square wave on the output. Figure 13.7 shows a

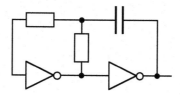

Figure 13.7 Relaxation oscillator formed from two inverters.[14,39]

simple relaxation oscillator based on two CMOS inverters. Notice that when designing with discrete components, it is possible to use current sources and current mirrors to control the current flowing into a capacitor and hence the frequency of the oscillator. Once a square wave signal has been generated, it can be divided down to create the necessary clock phases. Inclusion of inverter chains in this logic can ensure that edges do not overlap.

Given the caveats regarding implementation of capacitors in integrated circuits, it is apparent that such circuits cannot be relied on to oscillate at precisely the desired frequency nor to be as stable as those that incorporate quartz or similar resonators. Thus, these circuits cannot be relied on for precise timing. One use for such oscillators has been the control of analog multiplexers for multichannel recording probes.[13,51,58,59] Here the clock signal of the oscillator has been regenerated off-chip in order to decode and demultiplex the recorded signals. Thus, frequency differences between individual oscillators could be accounted for if necessary. Peeters et al.[59] implemented a two-wire system for recording probes, which obviously required an on-chip oscillator for multiplexing.

Recent designs have, however, tended to rely on off chip clock signals,[60-62] and Ji's design[51] implemented a serial communications protocol that enabled the clock to be regenerated from the data received. This was achieved by using multilevel logic signals (5 V on the I/O line toggled the direction; 4 V was a logic 1, and 2V a logic 0; and bits were separated by a 0-V period, so that the logic received a series of pulses that could be used to regenerate the clock). Injectable stimulators[31,32] can use a similar approach, whereby the RF carrier (of a few megahertz) used to power and communicate with the device provides it with its primary clock signal.

The main problem with clocked logic, notably in recording applications, is that of clock coupling through to the analog section. For example, Bai et al.[52] reported about 40 µV of noise on recovered demultiplexed signals, compared to 20 µV of noise with nonmultiplexed signals. Coupling can be through stray capacitance or through power line glitches; it is difficult to separate digital and analog sections of circuitry on ICs in the manner that is common for discrete circuit design, and decoupling capacitors are not available. Various strategies are available for ameliorating this problem, the most basic being good circuit design, care when laying out the circuit, and good use of extraction and simulation from the layout.[64]

13.2.2.2   *Power-on reset and mode control*

It is recommended that logic circuits be supplied with a reset signal to ensure that they can be put into a known state when desired. This is normally performed by supplying an additional reset line to the circuit, but when concerned about the number of wires attached to an implant it would be desirable to eliminate a dedicated reset connection if possible. The first way to do this is to ensure that a reset signal is supplied on power-up. This is normally performed using a capacitor and flip-flop circuit. A transistor or low resistance path to ground is placed so as to ensure that the flip-flop powers up in one particular state (reset signal active). When power is supplied to the circuit, the capacitor will begin to charge, thus providing a delayed signal to change the state of the flip-flop, inactivating the reset signal.[14,51] A few caveats regarding this particular approach should be noted. The first is that the power supply must ramp up to maximum much more rapidly than the capacitor charges. This could be a problem with implanted integrated circuits where not only is the size of the capacitor limited by available silicon real estate, but the current available from an implanted power source may also be limited.

Where an external clock signal has been made available, designs have incorporated this into the power-on reset circuitry such that it is the first pulse on the clock line that inactivates the reset signal.[60] Designs for RF-powered injectable stimulators and retina prostheses[31,35,36,65] generally appear to incorporate a synchronization sequence within the data transmission protocol that effectively resets the device. This approach can be effective when function is relatively limited; however, state machine design can become burdensome and demanding on silicon real estate. Additionally, without careful implementation of power-up reset circuitry, one risks the possibility that the device will power up in an unsafe state, and the safe power cycling of stimulation circuits must be considered (see, for example, Loeb et al.[31]).

Because CMOS circuits, unlike bipolar circuits, can operate over a range of supply voltages, this provides the designer with further options in resetting or changing the operating mode of the circuit. Self-test features have been implemented[13,58] that are activated by raising the supply voltage above normal operating voltage. Once again, this can be implemented by a flip-flop activated by a transistor attached to the supply rail in such a way that it does not reach threshold and turn on until the supply rail is raised to a specific voltage. Ji and Wise[58] also introduced a multilevel logic system whereby a comparator was used to detect a 5-V signal on the signal I/O line. This enabled the design to toggle between send and receive modes without requiring an additional signal line.

13.2.2.3 Integrators

The integrator is a useful circuit element that functions as the name suggests by integrating an input voltage signal. Electronic integrators are not perfect, being limited by finite supply rail voltages, currents, and leakage; they are, in fact, a form of low-pass filter. The designer has two basic forms to choose from, both of which use a capacitor as the integrating element.[40]

A transconductance integrator involves conversion of the input voltage to a current, usually through transistor action, which is used to charge a capacitor. The output voltage is measured across the capacitor. A disadvantage of this circuit is the need to place additional circuit elements in the output circuit of the converter to provide feedback control, which can affect noise performance or dynamic range. For this reason, operational amplifier integrators are often preferred for such applications; this is essentially an operational amplifier circuit with a capacitor in the feedback path.

Beyond filtering, the principle application for integrators is to provide sawtooth signals and reference slopes (as elements within relaxation oscillators, for instance). These may be incorporated into analog-to-digital converters or voltage-controlled oscillators for use in phase-locked loops.

13.2.2.4 Phase-locked loops

The phase-locked loop (PLL) is used to synchronize a system clock signal with an incoming clock signal for which only the approximate frequency range is known. They can also be used as frequency multipliers to synthesize high-frequency clocks from low-frequency reference signals. The principle use of PLLs in neurotechnology has been in decoding the complex multiplexed data signals generated by recording devices.[66] The PLL (Figure 13.8) consists of a voltage-controlled oscillator (VCO), the control being provided through an integrator element which ramps up at a rate determined by the input voltage, before being reset. The phase of the shaped (square wave) output of the VCO is compared with input signal, and the difference is fed back as a voltage through a low-pass filter to the controlling input of the VCO. The output of the PLL locks onto the reference signal after a period of time determined by the filter in the feedback path; thereafter, synchronization can be maintained even if some transitions of the reference signal are missed.

13.2.3 Recording

Further to the general circuit elements discussed in the proceeding section, a number of elements have more specific application in the recording of neural signals through MEMS-based microelectrode devices. These are, in particular, preamplifiers, filters, switched-capacitor filters, and analog-to-digital converters.

13.2.3.1 Signal preamplifiers

One of the earliest goals of MEMS-based electrodes for neurophysiological applications was to be able to integrate signal preamplifiers as close to the recording site as possible, thus reducing as far as possible the problems of externally induced noise.[4,43] Beyond the general requirements of good low noise design, good power-supply rejection ratio (PSRR, a measure of the immunity of the amplifier from noise on the power-supply lines), and low power consumption, the designer is faced with additional problems not

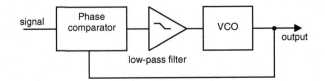

Figure 13.8 Components of a phase-locked loop (PLL).

normally encountered in amplifier design — specifically, baseline (DC) drift of microelectrode recording sites, polarization of the recording site, and charge build-up on the MOSFET gate.

It is known that even the passage of very small (fA) DC currents across microelectrode sites can result in offset signals that are orders of magnitude larger than the signals of interest.[14] This generally precludes the use of low-noise BJTs or JFETs as the input stage of preamplifiers, because the offset voltages induced would tend to saturate the input stage of the amplifier, and current flow may have a deleterious effect on the electrode site and surrounding tissue. The FET solution then suffers from the possibility of charge build up on the gate of the transistor. An additional problem, under certain circumstances, is optically induced currents.[67] The designer must compromise the need for a high-input impedance, which must be at least one order of magnitude greater than that of the site, against the requirements of low gain at DC and a path of known impedance for bias currents and discharge of charge build up. In discrete circuits this can be solved using a high-value resistor and a relatively high supply voltage (e.g., ±12 V) to avoid saturation of the first amplifier stage; more severe filtering would be performed by subsequent circuits.[68]

A biased diode can be employed to provide the appropriate input resistance required,[13,69] often in conjunction with a transmission gate that can be turned on to provide a discharge path to ground, if necessary. At least one early design implemented a capacitor to decouple the recording site.[67] This, however, required considerable silicon real estate for the 30 pF capacitors that were required, compared to capacitors of 3 pF or less employed in DC-coupled designs with filters.[13,58] More recent work by Bai and Wise[69] has reviewed a number of possibilities for DC baseline stabilization for a closed-loop amplifier without any filtering. In this case, a closed-loop amplifier was desired, as it was seen as providing more stable gain characteristics.

Furthermore, it must be recognized that the high-frequency cut-off is equally of importance when considering the noise performance of the system. Normally, however, this is less problematic than the low-frequency cut-off, as it can be implemented with a relatively small loading capacitor in the output stage (see, for example, Ji and Wise[58]) or through design (for example, Takahashi and Matsuo[67] used the gate capacitance of the input transistor to achieve a high-frequency cut-off). Further studies on noise performance for integrated neural signal preamplifiers have recently been published.[69–71]

13.2.3.2   Filtering

As previously stated, filter design is compromised by the problems of conserving silicon real estate and fabricating low-tolerance capacitors on-chip; in such cases, the designer may find it desirable to avoid continuous time active filters where possible. Takahashi and Matsuo,[67] however, used large decoupling capacitors between recording site and amplifier, whereas Bai and Wise[69] have attempted to obviate the need to attenuate low-frequency components in the preamplifier by careful input stage design. Otherwise, the

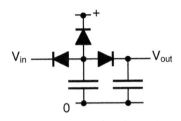

Figure 13.9 Diode-capacitor lowpass filter that has been implemented in several integrated neural signal preamplifier circuits.[58,60,72]

preferred approach in preamplifier design has been to implement a diode-capacitor filter[58,60,72] (Figure 13.9); the equivalent resistance of the diodes is relatively high, enabling the designer to use correspondingly small capacitors to achieve the desired low-frequency response. This filter design would normally provide negative feedback from the output stage of the amplifier to the input stage.

The principle alternative to continuous time filtering is the switched capacitor filter.[39] This is implemented with CMOS switches, as indicated in Figure 13.10. The switches are operated at a frequency much greater than the highest frequency in the signal, and the circuit acts as an integrator ($V_{out} = f_0 c_1/c_2 \int V_{in}dt$). It can therefore be used to substitute continuous time integrators in filter circuits. Because the transfer function of the circuit depends on the ratio of the two capacitors, rather than their absolute values, it lends itself to IC fabrication where the relative values of two components can be controlled to a much greater degree than their absolute values. It is also possible to tune the filters thus realized through the clock frequency.

Switched capacitor filters do, however, have a number of disadvantages for the present application. First, there is the problem of clock feedthrough, where transients appear in the output signal at the frequency of the clock signal used. Furthermore, any noise in the input signal at or near the clock frequency will be aliased down and appear in the output signal within the pass-band. Finally, in addition to a generally reduced dynamic range compared to continuous time-active filters, the switched-capacitor filter will typically consume more power. This is due to the fact that power dissipation in CMOS logic circuits is strongly related to clock frequency. For these rea-

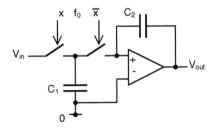

Figure 13.10 Principle of the switched capacitance filter.

sons, the switched capacitor filter is probably not appropriate in the early stages of a recording system; however, it could potentially be used to good effect in some applications.

13.2.3.3 Analog-to-digital converters

The principle problem with multiplexing analog signals onto a single wire for transmission to a centralized control unit of a neuroprosthesis is that the signals, although amplified, are still susceptible to interference. For this reason, conversion of the signals to digital format is desirable, and the designer has a number of options by which to implement this.

At this point it will be assumed that a full reproduction of the recorded single unit action potential is required. Normally one would use an upper frequency of interest of 10 kHz, but for a best case when considering converter design an upper frequency of 3 kHz will be assumed here.[73,74] In this case, the converter would have to sample the signal at least every 0.1 msec (10-kHz sampling frequency). For 100 recording sites, this would require a 1-MHz converter, or several converters operating at lower frequencies. The problem is not, therefore, as simple as it first appears to be. A variety of analog-to-digital converter (ADC) designs can be selected for this purpose, the most common and promising being the successive approximation ADC, the flash converter, the sigma–delta converter, and integrating converters. The function of these being outlined below. These converters will generally need to be combined with sample and hold circuits to stabilize the input signal during the conversion process. These contain capacitors that are charged from the input signal through a digitally controlled switch.

13.2.3.3.1 Successive approximation ADC. The operation of the successive approximation ADC is illustrated diagrammatically in Figure 13.11. The

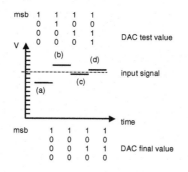

Figure 13.11 Illustration of the operation of a four-bit ADC. The input signal is shown as a dashed line, and the test value of the DAC is shown as a heavy line. In the first cycle (a), the most significant bit (msb) of the DAC is set and the output compared to the input signal. The input signal is greater; therefore, the bit remains set at the end of the step. Then (b) the next bit of the DAC is set. The input signal is less than the output of the DAC so this bit is reset. This is repeated in steps (c) and (d) to give a final value of 1010 after four steps.

converter is comprised of a comparator and a digital-to-analog converter (DAC), the output of which is compared to the analog input signal. The first bit of the DAC is set and the output compared to the input signal. If the input is less than the output of the DAC then the most significant bit is reset. This sequence is then repeated for the next most significant bit and the remaining bits in sequence. The number represented on the inputs to the DAC is then the output of the ADC. This is a clocked circuit, and power dissipation will be affected by clock frequency; this in turn will depend on the number of bits used in the converter and the desired conversion time.

13.2.3.3   Flash ADC.   The conversion time of the flash ADC is limited only by signal delays through the circuit and the frequency response of the analog components. It consists of a reference resistor chain and a number of comparators related to the number of quantization levels desired. If an eight-bit result is required, then 255 separate reference voltages are generated and 255 comparators are used to compare the input signal to each of these. The output of the comparators is then converted to an eight-bit number using conventional combinatorial digital logic. The power dissipation of the converter and area of the circuit are related to the number of bits required in the output. The converter will also be relatively inaccurate due to process variations. There are alternatives to full-flash converters that incorporate partial flash conversion of the signal. These designs trade reduced component count for speed.

13.2.3.4   Sigma–delta ADC.   The sigma–delta converter has recently gained popularity due to the increasing speed of CMOS ICs. There are various designs, but they essentially consist of a single comparator and an integrator. In the simplest form, a number of pulses are supplied to the input of the integrator over a specific time period. When the output of the integrator reaches the value of the input signal then the number of pulses supplied provides the output of the converter. This is the opposite of the flash converter. For an eight-bit converter, for instance, 255 clock pulses are required to effect the conversion. Sigma–delta converters are generally the most accurate converters available; however, the conversion times available and high clock frequencies suggest that they are not suitable for the present application.

13.2.3.5   Integrating ADCs.   Integrating ADCs are, perhaps, the most useful form of ADC for the present application. The simplest form of integrating ADC consists of a comparator, counter, current source, and capacitor. The counter is started at the same time that the current source is applied to the discharged capacitor. When the voltage across the capacitor is equal to the input signal, the counter is stopped and the number it contains represents the input signal level. The advantage of this approach is that the two halves of the converter can be separated. The capacitor and comparator can be implemented near the recording site, and the counter can be implemented on the central controller, which may not have so many design restrictions as remote components of the system. Signals from several different sites can be multiplexed together as pulse-width modulated (PWM) signals for transmission via a single connecting wire. The speed of the converter is essentially limited by the clock frequency, although the analog circuitry will itself play a part at high clock frequencies. Process variations can also limit the accuracy of the converter. Use of dual slope converters can help overcome this problem. In this case, the capacitor is charged for a fixed period of time from a current source that is proportional to the input voltage. The capacitor is then discharged using a fixed current source; the discharge time will be proportional to the input voltage.

13.2.4   Stimulation

Following this review, only one major circuit element remains that is specific to MEMS-based stimulators, and that is the stimulus delivery circuit and associated digital-to-analog converter. The need for a stable power supply must be kept in mind, however.

13.2.4.1   Voltage-output DACs
The output stage of a stimulator is normally designed to deliver biphasic charge balanced pulses. For this purpose, current output DACs, as described in the following section, are most appropriate. Applications for voltage-output DACs would include ADCs, as described in the previous section, or the control of current sources. Note that current output DACs can be used in these applications with an appropriate current to voltage output stage. Two common voltage-output DACs are shown in Figure 13.12, one based on a summing amplifier and one on the R-2R ladder network.

13.2.4.2   Current output DACs
The conventional form of the current output DAC is shown in Figure 13.13. It consists of a series of scaled current mirrors, with MOS switches to connect them to the electrode site. In Figure 13.13 both the source and sink halves of an eight-bit converter are shown, although not all the bits are shown. This format has been implemented for a number of different designs.[37,60,61] Normally, only one or a few converters would be implemented, with the outputs multiplexed onto different electrode sites.

An interesting bi-CMOS version has been implemented by Kim and Wise.[75] In this version, scaled transistors supply current to the base connection of a BJT. The current flowing through the BJT is proportional to the base current, thus a converter was realized with a much reduced transistor count, the current mirroring transistors being rendered unnecessary. In addition, the circuit was designed for reduced power operation. Tanghe and Wise[61] also published an interesting circuit, where the same current source path was used for both source and sink current mirror circuits. This should ensure closer matching between source and sink currents.

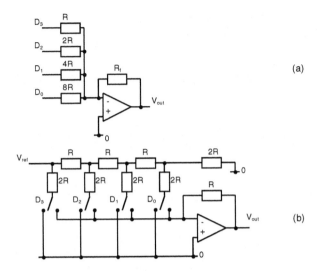

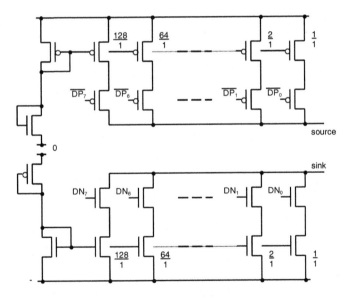

Figure 13.12 Four-bit DACs: (a) based on a summing amplifier, and (b) based on the R-2R ladder network. Both examples show inverting amplifiers so the outputs will be negative. D_3 is the msb; D_0 the lsb.

Figure 13.13 Eight-bit current output DAC with source and sink outputs. Transistor aspect ratios are for illustration only. Note that the data bits to the source transistors (DP_x) are inverted. Not all of the transistors have been shown.

Another option to ensure balanced charge operation has been to use a single-current source DAC and switch this between active and reference electrodes using a H-bridge style circuit.[36,76] Suaning and Lovell[36] also implement decoupling capacitors on the output of the bridge circuit to isolate the electrode sites from DC bias.

An alternative approach has been described for injectable stimulators.[31,32] In the system described by Loeb et al.,[31] a steady-state polarization is maintained between capacitive working and reference electrodes. Stimulation is by way of a current pulse (1.5 or 15 mA), which discharges the capacitor. A smaller current (0.01 or 0.1 mA) then recharges the system. Stimulation is principally effected by the high-amplitude pulse, and charge balance is maintained due to the longer recharge period at lower current.

13.2.5 Elements specific to CNS prostheses

The design constraints on prostheses located within the skull are a little more relaxed than for prostheses that have components distributed throughout the PNS. For this reason power regulation and communication are treated as elements specific to PNS prostheses, because greater problems are anticipated for PNS than CNS prostheses in this respect. As to elements specific to CNS prostheses, the more complex and resource-hungry circuits discussed are more likely to be implemented in CNS prostheses than in PNS prostheses. At present, technologies are not sufficiently advanced to be able to pick out particular circuit elements that are specifically applicable to CNS prostheses; those discussed tend to represent elements that are desirable for both CNS and PNS applications. More specific elements will appear as specialized signal processing elements, as MEMS-based neuroprosthetic systems become more application orientated.

13.2.6 Elements specific to PNS prostheses

Unlike components located within the skull, it is desirable that MEMS-based components located in the PNS should have as few leads attached as possible. This leads to one aspect more peculiar to PNS prostheses and retina prostheses than to CNS prostheses — specifically, radiofrequency (RF) communication and power transmission. This has been a particular factor in injectable stimulators[31,32] and retinal prostheses. The unique problems associated with this approach stem from its combination with MEMS and the resulting restrictions on size and location, particularly with injectable stimulators.[36,65,77,78] Within this section, however, the aspects addressed will be those of circuit design, particularly power regulation and approaches to communication.

13.2.6.1 Voltage regulation

Voltage regulation of some degree may be required for implants with wire connections, particularly if these incorporate sensitive circuitry and some signaling is performed by voltage changes on the power supply lines.[13,58] RF-coupled power[31,36,37,63] will be rectified and a large smoothing capacitor will used across the input of the regulator. Voltage regulators

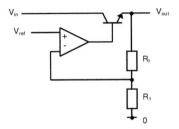

Figure 13.14 Principle of a voltage regulator based on an amplifier and reference source. Note that the output will be a multiple of the reference (V_{ref}) and that the input voltage must be smoothed sufficiently to ensure that the reference is stable and the amplifier can drive to at least $V_{out} + 0.7$ V.

operate from a reference voltage that will usually be in the form of a zener diode. Ziaie et al.[32] used zener diodes with BJTs, where the forward voltage drop of the base-emitter junction is known (0.7 V) to form a basic voltage regulator. More complex voltage regulators incorporate an amplifier of some sort with a high current output stage. The basic principle is illustrated in Figure 13.14. This amplifier could be as simple as a long-tailed pair, but note that the input supply has to be smoothed sufficiently by the capacitor to ensure operation.

Voltage regulators are going to be temperature hot spots on any die, because they will incorporate one or more components that have a large voltage drop across them with a large current flowing. This means that design and location of power regulation circuits needs to be carefully considered. Not all parts of the circuit may require voltage regulation; where stimulation is current controlled, for example, the rectified and smoothed signal could be used (see, for example, Loeb et al.[31]).

Recently, an alternative to powering via RF coupling or implanted batteries has been proposed: transcutaneous optical powering.[79–81] Optical powering and even data may be supplied by an infrared (IR) LED array or laser and converted to electrical power through PIN photodiode arrays. Edell[79] also presented a scheme by which recorded signals would be transmitted optically from the implant. Obviously this scheme has also received particular attention in retinal prosthetic applications.[33,34]

13.2.6.2 Communication schemes

Although IR powering and data transmission has been suggested for wireless implants,[33–35,79–81] RF approaches appear to be more developed, probably due to the history of this technique in prosthetic applications. A favored approach to data transmission has been the use of an amplitude-modulated (AM) carrier.[31,35,65] Fast switching of this carrier is achieved by frequency modulating the original data and delivering this to a tuned (resonant) antenna circuit. When the frequency of the signal deviates from the resonant frequency of the circuit, a reduction in amplitude of the transmitted signal results.

Demodulation of the transmitted signal can be achieved by filtering to remove the carrier frequency, and the subsequent use of a comparator or Schmitt trigger.[32,35] This is essentially a relatively low open-loop gain amplifier with some positive feedback. The positive feedback induces some hysteresis in the circuit, making it noise tolerant to some degree. The injectable devices described by Loeb et al.[31] demodulate the transmitted data by comparing short-term fluctuations in the received signal with a longer term average.

Some groups have used pulse-width modulation (PWM) techniques, or variations thereon, to encode the data.[37,65] In this situation, the ratio of the high (mark) period to the low (space) period of the AM signal provides the indication of a logic 1 or 0. The usual way to decode these signals is to implement a clock, possibly derived from the RF carrier frequency,[31] which drives a counter circuit. This would be used to count the time of a mark (or space) in the incoming signal, and hence determine the value of the corresponding bit.

Clements et al.[37,76] implemented a delay locked loop (DLL) and a variation on PWM. Three taps from a voltage-controlled delay line were used to determine whether or not the incoming signal represented a 1 or a 0. The advantage of this approach was that data transmission would have relatively little effect on the power level received by the implant, and a clock signal could be derived from the incoming data. The DLL consisted of a string of inverters, whereby the current controlling them (and hence the delay through them) was independently controlled. Thus, the total delay could be adjusted to the frequency of the incoming data in a manner reminiscent to the operation of a PLL.

Manchester encoding has also been implemented.[31,35,82] Logic levels are represented by a transition halfway through the bit; a logic 0 is represented by a low to high transition, and a logic 1 by a high to low transition. Clock signals may be recovered from this encoding scheme, and it has the additional advantage that the AM signal is low for exactly half a bit period, and high for the other half.

The data transmission protocol for such implanted devices has generally been of a broadly similar format. An initial reset or synchronization signal would be followed by a binary address — the number of the device for which the subsequent data were intended, and a set of stimulation parameters encoded in binary. The stimulation period either would be part of these binary-encoded parameters or would be signaled in some manner by the transmitter (e.g., the time the signal remains high following the last word of data). Parameters would typically be programmed for every single pulse, as preprogrammed parameters could become corrupted. If incomplete data were to be received, this would be discarded until a new synchronization or reset signal were received and the process would begin again.

Suaning and Lovell[36] reviewed RF communications with neuroprostheses in relation to their application, a retinal prosthesis. They implemented an approach whereby parameters were set by counting down the number of pulses transmitted in a particular packet. Different timing mechanisms were also implemented to ensure that data were correctly received. Although electrode selection was implemented, device addressing was not. Suaning and Lovell also implemented a reverse telemetry system that provided an indication of power levels supplied to the IC and to the stimulation circuitry. Connecting the supply voltage to the receiver antenna would cause a pulse that could be detected by the transmitter during quiet periods in the transmission.

Following completion of stimulation, the voltages of interest would be connected to resistor–capacitor circuits. These would discharge, and, upon reaching a reference level, the telemetry pulse would be effected. Thus, the time period between completion of stimulation and reception of the reverse telemetry pulse would indicate the levels of these voltages. These could be used as diagnostics, not only to confirm operation of the implant but also as an indication the electrical characteristics of the tissue adjacent to the electrodes.

13.2.7   Elements specific to retina prostheses

Most of the elements required by a retina prosthesis have been discussed in the preceding sections. One unique approach to the retina prosthesis has been to utilize arrays of photodiodes coupled to electrodes in such a way that illumination of the photodiode would stimulate the adjacent tissue,[33,34,83] power for the stimulation being derived from the incident illumination. Although the practical application of this approach has been demonstrated, at least one group feels that the intensity of illumination required to effect stimulation would be too great for long-term application and has recommended an array combined with an external power source for stimulation.[34] This would have the added advantage of enabling the use of a higher resolution array.

13.2.8   Other elements

The use of flexible polyimide microelectrode devices[10,11,84,85] prevents the direct integration of electronic devices onto the microelectrodes themselves (a monolithic solution). Hybrid solutions, where different elements of a system are produced using different technologies and then assembled, have advantages in terms of cost and speed of solution. Schuettler et al.[12] have, for example, reported a hybrid system incorporating a micromachined polyimide electrode array and interconnection system, three bare dice of commercially available ICs, and five passive components. A similar hybrid approach, although with custom-designed ICs, has been used in the development of a high-density connector system,[86] and a flip-chip approach has been proposed for contacting pincushion arrays.[57,87] Another example of the hybrid approach is an RF transmitter developed for biotelemetry.[63] This incorporated a surface-mount RF transistor with MEMS components (inductor, metal–polysilicon capacitors, and a polysilicon resistor).

13.3 Future directions

It is clear that we need to know a lot more about the nervous system and likely tissue reactions to MEMS-based devices in order for the technology to achieve its full potential. One can expect that application of the current generation of technology can show the way in this respect (e.g., see References 9, 21, and 88). In addition, injectable stimulators and retinal prosthesis projects are showing considerable promise in providing direct benefit to patients from current technologies.

One of the clear implications for the future is that the low-power circuit design techniques that have resulted from the demand for portable equipment will be taken up and implemented in neurotechnology. It also seems likely that the widening access to bi-CMOS processes on a multiproject basis (e.g., through MOSIS in the United States and Europractice in Europe) will result in increasing numbers of bi-CMOS designs, which are able to incorporate of the advantages of both bipolar and CMOS technologies.

What is more difficult to predict is the form that future circuit designs will take: full custom, semicustom, monolithic with MEMS, hybrid, etc. The mechanical properties of silicon and the biocompatibility results achieved with polyimide substrate microelectrode structures by the Fraunhöfer Institute for Biomedical Engineering[10,11] has led to further development of polyimide-based devices, including penetrating arrays.[85] In cases where the electrode substrate is not silicon, the fabrication process for the IC is not constrained by enforced compatibility with the MEMS process.

The design problem appears challenging enough that full-custom designs will, for the short term at least, provide the best solutions. The designer is, however, increasingly provided for in terms of tools, technologies, and intellectual properties that can be brought to bear on the problem. The current industrial trend for System-on-Chip (SoC) designs that incorporate analog and digital components on a single die should spin-off increasing knowledge and improved technologies that can be taken up in neuroprosthetics.

Another area of investigation is powering and control of implants. RF has tended to be dominant in the past, with IR a relative newcomer. New and innovative approaches to this problem would undoubtedly be welcome.

As mentioned, increased understanding of the operation of the nervous system is essential. This is particularly important when the potential data-gathering capacity of MEMS-based multimicroelectrode devices is considered (e.g., one device with 100 recording sites, each sampling extracellular spikes with information content to 10 kHz and eight-bit precision, implies a minimal data rate of 16 Mbps and the necessary processing power to make sense of the incoming data). Obviously, data compression and interpretation have to occur near the recording site; the best way to approach this problem will be informed by experimental results arising from the existing technology.

In 1991, Peeters et al.[59] presented a design that illustrates some of the ingenuity that has gone into solving the problems presented. A probe with ten recording sites was interfaced using only two wires: power and ground. The signal arising at each recording site was amplified and these were then multiplexed onto the input of a second amplification stage. An on-chip oscillator was used to drive the multiplexer. The output of this stage was compared with a reference voltage; any signals above a preset threshold indicated neural activity at the site. This was signaled through a current source that was connected across the power lines. The signal was pulse-width modulated to indicate the active channel. Thus, activity on channel 1 would cause a current pulse of 10-μsec duration to be drawn from the power supply, and activity on channel 10 would cause a 100-μs pulse; these pulses were detectable by the external circuitry. The disadvantages of such a system are quite apparent; however, the description provides considerable inspiration when dealing with the problems posed to neurotechnology.

Acknowledgments

The author would like to thank K. Bustamante, M. Hughes, D. Ewins, and the Biomedical Engineering Group at the University of Surrey for access to facilities and publications.

References

1. Stamford, J.A., Ed., *Monitoring Neuronal Activity: A Practical Approach*, IRL Press at Oxford University Press, Oxford, 1992.
2. Llinás, R., Nicholson, C., and Johnson, K., Implantable monolithic wafer recording electrodes for neurophysiology, in *Brain Unit Activity During Behaviour*, Philips, M.I., Ed., Charles C Thomas, Springfield, IL, 1973, chap. VII.
3. Wise, K.D., Angell, K.B., and Starr, A., An integrated circuit approach to extracellular microelectrodes, *IEEE Trans. Biomed. Eng.*, BME-17(3), 238, 1970.
4. Wise, K.D. and Angell, J.B., A low-capacitance multielectrode probe for use in extracellular neurophysiology, *IEEE Trans. Biomed. Eng.*, BME-22(3), 212, 1975.
5. Normann, R.A., Maynard, E.M., Guillory, K.S., and Warren, D.J., Cortical implants for the blind, *IEEE Spectrum*, May, 54, 1996.
6. Loeb, G.E., Walmsley, B., and Duysens, J., Obtaining proprioceptive information from natural limbs: implantable sensors vs. somatosensory neuron recordings, in *Physical Sensors for Biomedical Applications*, CRC Press, Boca Raton, FL, 1980, chap. 10.
7. Prochazka, A., Westerman, R.A., and Ziccone, S.P., Discharges of single hindlimb afferents in the freely moving cat, *J. Neurophysiol.*, 39, 1090, 1976.
8. Prohaska, O.J., Olcaytug, F., Pfundner, P., and Dragaun, H., Thin-film multiple electrode probes: possibilities and limitations, *IEEE Trans. Biomed. Eng.*, BME-33(3), 223, 1986.
9. Meister, M., Pine, J., and Baylor, D.A., Multi-neuronal signals from the retina: acquisition and analysis, *J. Neurosci. Meth.*, 51, 95, 1994.
10. Navarro, X., Calved, S., Rodríguez, F.J., Stieglitz, T., Blau, C., Butí, M., Valderrama, E., and Meyer, J-U., Stimulation and recording from regenerated peripheral nerves through polyimide sieve electrodes, *J. Periph. Nervous Syst.*, 3(2), 91, 1998.

11. Stieglitz, T., Beutel, H., Schuettler, M., and Meyer, J.-U., Micromachined, poly-imide based devices for flexible neural interfaces, *Biomed. Microdevices*, 2(4), 283, 2000.

12. Schuettler, M., Kock, K.P., Stieglitz, T., Scholz, O., Haberer, W., Keller, R., and Meyer, J.-U., Multichannel neural cuff electrodes with integrated multiplexer circuit, in *Proc. 1st Ann. Int. IEEE-EMBS Special Topic Conf. on Microtechnologies in Medicine and Biology*, Lyon, October 12–14, 2000, p. 624.

13. Najafi, K. and Wise, K.D., An implantable multielectrode array with on-chip signal processing, *IEEE J. Solid-State Circuits*, SC-21(6), 1035, 1986.

14. Najafi, K., Multielectrode Intracortical Recording Arrays with On-Chip Signal Processing, Ph.D. thesis, University of Michigan, Ann Arbor, 1986.

15. Edell, D.J., Clark, L.D., and McNeil, V.M., Optimization of electrode structure for chronic transduction of electrical neural signals, in *Proc. 8th Ann. Int. Conf. IEEE Eng. Med. Biol. Soc.*, Houston, TX, 1986, p. 1626.

16. Najafi, K., Ji, J., and Wise, K.D., Scaling limitations of silicon multichannel microprobes, *IEEE Trans. Biomed. Eng.*, 37(5), 474, 1990.

17. Normann, R.A., Visual neuroprosthetics: functional vision for the blind, *IEEE Eng. Med. Biol. Mag.*, January/February, 77, 1995.

18. Dobelle, W.H., Artificial vision for the blind by connecting a television camera to the visual cortex, *ASAIO J.*, 46, 3, 2000.

19. Anderson, D.J., Najafi, K., Tanghe, S.J., Evans, D.A.. Levy, K.L., Hetke, J.F., Xue, X., Zappa, J.J., and Wise, K.D., Batch-fabricated thin-film electrodes for stimu-lation of the central auditory system, *IEEE Trans. Biomed. Eng.*, 36(7), 870, 1989.

20. Rauschecker, J.P. and Shannon, R.V., Sending sound to the brain, *Science*, 295, 1025, 2002.

21. Serruya, M.D., Hatsopoulos, N.G., Paninski, L., Fellows, M.R., and Donoghue, J.P., Instant neural control of a movement signal, *Nature*, 416, 141, 2002.

22. Kennedy, P.R., Bakay, R.A., Moore, M.M., Adams, K., and Goldwaithe, J., Direct control of a computer from the human central nervous system, *IEEE Trans. Rehabil. Eng.*, 8(2), 198, 2002.

23. Najafi, K., Micromachined systems for neurophysiological applications, in *Handbook of Microlithography, Micromachining, and Microfabrication*, Vol. 2. *Micromachining and Microfabrication*, Rai-Choudhury, P., Ed., SPIE Optical Engi-neering Press, Washington, D.C., and IEE, London, 1997, chap.9.

24. Bogart, T.F., *Electronic Devices and Circuits*, Merrill Publishing, Columbus, OH, 1986.

25. Andrews, B.J. and Rushton, D.N., Lower limb functional neurostimulation, *Baillière's Clin. Neurol. Int. Practice Res.: Neuroprostheses*, 4(1), 35, 1995.

26. Troyk, P.R. and Donaldson, N. de N., Implantable FES stimulation systems: what is needed?, in *Proc. IFESS 2001, the 6th Ann. Conf. of the Int. FES Society*, Cleveland, OH, June 16–20, 2001.

27. Kovacs, G.T.A., Storment, C.W., and Rosen, J.M., Regeneration microelectrode array for peripheral nerve recording and stimulation, *IEEE Trans. Biomed. Eng.*, 39(9), 893, 1992.

28. Edell, D.J., A peripheral nerve information transducer for amputees: long-term multichannel recordings from rabbit peripheral nerves, *IEEE Trans. Biomed. Eng.*, BME-33(2), 203, 1986.

29. Rutten, W.L.C., Multi-microelectrode devices for intrafascicular use in periph-eral nerve, in *Proc. 18th Ann. Int. Conf. IEEE Eng. Med. Biol. Soc.*, October 31 – November 3, Amsterdam, 1996.

30. Banks, D.J., Ewins, D.J., Balachandran, W., and Richards, P.R., Development of an insertable neural signal transducer for peripheral nerve, in *Proc. BES Symp. on Electrical Stimulation: Clinical Systems*, April 6–7, University of Strathclyde, Glasgow, 1995, p. S6.4.

31. Loeb, G.E., Zamin, C.J., Schulman, J.H., and Troyk, P.R., Injectable microstimulator for functional electrical stimulation, *Med. Biol. Eng. Comput.*, 29, NS13, 1991.

32. Ziaie, B., Nardin, M.D., Coghlan, A.R., and Najafi, K., A single-channel implantable microstimulator for functional neuromuscular stimulation, *IEEE Trans. Biomed. Eng.*, 44(10), 909, 1997.

33. Peachey, N.S. and Chow, A.Y., Subretinal implantation of semiconductor-based photodiodes: progress and challenges, *J. Rehab. Res. Devel.*, 36(4), 371, 1999.

34. Stelzle, M., Stett, A., Brunner, B., Graf, M., and Nisch, W., Electrical properties of micro-photodiode arrays for use as artificial retina implant, *Biomed. Microdevices*, 3(2), 133, 2001.

35. Schwarz, M., Ewe, L., Hijazi, N., Hosticka, B.J., Huppertz, J., Kolnsberg, S., Mokwa, W., and Trieu, H.K., Micro implantable visual prostheses, in *Proc. 1st Ann. Int. IEEE-EMBS Special Topic Conf. on Microtechnologies in Medicine and Biology*, October 12–14, Lyon, 2000, p. 461.

36. Suaning, G.J. and Lovell, N.H., CMOS neurostimulation ASIC with 100 channels, scaleable output, and bi-directional radio-frequency telemetry, *IEEE Trans. Biomed. Eng.*, 48(2), 248, 2001.

37. Clements, M., Vichienchom, K., Liu, W., Hughes, C., McGucken, E., DeMarco, C., Mueller, J., Humayun, M., De Juan, E., Weiland, J., and Greenberg, R., An implantable power and data receiver and neuro-stimulus chip for a retinal prosthesis system, in *Proc. 1999 IEEE Int. Symp. on Circuits and Systems*, Vol. 1, 1999, p. 194.

38. Brock, K., Combining high speed and low power in 0.13 µm SoC designs, *Electron. Eng. Des.*, May, 20, 2002.

39. Horowitz, P. and Hill, W., *The Art of Electronics*, 2nd ed., Cambridge University Press, Cambridge, U.K., 1989.

40. Ismail, M. and Fiez, T., Eds., *Analog VLSI Signal and Information Processing*, McGraw-Hill, New York, 1994.

41. Geiger, R.L., Allen, P.E., and Strader, N.R., *VLSI Design Techniques for Analog and Digital Circuits*, McGraw-Hill, New York, 1990.

42. Soclof, S., *Design and Applications of Analog Integrated Circuits*, Prentice-Hall, Englewood Cliffs, NJ, 1991.

43. Jobling, D.T., Smith, J.G., and Wheal, H.V., Active microelectrode array to record from the mammalian central nervous system *in vitro*, *Med. Biol. Eng. Comput.*, 19, 553, 1981.

44. Fromherz, P., Offenhäusser, A., Vetter, T., and Weis, J., A neuron-silicon junction: a Retzius cell of the Leech on an insulated-gate field-effect transistor, *Science*, 252, 1290, 1991.

45. Fromherz, P. and Stett, A., Silicon-neuron junction: capacitive stimulation of an individual neuron on a silicon chip, *Phys. Rev. Lett.*, 75(8), 1670, 1995.

46. Offenhäusser, A., Sprössler, C., Matsuzara, M., and Knoll, W., Field-effect transistor array for monitoring electrical activity from mammalian neurons in culture, *Biosens. Bioelect.*, 12(8), 819, 1997.

47. Sprössler, C., Richter, D., Denyer, M., and Offenhäusser, A., Long-term recording system based on field-effect transistor arrays for monitoring electrogenic cells in culture, *Biosens. Bioelect.*, 13, 613, 1998.
48. Fromherz, P., Electrical interfacing of nerve cells and semiconductor chips, *Chem. Phys. Chem.*, 3, 276, 2002.
49. Fromherz, P., Müller, C.O., and Weis, R., Neuron transistor: electrical transfer function measured by the patch-clamp technique, *Phys. Rev. Lett.*, 71(24), 4079, 1993.
50. Grattarola, M. and Martinioia, S., Modeling the neuron-transistor junction: from extracellular to patch recording, *IEEE Trans. Biomed. Eng.*, 40(1), 35, 1993.
51. Ji, J., A Scaled Electrically Configurable CMOS Multichannel Intracortical Recording Array, Ph.D. thesis, University of Michigan, Ann Arbor, 1990.
52. Bai, Q., Gingerich, M.D., and Wise, K.D., An active three-dimensional microelectrode array for intracortical recording, in *Proc. Solid-State Sensor and Actuator Workshop*, June 8 – 11, Hilton Head Island, SC, 1998, p. 15.
53. Edell, D.J., Toi, V.V., MvNiel, V.M., and Clark, L.D., Factors influencing the biocompatibility of insertable silicon microshafts in cerebral cortex, *IEEE Trans. Biomed. Eng.*, 39(6), 635, 1992.
54. Drake, K.L., Wise, K.D., Farraye, J., Anderson, D.J., and BeMent, S.L., Performance of planar multisite microprobes in recording extracellular single-unit intracortical activity, *IEEE Trans. Biomed. Eng.*, 35(9), 719, 1988.
55. Rutten, W.L. C., van Wier, H.J., Put, J.H.M., Rutgers, R., and De Vos, R.A. I., Sensitivity, selectivity, and bioacceptance of an intraneural multi electrode stimulation device in silicon technology, *Electrophysiol. Kinesiol.*, 135, 1998.
56. Buitenweg, J.R., Rutten, W.L.C., and Marani, E., Finite element modeling of the neuron-electrode interface, *IEEE Eng. Med. Biol. Mag.*, 19(6), 46, 2000.
57. Frieswijk, T.A., Bielen, J.A., and Rutten, W.L.C., Development of a solder bump technique for contacting a three-dimensional multi electrode array, in *Proc. 18th Ann. Int. Conf. IEEE Eng. Med. Biol. Soc.*, October 31 – November 3, Amsterdam, 1996.
58. Ji, J. and Wise, K.D., An implantable CMOS circuit interface for multiplexed microelectrode recording arrays, *IEEE J. Solid-State Circuits*, 27(3), 433, 1992.
59. Peeters, E., Puers, B., and Sansen, W., A two-wire, digital output multichannel microprobe for recording single-unit neural activity, *Sensors Actuators B*, 4, 217, 1991.
60. Kim, C. and Wise, K.D., A 64-site multishank CMOS low-profile neural stimulating probe, *IEEE J. Solid-State Circuits*, 31(9), 1230, 1996.
61. Tanghe, S.J. and Wise, K.D., A 16-channel CMOS neural stimulating array, *IEEE J. Solid-State Circuits*, 27(12), 1819, 1992.
62. Gingerich, M.D. and Wise, K.D., An active microelectrode array for multipoint stimulation and recording in the central nervous system, in *Proc. Transducers '99*, June 7–10, Sendai, Japan, 1999, p. 280.
63. Ziaie, B., Najafi, K., and Anderson, D.J., A low-power miniature transmitter using a low-loss silicon platform for biotelemetry, in *Proc. 19th Ann. Int. Conf. IEEE Eng. Med. Biol. Soc.*, October 30 – November 2, Chicago, 1997, p. 2221.
64. Gatti, U. and Maloberti, F., Analog and mixed analog-digital layout, in *Analog VLSI Signal and Information Processing*, Ismail, M. and Fiez, T., Eds, McGraw-Hill, New York, 1994, chap. 16.

65. Akin, T., Ziaie, B., and Najafi, K., RF telemetry powering and control of hermetically sealed integrated sensors and actuators, in *Proc. IEEE Solid-State Sensors and Actuators Workshop*, 1990, p. 145.

66. Ji, J., Najafi, K., and Wise, K.D., A low-noise demultiplexing system for active multichannel microelectrode arrays, *IEEE Trans. Biomed. Eng.*, 38(1), 75, 1991.

67. Takahashi, K. and Matsuo, T., Integration of multi-microelectrode and interface circuits by silicon planar and three-dimensional fabrication technology, *Sensors Actuators*, 5, 89, 1984.

68. Banks, D.J., Balachandran, W., Richards, P.R., and Ewins, D., Instrumentation to evaluate neural signal recording properties of micromachined microelectrodes inserted in invertebrate nerve, *Physiol. Meas.*, 23, 437, 2002.

69. Bai, Q. and Wise, K.D., Single-unit neural recording with active microelectrode arrays, *IEEE Trans. Biomed. Eng.*, 48(8), 911, 2001.

70. Kim, K.H. and Kim, S.J., Noise characteristic design of CMOS source follower voltage amplifier for active semiconductor neural signal recording, *Med. Biol. Eng. Comput.*, 38(4), 469, 2000.

71. Kim, K.H. and Kim, S.J., Noise performance design of CMOS preamplifier for the active semiconductor neural probe, *IEEE Trans. Biomed. Eng.*, 47(8), 1097, 2000.

72. Dorman, M.G., Prisbe, M.A., and Meindl, J.D., A monolithic signal processor for a neurophysiological telemetry system, *IEEE J. Solid-State Circuits*, SC-20(6), 1185, 1985.

73. Maher, M.P., Pine, J., Wright, J., and Tai, Y-C., The neurochip: a new multielectrode device for stimulating and recording from cultured neurons, *J. Neurosci. Meth.*, 87, 45, 1999.

74. Banks, D.J., Ewins, D.J., and Balachandran, W., The effects of high pass filtering extracellularly recorded action potentials predicted by computer models, *Biomed. Eng. Appl. Basis Commun.*, 9(6), 363, 1997.

75. Kim, C. and Wise, K.D., Low-voltage electronics for the stimulation of biological neural networks using fully complementary BiCMOS circuits, *IEEE J. Solid-State Circuits*, 32(10), 1483, 1997.

76. Clements, M., Vichienchom, K., Liu, W., Hughes, C., McGucken, E., DeMarco, C., Mueller, J., Humayun, M., De Juan, E., Weiland, J., and Greenberg, R., An implantable neuro-stimulator device for a retinal prosthesis, in *Digest of Technical Papers, IEEE Int. Solid-State Circuits Conf.*, 1999, p. 216.

77. Heetderks, W.J., RF powering of millimeter and submillimeter-sized neural prosthetic implants, *IEEE Trans. Biomed. Eng.*, 35(5), 323, 1988.

78. Shah, M.R., Philips, R.P., and Normann, R.A., A study of printed spiral coils for neuroprosthetic transcranial telemetry applications, *IEEE Trans. Biomed. Eng.*, 45(7), 867, 1998.

79. Edell, D.J., Chronically implantable neural information transducers, in *Proc. 18th Ann. Int. Conf. IEEE Eng. Med. Biol. Soc.*, October 31 – November 3, Amsterdam, 1996.

80. Murakawa, K., Kobayashi, M., Nakamura, O., and Kawata, S., A wireless near-infrared energy system for medicinal implants, *IEEE Eng. Med. Biol. Mag.*, Nov./Dec., 70, 1999.

81. Goto, K., Nakagawa, T., Nakamura, O., and Kawata, S., An implantable power supply with an optically rechargeable lithium battery, *IEEE Trans. Biomed. Eng.*, 48(7), 830, 2001.

82. Cameron, T., Loeb, G.E., Peck, R.A., Schulman, J.H., Strojnik, P., and Troyk, P.R., Micromodular implants to provide electrical stimulation of paralyzed muscles and limbs, *IEEE Trans. Biomed. Eng.*, 44(9), 781, 1997.

83. Schlosshauer, B., Hoff, A., Guenther, E., Zrenner, E., and Hämmerle, H., Towards a retina prosthesis model: neurons on microphotodiode arrays *in vitro*, *Biomed. Microdevices*, 2(1), 61, 1999.

84. Stieglitz, T., Beutel, H., and Meyer, J.-U., A flexible, light-weight multichannel sieve electrode with integrated cables for interfacing regenerating peripheral nerves, *Sensors Actuators A*, 60, 240, 1997.

85. Rousche, P.J., Pellinen, D.S., Pivin, D.P., Williams, J.C., Vetter, R.J., and Kipke, D.R., Flexible polyimide-based intracortical electrode arrays with bioactive capability, *IEEE Trans. Biomed. Eng.*, 48(3), 361, 2001.

86. Akin, T., Ziaie, B., Nikles, S.A., and Najafi, K., A modular micromachined high-density connector system for biomedical applications, *IEEE Trans. Biomed. Eng.*, 46(4), 471, 1999.

87. Jones, K.E. and Normann, R.A., An advanced demultiplexing system for physiological stimulation, *IEEE Trans. Biomed. Eng.*, 44(12), 1211, 1997.

88. Stanley, G.B., Li, F.F., and Dan, Y., Reconstruction of natural scenes from ensemble responses in the lateral geniculate nucleus, *J. Neurosci.*, 19(18), 8036, 1999.

chapter fourteen

Molecular and nanoscale electronics

Michael C. Petty and C. Pearson

Contents

14.1 Introduction

The past 30 years have witnessed the emergence of *molecular electronics* as an important technology for the 21st century.[1,2] While modern electronics is based largely on silicon, molecular electronics is concerned with the exploitation of organic and biological materials in electronic and opto-

0-8493-1100-4/03/$0.00+$1.50
© 2003 by CRC Press LLC

electronic devices. Serious research started in the United States in the 1970s with pioneering work by Aviram, at IBM, who proposed a structure for a molecular rectifier.[3] Impetus was provided by the enthusiasm of Carter working in the U.S. Naval Research Laboratory,[4] and the field was epitomized by the elegant experiments on monolayer films undertaken by Kuhn in Göttingen.[5] However, the term *molecular electronics* dates further back, to an ambitious project instigated by the U.S. Air Force in the 1950s in an attempt to develop the integrated circuit.[6] The idea, which was radically different from other approaches to microelectronics, was to build a circuit in a solid without reproducing the individual component functions. Although analogies with the biological world are evident, the project required technology that was not available at the time (and perhaps is only now beginning to emerge), and little progress was made.

The subject of molecular electronics, as it has matured, can broadly be divided into two themes, as indicated in Figure 14.1, although there is substantial overlap. The first concerns the development of electronic and opto-electronic devices using the unique macroscopic properties of organic compounds (*molecular materials for electronics*). This division of molecular electronics is already making a technological impact, the best-known example being the liquid crystal display. Other areas in which organic compounds are becoming increasingly important are organic light-emitting displays, pyroelectric plastics for infrared imaging, and chemical and biochemical sensors.

The second strand to molecular electronics (*molecular scale electronics*) recognizes the dramatic size reduction in the individual processing elements

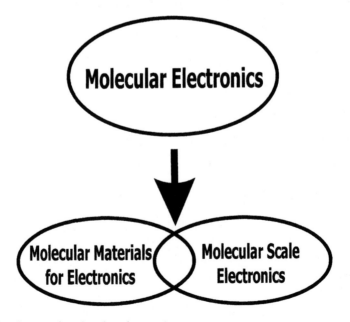

Figure 14.1 Scope of molecular electronics.

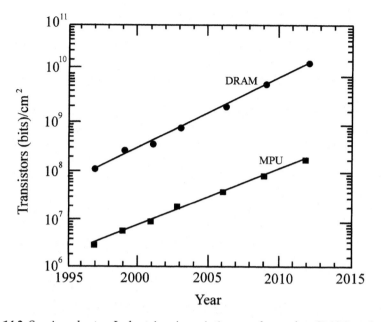

Figure 14.2 Semiconductor Industries Association roadmap for CMOS technology showing the predicted growth in density of the transistors in the microprocessor unit (MPU) and the dynamic random access memory (DRAM).[7]

in integrated circuits over recent years. Moore's law — namely, the functions per chip double every 1.5 years — will probably describe developments over at least the next decade. The semiconductor industries have produced an International Technology Roadmap on the future of complementary metal-oxide semiconductor (CMOS) technology.[7] Figure 14.2 shows the expected growth in the density of the transistors in both the microprocessor unit (MPU) and the dynamic random access memory (DRAM) of a CMOS chip over the next decade. The prediction is for a 30-nm minimum feature size (gate length) for the MPU and 10^{10} transistors per cm² in the case of the memory by the year 2012. These figures are regularly updated.

14.2 Conductive organic materials

The electronic and opto-electronic properties of silicon and gallium arsenide derive from the covalent bonds formed between the individual atoms. However, semiconductive behavior is not restricted to inorganic materials. Certain organic compounds possess significant electrical conductivity.[8,9] The physical explanation can be found in the band theory for solids, one of the great success stories of modern physics.

Whenever two identical atoms are brought close together, the electron orbitals overlap and the energy level associated with each electron in the separated atoms is split into two new levels, with one above and one below the original level. Consider, for example, the formation of a hydrogen molecule from two separated atoms. When the two atoms approach each other so that their 1s electron orbitals overlap, two new σ-electronic orbitals are formed around the atoms, symmetric with respect to the interatomic axis. In one orbital, the bonding orbital, the electron has a lower energy than in the isolated atom orbital and, in the other, the antibonding orbital, an electron has a higher energy.

In an extended solid many atoms can interact and many similar splittings of energy levels occur. For a solid containing approximately 10^{26} (Avogadro's number) atoms, each energy level splits, but the energies between these split levels are very small, and continuous ranges or bands of energy are formed. Two such important bands are the valence band and the conduction band, analogous to the bonding and antibonding levels of the two-atom model. The energy gap, or band gap, between them is a forbidden zone for electrons. Electrical conduction takes place by electrons moving under the influence of an applied electric field in the conduction band and/or holes moving in the valence band. Holes are really vacancies in a band but, for convenience, they may be regarded as positively charged carriers.

An important feature of the band model is that the electrons are delocalized or spread over the lattice. The strength of the interaction between the overlapping orbitals determines the extent of delocalization that is possible for a given system. An important material parameter is the mobility of the charge carriers. This is defined as the carrier velocity divided by the electric field and provides an indication of how quickly the carriers react to the field (i.e., the frequency response of the material). The greater the degree of electron delocalization, the larger the width of the bands (in energy terms) and the higher the mobility of the carriers within the band. For many polymeric organic materials, the molecular orbitals responsible for bonding the carbon atoms of the chain together are the sp^3 hybridized σ-orbitals that do not give rise to extensive overlapping. The resulting band gap is large, as the electrons involved in the bonding are strongly localized on the carbon atoms and cannot contribute to the conduction process. This is why a simple saturated polymer such as polyethylene, $-(CH_2)_n-$, is an electrical insulator.

A significant increase in the degree of electron delocalization may be found in unsaturated polymers (i.e., those containing double and triple carbon–carbon bonds). If each carbon atom along the chain has only one other atom (e.g., hydrogen) attached to it, the spare electron in a p_z orbital of the carbon atom overlaps with those of carbon atoms on either side, forming delocalized molecular orbitals of π-symmetry. For a simple lattice of length $L = Na$, where N is the total number of atoms and a is the spacing between them, it can be shown that the total number of electron states in the lowest energy band is exactly equal to N. This result is true for every energy band in the system and applies to three-dimensional lattices. Allowing for two spin orientations of an electron, the Pauli exclusion principle requires that there will be room for two electrons per cell of the lattice in an energy band. If each atom contributes one bonding electron, the valence band will be only half filled.

From the above it might be expected that a linear polymer backbone consisting of many strongly interacting coplanar p_z orbitals, each of which contributes one electron to the resultant continuous π-electron system, would behave as a one-dimensional metal with a half-filled conduction band. In chemical terms, this is a conjugated chain and may be represented by a system of alternating single and double bonds. It turns out that, for one-dimensional systems, such a chain can more efficiently lower its energy by introducing bond alternation (alternating short and long bonds). This limits the extent of electronic delocalization that can take place along the backbone. The effect is to open an energy gap in the electronic structure of the polymer. All conjugated polymers are large band-gap semiconductors, with band gaps more than about 1.5 eV, rather than metals (the band gap in silicon is 1.1 eV at room temperature).

Figure 14.3 depicts the bond formation and electronic band structure of a simple conjugated polymer, polyacetylene. The overlapping of the p_z orbitals to form π bonds is shown in Figure 14.3a and the resulting band structure in Figure 14.3b. The filled π band is the valence band, while the empty π^* band is the conduction band. Like silicon, the conductivity of polyacetylene can be changed by the addition of impurity atoms. However, the term *doping* is a misnomer, as it tends to imply the use of minute quantities, parts per million or less, of impurities introduced into a crystal lattice. In the case of conductive polymers, typically 1 to 50% by weight of chemically oxidizing (electron withdrawing) or reducing (electron donating) agents are used to physically alter the number of π electrons on the polymer backbone, leaving oppositely charged counter ions alongside the polymer chain. These processes are redox chemistry. The doping effect can be achieved because a π electron can be removed (added) without destroying the σ backbone of the polymer so that the charged polymer remains intact. The increase in conductivity can be as much as 11 orders of magnitude.

Derivatives of the "basic" conductive polymers have been synthesized to provide certain electronic features (e.g., band gap, electron affinity) and to increase their processability. Figure 14.4 shows a few examples: polypyrrole, polyparaphenylene, and polyphenylenevinylene. The monomer repeat units are based on five-membered or six-membered (benzene) carbon ring systems.

The electrical properties of organic polymers such as those described above are not directly comparable to those of silicon. In the latter, the three-dimensional crystallographic structure provides for extensive carrier delocalization throughout the solid, resulting in a relatively high mobility. Electrical conduction in polymers not only requires carrier transport along the polymer chains but also "hopping" between these chains, which tend to lie tangled up like a plate of spaghetti. Although some improvement in the carrier mobility can be achieved by both increasing the order of the polymer chains and by improving the purity of the material, the charge carrier mobilities in organic compounds are usually quite low, making it difficult to produce very high-speed electronic computational devices that

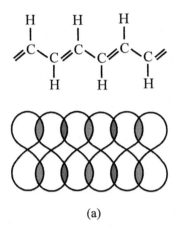

(a)

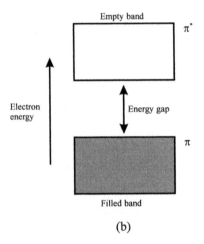

(b)

Figure 14.3 (a) Chemical (π) bonding in polyacteylene. (b) Electronic band structure showing the normally empty π* band (conduction band) and the normally filled π band (valence band).

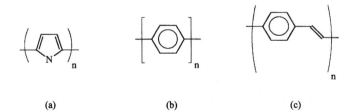

(a)   (b)   (c)

Figure 14.4 Chemical structures of conductive polymers: (a) polypyrrole, (b) poly-paraphenylene, and (c) polyphenylenevinylene.

are competitive with those based on silicon and gallium arsenide. None-theless, other features make organic compounds attractive for certain types of electronic device, as indicated in the following sections.

14.3 Thin-film architectures

Technological applications of new materials often require them to be in the form of thin films. Many thin-film-processing techniques have already been developed for the fabrication of electronic and opto-electronic components. Well-established methods of organic film deposition include electrodeposition, thermal evaporation, and spinning.[2] The Langmuir–Blodgett (LB) technique, self-assembly, and layer-by-layer electrostatic deposition are further means of producing layers of organic materials; these methods allow ultra-thin-film assemblies of organic molecules to be engineered at the molecular level.[10,11]

14.3.1 Langmuir–Blodgett technique

Langmuir–Blodgett films are prepared by first depositing a small quantity of an amphiphilic compound (i.e., one containing both polar and nonpolar groups) dissolved in a volatile solvent onto the surface of purified water.[11] The classical materials are long-chain fatty acids, such as *n*-octadecanoic acid (stearic acid), as shown in Figure 14.5. When the solvent has evaporated, the

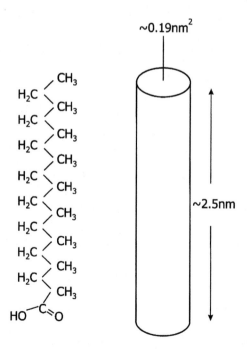

Figure 14.5 Chemical formula for *n*-octadecanoic acid (stearic acid). The approximate geometrical shape and dimensions of the molecule are shown on the right.

organic molecules may be organized into a floating two-dimensional "crystal" by compression on the water surface. As the area available to the organic molecules is reduced, the floating film will undergo several phase transformations. These are, to a first approximation, analogous to three-dimensional gas, liquid, and solid phases. The phase changes may readily be identified by monitoring the surface pressure as a function of the area occupied by the molecules in the film. This is the two-dimensional equivalent to the pressure vs. volume isotherm for a gas. Figure 14.6 shows such a plot for a hypothetical long-chain organic monolayer material (e.g., a long-chain fatty acid). In the gaseous state (G in Figure 14.6) the molecules are far enough apart on the water surface that they exert little force on one another. As the surface area of the monolayer is reduced, the hydrocarbon chains will begin to interact. The liquid state that is formed is generally called the expanded monolayer phase (E). The hydrocarbon chains of the molecules in such a film are in a random, rather than regular, orientation with their polar groups in contact with the subphase. As the molecular area is progressively reduced, condensed (C) phases may appear. There may be more than one of these, and the emergence of each condensed phase can be accompanied by constant pressure regions of the isotherm, as observed in the cases of a gas condensing to a liquid and a liquid solidifying. In the condensed monolayer states, the molecules are closely packed and are oriented with their hydrocarbon chain pointing away from the water surface. The area per molecule in such a state

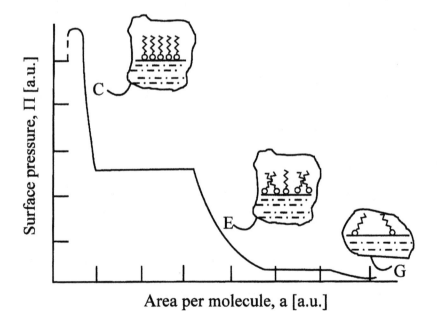

Figure 14.6 Surface pressure vs. area per molecule isotherm for a long-chain organic compound showing the gaseous (G), expanded (E), and condensed (C) phases. The surface pressure Π and area *a* are in arbitrary units (a.u.).

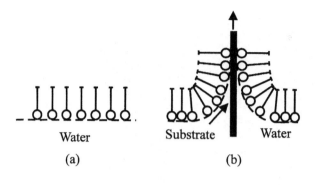

Water

(a)

Substrate   Water

(b)

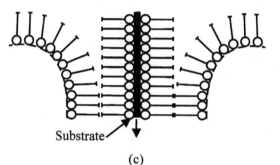

Substrate

(c)

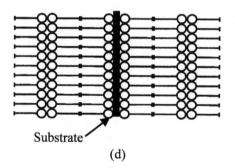

Substrate

(d)

Figure 14.7 Langmuir–Blodgett film deposition showing the transfer of amphiphilic molecules from the surface of water onto a solid substrate.

will be similar to the cross-sectional area of the hydrocarbon chain (i.e., ≈ 0.19 nm^2 molecule^{-1}).

If the surface pressure is held constant in one of the condensed phases, then the film may be transferred from the water surface onto a suitable solid substrate simply by raising and lowering the latter through the mono-layer–air interface. This technique, introduced by Langmuir and Blodgett,[11] is illustrated in Figure 14.7. In this example, the substrate is hydrophilic and the floating condensed monolayer (Figure 14.7a) is transferred, like a carpet, as the substrate is raised through the water (Figure 14.7b). Subsequently, a

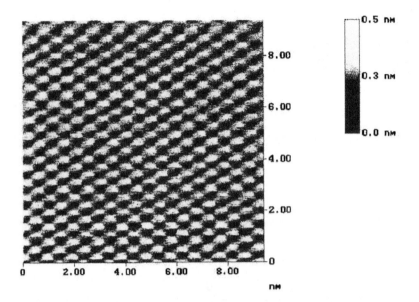

Figure 14.8 Atomic force micrograph of a 12-layer, *n*-eicosanoic acid Langmuir–Blodgett film deposited onto silicon. Unfiltered data obtained with a Digital Instruments Nanoscope III run in contact mode.[12]

monolayer is deposited on each traversal of the monolayer–air interface (Figure 14.7c). As shown in the figure, these stack in a head-to-head and tail-to-tail pattern (Figure 14.7d); this deposition mode is called *Y-type* and is that most frequently encountered. Instances in which the floating monolayer is only transferred to the substrate as it is being inserted into the subphase, or only as it is being removed, are also observed. These deposition modes are called *X-type* and *Z-type*, respectively.

The organization of the organic molecules in LB assemblies may be investigated by a number of analytical techniques including x-ray and neutron reflection, electron diffraction, and infrared spectroscopy.[11] Figure 14.8 shows an example of an atomic force micrograph (AFM) of a 12-layer, *n*-eicosanoic acid LB film deposited onto single-crystal silicon.[12] Lines of individual molecules are evident at the magnification shown.

Monolayer and multilayer films bear a striking resemblance to the naturally occurring biological membrane, as illustrated in Figure 14.9. The basis for this structure is a bilayer of amphiphilic phospholipid molecules. Proteins, shown as large globular molecules in the figure, are partially embedded in and protruding from this layer. Many components of cell membranes form condensed layers at the air–water interface, and some can be assembled into multilayer films.[13] Phospholipids, such as phosphatidic acid,[14] are good examples. However, chlorophyll-a, the green pigment in higher plants; vitamins A, E, and K; and cholesterol can also form condensed floating monomolecular layers. This has led to some work on LB biomimetic systems, in particular, the development of biological sensors.[15]

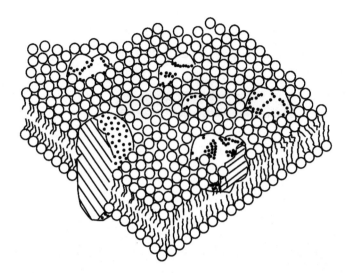

Figure 14.9 Schematic diagram showing the structure of a biological membrane.

14.3.2 Self-assembly

Self-assembly is a much simpler process than that of LB deposition. Mono-molecular layers are formed by the immersion of an appropriate substrate into a solution of the organic material (Figure 14.10a). The best known examples of self-assembled systems are organosilicon on hydroxylated surfaces (SiO_2, Al_2O_3, glass, etc.) and alkanethiols on gold, silver, and copper.[10] However, other combinations include dialkyl sulfides on gold; dialkyl disulfides on gold; alcohols and amines on platinum; and carboxylic acids on aluminum oxide and silver. The self-assembly process is driven by the interactions between the head group of the self-assembling molecule and the substrate, resulting in a strong chemical bond between the head group and a specific surface site (e.g., a covalent Si–O bond for alkyltrichlorosilanes on hydroxylated surfaces).

The combination of the self-assembly process with molecular recognition offers a powerful route to the development of nanoscale systems that may have technological applications as sensors, devices, and switches. For instance, the complexation of a neutral or ionic guest at one site in a molecule may induce a change in the optical or redox properties of the system. Figure 14.10b shows the proposed orientation of a derivative of the electroactive molecule tetrathiafulvalene (TTF) on a gold surface.[16] The incorporation of the metal-binding macrocyclic structure is to enable the molecule to function as a metal-cation sensor. Monolayers assembled onto platinum have been shown to exhibit electrochemical recognition to Ag^+ ions.

The self-assembly process, as described above, is usually restricted to the deposition of a single molecular layer on a solid substrate; however, chemical means can be exploited to build up multilayer organic films. A method pioneered by Sagiv is based on the successive absorption and

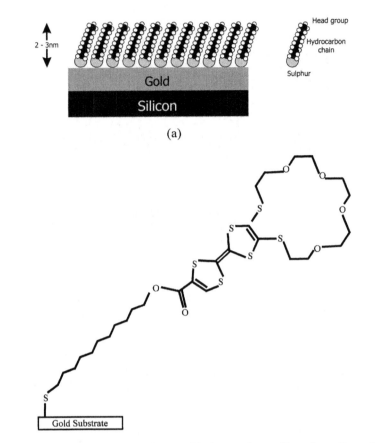

Figure 14.10 Self-assembly: (a) self-assembled monolayer film of an alkanethiol on a gold-coated substrate, and (b) possible orientation of a self-assembled layer of a cation-sensitive, redox-active molecule.[16]

reaction of appropriate molecules.[17,18] The headgroups react with the substrate to give a permanent chemical attachment, and each subsequent layer is chemically attached to the one before in a very similar way to that used in systems for supported synthesis of proteins.

14.3.3　*Electrostatic layer-by-layer deposition*

Another technique for building up thin films of organic molecules is driven by the ionic attraction between opposite charges in two different polyelectrolytes. Figure 14.11 shows a schematic diagram of this layer-by-layer assembly technique.[19,20] A solid substrate with a positively charged planar surface is immersed in a solution containing an anionic polyelectrolyte and a monolayer of polyanion is adsorbed (Figure 14.11a). Because the adsorption is carried out at relatively high concentrations of the polyelectrolyte, most of the ionic groups remain exposed to the interface with the solution,

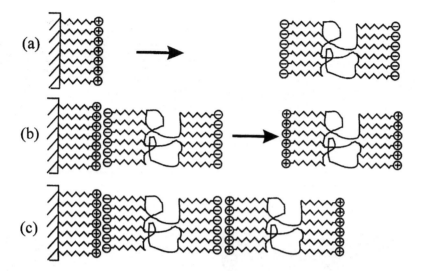

Figure 14.11 Schematic representation of the assembly of a multilayer architecture by consecutive adsorption of anionic and cationic polyelectrolytes.[19]

thus the surface charge is reversed. After rinsing in pure water, the substrate is immersed in a solution containing the cationic polyelectrolyte. Again, a monolayer is adsorbed but now the original surface charge is restored (Figure 14.11b), thus resulting in the formation of a multilayer assembly of both polymers (Figure 14.11c). It is possible to use a sensitive optical technique, such as surface plasmon resonance, to monitor, *in situ,* the growth of such electrostatically assembled films.[21] The layer-by-layer method has been used to build up layers of conductive polymers (e.g., partially doped polyaniline and a polystyrene polyanion).[22] Biocompatible surfaces consisting of alternate-layers of charged polysaccharides and oppositely charged synthetic polymers can also be deposited in this way.[23] A related, but alternative, approach uses layer-by-layer adsorption driven by hydrogen bonding interactions.[24]

14.3.4 Patterning

Patterning technologies are essential in the fabrication of most microelectronic device structures. Planar components are normally patterned using photolithography. Here, a surface is first covered with a light-sensitive photoresist, which is exposed to ultraviolet light through a contact mask. Either the exposed photoresist (positive resist) or the unexposed regions (negative resist) can then be washed away to leave a positive or negative image of the mask on the surface. This approach is routinely used in the fabrication of devices based on inorganic semiconductors. However, difficulties can be encountered when used with organic films, as the photoresists themselves are based on organic compounds.

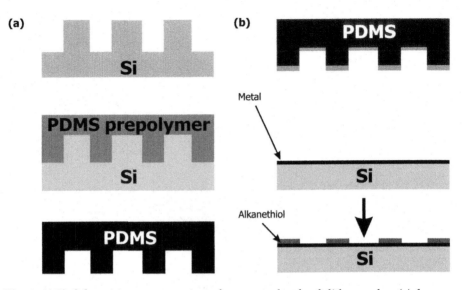

Figure 14.12 Schematic representation of an example of soft lithography: (a) formation of an elastomeric stamp, and (b) transfer of chemical "ink" to a surface.[25]

Brittain et al.[25] describe a series of "soft" lithographic methods that may be better suited to the patterning of organic layers. Pouring a liquid polymer, such as polydimethylsiloxane (PDMS) onto a "master" made from silicon forms a pattern-transfer element, as shown in Figure 14.12. The polymer is allowed to cure to form an elastomer, which can then be removed from the master (Figure 14.12a). This replica can subsequently be used as a stamp to transfer chemical ink, such as a solution of an alkanethiol, to a surface (Figure 14.12b).

Scanning microscopy methods offer a powerful means of manipulating molecules. Careful control of an AFM tip can allow patterns to be drawn in an organic film.[26] Such techniques can also be used to reposition molecules, such as the fullerene C60, on surfaces and to break up an individual molecule.[27] A further approach that has recently been developed at Northwestern University is called dip-pen nanolithography (DPN).[28] This technique, illustrated in the Figure 14.13, is able to deliver organic molecules in a positive printing mode. An AFM tip is used to "write" alkanethiols on a gold thin film in a manner analogous to that of a fountain pen. Molecules flow from the AFM tip to a solid substrate ("paper") via capillary transport, making DPN a potentially useful tool for assembling nanoscale devices.

The chemisorption of the "ink" is the driving force that moves the ink from the AFM tip through the water to the substrate as the tip is scanned across this surface. Adjusting the scan rate and relative humidity can control line widths. The current line-width resolution is 15 nm and spatial resolution is less than 10 nm. These parameters are limited by the tip radii of curvature associated with conventional AFM tips but in principle can

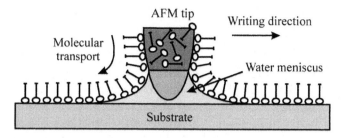

Figure 14.13 Dip-pen patterning technique showing the transfer of an "ink" onto "paper" using the tip of an atomic force microscope (AFM).[28]

be decreased through improvements in tip fabrication technology. DPN is a direct-write nanolithographic process where one can pattern and image with the same tool.

Recent developments of DPN have included an overwriting capability that allows one nanostructure to be generated and the areas surrounding that nanostructure to be filled with a second type of ink.[29] Perhaps the greatest limitation in using scanning probe methodologies for ultra-high-resolution nanolithography over large areas derives from the serial nature of most techniques; however, an eight-pen nanoplotter capable of doing parallel lithography has been reported.[30] The DPN method has also been used to deposit magnetic nanostructures[31] and arrays of protein molecules.[32]

There is also an intriguing link between the DPN process and LB deposition. For example, if the DPN experimental arrangement shown in Figure 14.13 is turned by 90°, a nanometer-scale LB trough, similar to those reported that use a moving subphase to compress the monolayer film is evident.[33] As the amphiphilic molecules flow down the AFM tip, supported by a thin layer of water, their surface pressure will rise. The resulting condensed monomolecular film is then transferred to a solid substrate; however, in the case of DPN, the substrate is stationary while the "nano-trough" moves in relation to it.

14.4   Molecular materials for electronics

14.4.1   Plastic electronics

Since the discovery of semiconducting behavior in organic materials, considerable research effort has been aimed at exploiting these properties in electronic and opto-electronic devices. Organic semiconductors can have significant advantages over their inorganic counterparts. For example, thin layers of polymers can easily be made by low-cost methods such as spin coating. High-temperature deposition from vapor reactants is generally needed for inorganic semiconductors. Synthetic organic chemistry also offers the possibility of designing new materials with different bandgaps. As noted in Section 14.2, the mobilities of the charge carriers in organic field effect transistors are low. Nevertheless, the simple fabrication techniques for polymers have attracted several companies to work on polymer transistor applications such as data storage and thin-film device arrays to address liquid crystal displays.[34]

Semiconducting organic films have been used similarly to inorganic semiconductors (e.g., Si, GaAs) in metal/semiconductor/metal structures. Perhaps the simplest example is that of a diode. Here, the semiconductor is sandwiched between metals of different work functions. In the ideal case, an n-type semiconductor should make an ohmic contact to a low work function metal and a rectifying Schottky barrier to a high work function metal.[35] An early example is that of a phthalocyanine LB film sandwiched between aluminum and indium–tin–oxide electrodes.[36]

Of particular interest is the possibility of observing molecular rectification using monolayer or multilayer films. This follows the early prediction of Aviram and Ratner (noted in the Introduction) that an asymmetric organic molecule containing a donor and an acceptor group separated by a short σ-bonded bridge, allowing quantum mechanical tunneling, should exhibit diode characteristics.[3] Many attempts have been made to demonstrate this effect in the laboratory, particularly in LB film systems.[37,38] Asymmetric current vs. voltage behavior has certainly been recorded for many LB film metal/insulator/metal structures, although some of these results are open to several interpretations as a result of the asymmetry of the electrode configuration.

Organic thin films have also been used as the semiconducting layer in field effect transistor (FET) devices.[39,40] These are three-terminal structures; a voltage applied to a metallic gate affects an electric current flowing between source and drain electrodes. For transistor operation, the charge must be easily injected from the source electrode into the organic semiconductor and the carrier mobility should be high enough to allow useful quantities of source–drain current to flow. The organic semiconductor and other materials with which it is in contact must also withstand the operating conditions without thermal, electrochemical, or photochemical degradation. Two performance parameters to be optimized in organic field effect transistors are the field effect mobility and on/off ratio.[40]

Figure 14.14 shows a schematic diagram for the possible structure of an organic thin-film FET. In this arrangement, the organic film is deposited in the final stages of FET fabrication. It is therefore not necessary for this layer to withstand any post deposition, chemical, and thermal processing. Figure 14.15 shows how a series of transistor devices can be made on the same silicon substrate incorporating different thicknesses of LB films. The silicon serves as the gate electrode, while silicon dioxide forms the insulator. Experimental data for such a device, using an organometallic complex as the semiconductive layer, are shown in Figure 14.16.[41,42] The graph shows the dependence of the saturated source–drain current vs. gate bias voltage for a device incorporating a film consisting of 59 LB layers of the iodine-doped complex on top of the interdigitated source and drain electrodes.[42] The slope

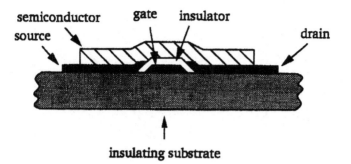

semiconductor   gate   insulator

source   drain

↑

insulating substrate

Figure 14.14 Schematic diagram of a field effect transistor (FET) structure.

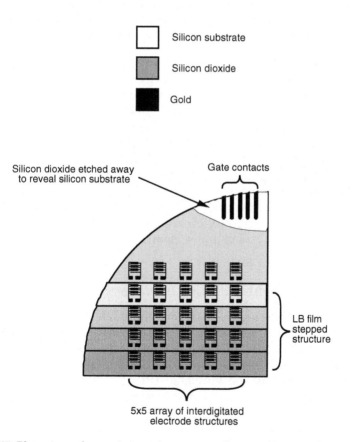

☐ Silicon substrate

▨ Silicon dioxide

■ Gold

Silicon dioxide etched away   Gate contacts
to reveal silicon substrate

LB film
stepped
structure

5x5 array of interdigitated
electrode structures

Figure 14.15 Plan view of organic transistors on a silicon substrate. The structure is that shown in Figure 14.14. The gate is silicon. Silicon dioxide forms the gate insulator, while the semiconductor is a Langmuir–Blodgett film. A stepped structure provides different thicknesses of semiconductive film.[42]

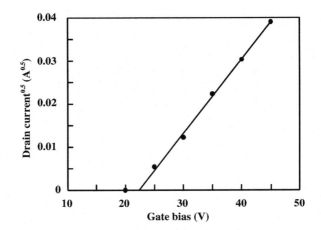

Figure 14.16 (Saturated drain-source current)$^{0.5}$ vs. gate bias voltage for a thin-film transistor incorporating 59 Langmuir–Blodgett layers of an iodine-doped charge transfer complex on top of interdigitated source and drain electrodes.[41,42]

of the straight line gives a carrier mobility of 0.3 cm^2 V^{-1} sec,$^{-1}$ a relatively high figure for an organic transistor device.

The operating characteristics of organic transistors have improved markedly over recent years. This has been brought about by both improvements in the material synthesis and in the thin-film processing techniques.[43–48] State-of-the-art devices possess characteristics similar to those of devices prepared from hydrogenated amorphous silicon, with mobilities around 1 cm^2 V^{-1} sec^{-1} and on/off ratios greater than 10^6. The use of ambipolar thin-film devices should lead to a significant simplification of complementary logic circuits. Thin-film transistors based on organic semiconductors are likely to form key components of plastic circuitry for use as display drivers in portable computers and pagers, and as memory elements in transaction cards and identification tags.

14.4.2   *Organic light-emitting structures*

Reports of light emission from organic materials on the application of an electric field (electroluminescence) have been around for many years. However, there has been an upsurge in interest following the initial report of organic light-emitting devices (OLEDs) incorporating the conjugated polymer poly(p-phenylene vinylene) (PPV).[49] The simplest OLED is an organic semiconductor sandwiched between electrodes of high and low work function. On application of a voltage, electrons are injected from one of the electrodes and holes from the other; recombination of these carriers then results in the emission of light. Electron- and hole-transporting films may be deposited between the cathode and the emitting layer and the anode and the emitting layer, respectively, to improve and balance the injection of carriers. Figure 14.17 shows the structure of an OLED that

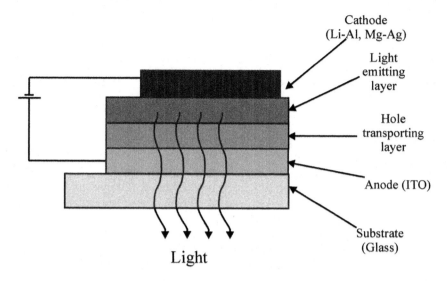

Cathode
(Li-Al, Mg-Ag)

Light
emitting
layer

Hole
transporting
layer

Anode (ITO)

Substrate
(Glass)

Light

Figure 14.17 Organic light-emitting device structure incorporating a light-emitting layer and a hole transport layer sandwiched between an indium–tin–oxide (ITO) anode and a metallic cathode.

uses a hole transport layer sandwiched between an indium–tin–oxide (ITO) anode and a light-emitting layer. Work is focused on the use of low-molecular-weight organic molecules and polymers and industrial interest in the application of such materials to various display technologies is considerable.[50]

Many techniques have been used in attempts to optimize the performance of OLEDs. For example, an inorganic insulating layer, such as LiF, may be inserted between the cathode and the emissive material.[50] The ITO anode electrode can be treated (e.g., with an oxygen plasma[51]) to reduce the turn-on voltage. Bilayer anodes such as ITO/polyaniline,[52] ITO/poly-(3,4-ethylene dioxythiophene),[53] and ITO/phthalocyanine[54] have also been shown to be beneficial.

The thin polymeric or oligomeric layers required for OLEDs are usually deposited by spin coating or thermal evaporation. The LB technique and layer-by-layer self-assembly offer alternative methods for building up ultra-thin organic films of nanometer dimensions. These approaches to thin-film deposition allow the thickness requirements of OLEDs to be fully explored in the laboratory. They may also offer certain advantages for the fabrication of devices containing ultrathin organic films and/or alternate-layer films. In a similar manner to the use of LiF noted above, improvements in device efficiency may be achieved by the insertion of an LB film "spacer" between the emissive film and the cathode.[55] The relatively ordered nature of LB arrays of molecules may be exploited in light-emitting structures. For example, the preferential alignment of the molecules can result in polarized light emission.[56,57] Langmuir–Blodgett deposition can also be used to control the positions of luminescent species within metal mirror microcavities, allowing the optical mode structure of the cavities to be probed by both the emission and absorption of the chromophore.[58]

14.4.3 *Chemical sensors*

The development of effective devices for the identification and quantification of chemical and biochemical substances for process control and environmental monitoring is a growing need.[59,60] Many sensors do not possess the specifications to conform to existing or forthcoming legislation; some systems are too bulky or expensive for use in the field. Inorganic materials such as the oxides of tin and zinc have traditionally been favored as the sensing element;[61] however, one disadvantage of sensors based on metallic oxides is that they usually have to be operated at elevated temperatures, limiting some applications. As an alternative, there has been considerable interest in trying to exploit the properties of organic materials. Many such substances (in particular, phthalocyanine derivatives) are known to exhibit a high sensitivity to gases.[62] Lessons can also be taken from the biological world; one household carbon monoxide detector currently being marketed is designed to simulate the reaction between CO and hemoglobin. A significant advantage of organic compounds is that their sensitivity and selectivity can be tailored to a particular application by modifications to their chemical structure. Moreover, thin-film technologies, such as self-assembly or layer-by-layer electrostatic deposition, enable ultra-thin layers of organic materials to be engineered at the molecular level.[63]

There are many physical principles upon which sensing systems might be based; changes in electrical resistance (chemiresistors), refractive index (fiberoptic sensors), and mass (quartz microbalance) have all been exploited in chemical sensing. The main challenges in the development of new sensors are in the production of inexpensive, reproducible, and reliable devices with adequate sensitivities and selectivities.

While many sensing devices may show adequate sensitivities, the selectivity can be poor. A semiconductive polymer may show a similar change in electrical resistance to a range of oxidizing (reducing) gases. To get around this difficulty, one approach that is being embraced enthusiastically by researchers is to use an array of sensing elements, rather than a single device. This is the method favored by nature. The human olfactory system, depicted in Figure 14.18a, has many receptors cells (sensors), which are individually nonspecific; signals from these are fed to the brain via a network of primary (glomeruli layer) and secondary (mitral layer) neurons for processing. It is generally believed that the selectivity of the olfactory system is a result of a high degree of parallel processing in the neural architecture. Artificial neural networks (Figure 14.18b) can, to some extent, emulate the connectivity of the olfactory neurons.[64] The electronic nose is an attempt to mimic the human olfactory system, and several companies now market such equipment.[60,65] Individual sensors can be based on polymer films. Each element is treated

Olfactory receptor cell

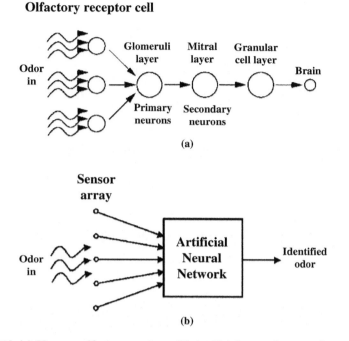

(a)

(b)

Figure 14.18 (a) Human olfactory system. (b) Artificial neural network.

in a slightly different way during deposition so that it responds uniquely on exposure to a particular gas or vapor. The pattern of resistance changes in the sensor array can then be used to fingerprint the vapor.

An alternative technique to using an array of sensing elements is to make multiple measurements on one sample. For example, discrimination between different vapors may be achieved by monitoring the complex electrical admittance of a sensing layer at several frequencies (admittance spectroscopy). Figure 14.19 shows the frequency behavior, at room temperature over the range of 10^{-1} to 10^3 kHz, of the capacitance of an LB film of a coordination polymer formed by the reaction of the bifunctional amphiphilic ligand, 5,5'methylenebis (N-hexadecylsalicylideneamine) and copper ions in an interfacial reaction at the water surface.[66,67] The measurements were undertaken in both nitrogen gas and ethanol vapor (3.3%). The inset shows the transient behavior measured at a fixed frequency (1 kHz) when the ethanol vapor was turned on and off. On exposure to other vapors, similar transient responses were noted. The percentage changes in the capacitance and conductance, normalized to the vapor concentrations, are shown in Figure 14.20.

It is evident that the capacitance increases when the device is exposed to ethanol and acetonitrile, with a greater fractional increase for the acetonitrile. This is consistent with the strongly polar nature of the latter solvent (dipole moment for acetonitrile = 3.92 Debye compared to 1.69 Debye for

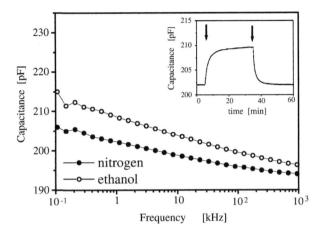

Figure 14.19 Capacitance vs. frequency for a coordination polymer Lang-muir–Blodgett film in nitrogen and exposed to ethanol vapor (3.3%). The inset shows the transient behavior measured at a fixed frequency (1 kHz) when the ethanol vapor was turned on and off (indicated by the arrows).[66]

ethanol). A decrease of capacitance is observed with the nonpolar benzene, almost certainly associated with the swelling of the film. From a practical sensing viewpoint, it is clear that monitoring the admittance at four frequencies provides a means of discrimination between the vapors studied.

14.5 Molecular scale electronics

The "bottom-up" approach to molecular electronics offers many intriguing prospects for manipulating materials on the nanometer scale, thereby providing opportunities to build up novel architectures with predetermined and unique physical and/or chemical properties. Two relatively simple examples are described below: the incorporation of an electric polarization into a multilayer array and the use of conductive nanoparticles to realize single electron devices. The possible exploitation of the electronic properties of DNA is also discussed.

14.5.1 Molecular superlattices

It is widely recognized that pyroelectric materials have considerable advantages over narrow bandgap semiconductors, such as mercury cadmium telluride (CMT), as detectors of infrared radiation. In addition to their broader spectral sensitivity, pyroelectric detectors possess the advantage of efficient operation at ambient temperatures, obviating the need for expensive cooling systems that are required for their CMT counterparts.[68]

For a material to be pyroelectric, it must possess a noncentrosymmetric crystal structure with a unique polar axis.[69] Furthermore, if the material is to form the basis of an efficient pyroelectric detector, it must be fabricated

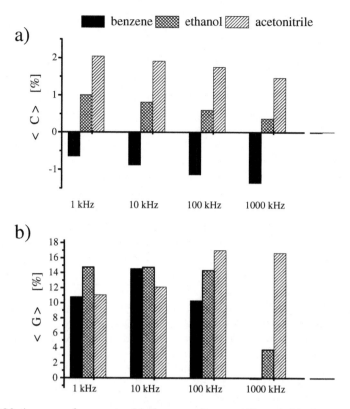

Figure 14.20 Average changes in (a) the capacitance, ΔC, and (b) the conductance, ΔG, for a coordination polymer Langmuir–Blodgett film, measured at four frequencies and for three different vapors.[66]

in thin-film form. Conventionally, two broad approaches have been taken to producing pyroelectric thin films: the first is to grow a single crystal of a material such as triglycine sulfate, which belongs to one of the polar space groups, and to thin the crystal by slicing, mechanical grinding, or etching; the second, and more satisfactory method is to grow a thin polycrystalline, nonpolar film and to render it pyroelectric by applying a large electric field ("poling"). This is particularly applicable to ceramics, which can be deposited by radio frequency sputtering, and to polymers, which can be spin-coated onto a suitable substrate.

Several advantages make LB films particularly attractive candidates for pyroelectric materials. The most important is that the sequential deposition of single monolayers enables the symmetry of the film to be precisely defined; in particular, layers of different materials can be built up to produce a highly polar structure. Second, the polarization of an LB film is "frozen-in" during deposition, and it is therefore not necessary to subject the film to a poling process. The third advantage is that the LB technique uses amphiphilic organic materials which possess low permittivities, ε, and the

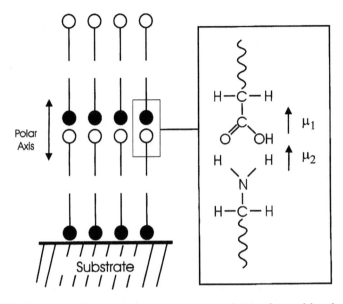

Figure 14.21 Schematic diagram of an organic superlattice formed by the alternate-layer Langmuir–Blodgett deposition of a long-chain acid and a long-chain amine. The dipole moments associated with the polar head groups are shown on the right.[68]

figure of merit for voltage responsivity p/ε (where p is the pyroelectric coefficient) is expected to be large. Finally, the LB method enables the preparation of much thinner films than are usually attainable by more conventional techniques.

Figure 14.21 illustrates the principle of a superlattice consisting of acid and amine molecules whose dipole moments are in opposite senses but when deposited in Y-type LB film form are aligned in the same direction. Infrared studies have indicated that the deposition results in a proton transfer from the acid head group to that of the amine, giving rise to an overall polarization component perpendicular to the multilayer plane.[70]

Many LB materials deposited as alternate-layer films or as X- or Z-type layers exhibit such behavior, including polymeric compounds and phospholipid materials.[71] The latter compounds can possess large dipole moments associated with their head groups. The pyroelectric coefficients of multilayer acid/amine LB films can be about 10 μC m^{-2} K^{-1} and depend on the thermal expansion coefficient of the substrate, indicating the presence of a significant secondary contribution (i.e., as a result of the piezoelectric effect) to the measured pyroelectric response.[68,70] Although these pyroelectric coefficients are still less (by about a factor of three) than those measured for polyvinylidene fluoride, a well-known pyroelectric polymer, the dielectric constants are also less, providing comparable figures of merit for infrared detection devices. Alternate-layer LB structures can also exhibit a significant second-order nonlinear optical response (e.g., second-harmonic generation).[11]

14.5.2 Single electron devices

Progress down the Moore's law curve (Figure 14.2) has enabled the design of electronic structures on a scale where quantum mechanical effects become important. The current flow in such devices can be determined by tunneling through energy barriers and may also exploit the fact that charge is quantized in units of e (1.6×10^{-19} C). Consequently, there has been much interest focused on single-electron devices, in which the addition or subtraction of a small number of electrons to or from an electrode can be controlled with one-electron precision.[72,73]

A single-electron tunneling device is a structure based on the principle of Coulomb blockade, where the number of electrons on an island (semiconducting or metallic dot) is controlled by a capacitatively coupled gate, as depicted in Figure 14.22a.

The associated change in Coulomb energy on addition of one electronic charge is conveniently expressed in terms of the capacitance, C, of the island. The extra charge changes the electrostatic potential by the charging energy, e^2/C. This charging energy becomes important when it exceeds the thermal energy $k_B T$, where k_B is the Boltzmann constant (1.38×10^{-23} JK^{-1}) and T is the absolute temperature.

A second condition is that the barrier between the electron reservoir and the island is sufficiently opaque that the electrons are either located on the island or in the reservoir. This means that quantum fluctuations in the number due to tunneling through the barriers is much less than one. This requirement translates to a lower value for the tunnel resistance, R_t, of the barrier, which should be much larger than the resistance quantum, $h/e^2 = 25.813$ kΩ in order for the energy uncertainty to be much smaller than the charging energy. The conditions to be fulfilled are:[73]

$$R_t \gg \frac{h}{e^2} \quad \text{and}$$

$$\frac{e^2}{C} \gg k_B T$$

The first criterion can be met by weakly coupling the island to the electron reservoir. The second condition is satisfied by making the island very small. Room-temperature operation requires structures of less than 10 nm, much smaller than present lithography resolutions and difficult to achieve.

Figure 14.22b shows the operation of the tunneling device. By applying a positive gate voltage, an electron is transferred to the dot at the critical gate voltage. At this moment, the potential in the dot region is decreased and blocks the transfer of other electrons (Coulomb blockade). If the nanoparticles/gate/reservoir form part of an FET, then this trapping of an electron in the dot shifts the threshold voltage of the transistor. Therefore, by sensing the current difference between the two states, the stored information can be read.

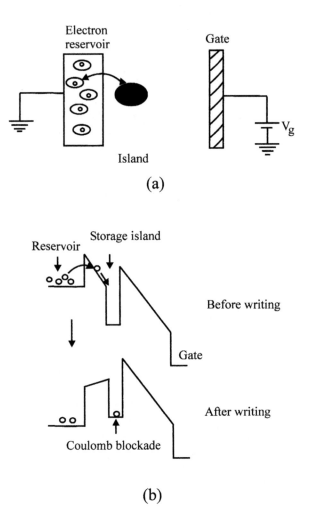

Figure 14.22 (a) Principle of Coulomb blockade; transfer of single electrons from an electrode (reservoir) to a conductive dot (island). (b) Schematic energy diagram for a single-electron memory showing the writing of one bit of data.[72]

Many approaches have been used to form the size of nanoparticle necessary for operation of such single-electron devices at room temperature. For example, iron oxide nanoparticles can be synthesized by the reduction of ferric chloride.[74] Langmuir–Blodgett layers of the cadmium salt of a long-chain fatty acid can be converted to arrays of CdS nanoclusters by exposure to H_2S gas.[75,76] Nanoparticles consisting of metallic particles capped with a short-chain oligomer can be manipulated directly by the LB method. Figure 14.23 shows a transmission electron micrograph of such particles deposited onto a carbon-coated microscope grid.[77] The individual particles, with a mean diameter of approximately 8 nm, are organized in a close-packed arrangement.

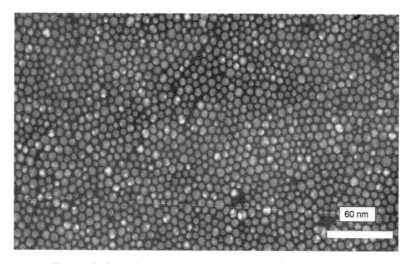

Figure 14.23 Transmission electron micrograph of a Langmuir–Blodgett layer of gold nanoparticles deposited onto a carbon-coated grid.[77]

14.5.3 DNA electronics

The study of the electronic behavior of organic compounds has led some scientists to work on the electrical properties of biological materials. Deoxyribonucleic acid (DNA) is arguably the most significant molecule in nature. Reports of the electronic properties of DNA have already generated controversy in the literature.[78–82] According to some, DNA is a molecular wire of very small resistance; others, however, find that DNA behaves as an insulator. These seemingly contradictory findings can probably be explained by the different DNA sequences and experimental conditions used to monitor the conductivity.[79]

The DNA strand in the double-helix arrangement consists of a long polymer backbone comprising repeating sugar molecules and phosphate groups. Each sugar group is attached to one of four bases, guanine (G), cytosine (C), adenine (A), and thymine (T). The chemical bonding is such that an A base only ever pairs with a T base, while a G always pairs with a C. Some of the electron orbitals belonging to the bases overlap quite well with each other along the long axis of the DNA. This provides a path for electron transfer along the molecule, in a similar fashion to the one-dimensional conduction seen in conjugated polymers (see Section 14.2).

Theoretical and experimental work now suggests that a hole (i.e., a positive charge) is more stable on a G–C base pair than on an A–T base pair.[79] The energy difference between these two pairs is substantially larger than the thermal energy of the charge carrier. Under these conditions, a hole will localize on a particular G–C base pair. Because the A–T base pairs have a higher energy, they act as a barrier to hole transfer. However, if the distance between two G–C base pairs is small enough, the hole can tunnel quantum mechanically from one pair to the next. In this way, charge carriers are able to shuttle along a single DNA molecule over a distance of a few nanometers.

"DNA chips" exploit the fact that short strands of DNA will bind to other segments of DNA that have complementary sequences and can therefore be used to probe whether certain genetic codes are present in a given specimen of DNA. Microfabricated chips with many parallel DNA probes are becoming widespread in analytical and medical applications. Currently, the chips are read out optically, but further miniaturization might require new read-out schemes, possibly involving the electron-transfer properties of DNA.

Computations by chemical or biological reactions overcome the problem of parallelism and interconnections in a classical system. If a string of DNA can be put together in the proper sequence, it can be used to solve combinational problems. The calculations are performed in test tubes filled with strands of DNA. Gene sequencing is used to obtain the result. For example, Adleman[83] calculated the "traveling salesman" problem to demonstrate the capabilities of DNA computing. DNA computing on parallel problems potentially provides 10^{14} MIPS (millions of instructions per second) and uses less energy and space than conventional supercomputers. While CMOS supercomputers operate 10^9 operations per Joule, a DNA computer could perform about 10^{19} operations per Joule. Data could potentially be stored on DNA in a density of approximately 1 bit per nm^3 while existing storage media such as DRAMs require 10^{12} nm^3 to store 1 bit.

14.6 Conclusions

Organic compounds possess a wide range of fascinating physical and chemical properties that make them attractive candidates for exploitation in electronic and opto-electronic devices. It is not, however, expected that these materials will displace silicon in the foreseeable future as the dominant material for fast signal processing. It is far more likely that organic materials will find uses in other niche areas of electronics, where silicon and other inorganic semiconductors cannot compete. Examples already exist, such as liquid crystal displays and certain chemical sensors. Organic light emitting structures are likely to make a major impact in the market place over the next ten years.

For commercialization, organic materials are usually required in the form of thin films. Simple processing methods, such as spin-coating and thermal evaporation, are already available. A number of other thin-film technologies offer the means to manipulate arrays of organic molecules (i.e., molecular engineering). Techniques such as Langmuir–Blodgett film deposition, self-assembly, and layer-by-layer electrostatic deposition allow intriguing molecular architectures to be built up on solid surfaces. Quantum mechanical effects can control the behavior of devices based on such assemblies of molecules.

Over the first decades of the 21st century, classical CMOS technology will come up against a number of technological challenges. The bottom-up approach to molecular electronics provides an alternative and attractive way forward and, as such, it is currently an area of exciting interdisciplinary activity. Analogies with the living world are seductive and much can be learnt from nature (for example, neural networks for chemical sensing). However, it is rather too early to tell whether sophisticated computing structures mimicking the human brain can be produced. It is probably not appropriate to emulate individual inorganic devices (e.g., Si MOSFETs) using organic, or even biological, compounds, and then to connect these together using the techniques developed by the microelectronics industry. The "wiring" problems could be insurmountable. Living systems assemble themselves from molecules and are extremely energetically efficient when compared with manmade computational devices. More radical approaches to materials fabrication and device design (e.g., exploiting self-organization) are almost certainly required if molecular-scale electronics is to be realized.

References

1. Petty, M.C., Bryce, M.R., and Bloor, D., Eds., *An Introduction to Molecular Electronics*, Edward Arnold, London, 1995.
2. Richardson, T.H., Ed., *Functional Organic and Polymeric Materials*, John Wiley & Sons, Chichester, 2000.
3. Aviram, A. and Ratner, M.A., *Chem. Phys. Lett.*, 29, 277, 1974.
4. Carter, F.L., *J. Vac. Sci. Technol. B*, 1, 959, 1983.
5. Kuhn, H., in *Molecular Electronics*, Hong, F.T., Ed., Plenum Press, New York, 1989.
6. Braun, E. and MacDonald, S., *Revolution in Miniature*, Cambridge University Press, Cambridge, U.K., 1978.
7. Semiconductor Industry Association Roadmap, http://public.itrs.net/.
8. Ferraro, J.R. and Williams, J.M., *Introduction to Synthetic Electrical Conductors*, Academic Press, Orlando, FL, 1987.
9. Roth, S., *One-Dimensional Metals*, VCH, Weinheim, 1995.
10. Ulman, A., *Ultrathin Organic Films*, Academic Press, San Diego, CA, 1991.
11. Petty, M.C., *Langmuir-Blodgett Films*, Cambridge University Press, Cambridge, U.K., 1996.
12. Evanson, S.A., Badyal, J.P.S., Pearson, C., and Petty, M.C., *J. Phys. Chem.*, 100, 11672, 1996.
13. Swart, R.M., in *Langmuir-Blodgett Films*, Roberts, G.G., Ed., Plenum Press, New York, 1990.
14. Cui, D.F., Howarth, V.A., Petty, M.C., Ancelin, H., and Yarwood, J., *Thin Solid Films*, 192, 391, 1990.
15. Nicolini, C., Ed., *Molecular Bioelectronics*, World Scientific, Singapore, 1996.
16. Moore, A.J., Goldenberg, L.M., Bryce, M.R., Petty, M.C., Monkman, A.P., Marenco, C., Yarwood, J., Joyce, M.J., and Port, S.N., *Adv. Mater.*, 10, 395, 1998.
17. Netzer, L. and Sagiv, J., *J. Am. Chem. Soc.*, 105, 674, 1983.
18. Netzer, L., Iscovici, R., and Sagiv, J., *Thin Solid Films*, 99, 235, 1983.
19. Decher, G., Hong, J.D., and Schmitt, J., *Thin Solid Films*, 210/211, 831, 1992.
20. Decher, G., Lvov, Y., and Schmitt, J., *Thin Solid Films*, 244, 772, 1994.
21. Pearson, C., Nagel, J., and Petty, M.C., *J. Phys. D: Appl. Phys.*, 34, 285, 2001.

22. Cheung, J.H., Stockton, W.B., and Rubner, M.F., *Macromolecules*, 30, 2712, 1997.
23. Lvov, Y., Onda, M., Ariga, K., and Kunitake, T., *J. Biomater. Sci., Polym. Ed.*, 9, 345, 1998.
24. Stockton, W.B. and Rubner, M.F., *Macromolecules*, 30, 2717, 1997.
25. Brittain, S., Paul, K., Zhao, X.-M., and Whitesides, G., *Phys. World*, 11, 31, 1998.
26. Chi, L.F., Eng, L.M., Graf, K. and Fuchs, H., *Langmuir*, 8, 2255, 1992.
27. Gimzewski, J., *Phys. World*, 11, 25, 1998.
28. Piner, R.D., Zhu, J., Xu, F., Hong, S., and Mirkin, C.A., *Science*, 283, 661, 1999.
29. Hong, S., Zhu, J., and Mirkin, C.A., *Science*, 286, 523, 1999.
30. Hong, S. and Mirkin, C.A., 288, 1808, 2000.
31. Liu, X.G., Fu, L., Hong, S.H., Dravid, V.P., and Mirkin, C.A., *Adv. Mater.*, 14, 231, 2002.
32. Lee, K.B., Park, S.J., Mirkin, C.A., Smith, J.C., and Mrksich, M., *Science*, 295, 1702, 2002.
33. Lu, Z., Qian, F., Zhu, Y., Yang, X., and Wei, Y., *Thin Solid Films*, 243, 371, 1994.
34. May, P., *Phys. World*, 8, 52, 1995.
35. Rhoderick, E.H., *Metal-Semiconductor Contacts*, Clarendon Press, Oxford, 1978.
36. Hua, Y.L., Petty, M.C., Roberts, G.G., Ahmad, M.M., Hanack, M., and Rein, M., *Thin Solid Films*, 149, 161, 1987.
37. Martin, A.S., Sambles, J.R., and Ashwell, G.J., *Phys. Rev. Lett.*, 70, 218, 1993.
38. Ashwell, G.J. and Gandolfo, D.S., *J. Mater. Chem.*, 11, 246, 2001.
39. Horowitz, G., *Adv. Mater.*, 2, 286, 1990.
40. Katz, H.E., *J. Mater. Chem.*, 7 369, 1997.
41. Pearson, C., Gibson, J.E., Moore, A.J., Bryce, M.R. and Petty, M.C., *Electron. Lett.*, 29, 1377, 1993.
42. Pearson, C., Langmuir–Blodgett Films of Organic Charge-Transfer Complexes, Ph.D. thesis, University of Durham, U.K., 1996.
43. Laquindanum, J.G., Katz, H.E., Lovinger, A.J., and Dodabalapur, A., *Chem. Mater.*, 8, 2542, 1996.
44. Bao, Z., Lovinger, A.J., and Dodabalapur, A., *Adv. Mater.*, 9, 42, 1997.
45. Dimitrakopoulos, C.D., Furman, B.K., Graham, T., Hedge, S., and Purushotha-man, S., *Synth. Met.*, 92, 47, 1998.
46. Brown, A.R., Pomp, A., Hart, C.M., and de Leeuw, D.M., *Science*, 270, 972, 1995.
47. Dimitrakopoulos, C.D., Purushothaman, S., Kymissis, J., Callegari, A., and Shaw, J.M., *Science*, 283, 822, 1999.
48. Sirringhaus, H., Kawase, T., Friend, R.H., Shimoda, T., Inbasekaran, M., Wu, W., and Woo, E.P., *Science*, 290, 2123, 2000.
49. Burroughes, J.H., Bradley, D.D.C., Brown, A.R., Marks, R.N., Mackay, K., Friend, R.H., Burns, P.L., and Holmes, A.B., *Nature*, 347, 359, 1990.
50. Hudson, A.J. and Weaver, M.S., in *Functional Organic and Polymeric Materials*, Richardson, T.H., Ed., John Wiley & Sons, Chichester, 2000, p. 365.
51. Kim, J.S., Granstrn, M., Friend, R.H., Johansson, N., Salaneck, W.R., Daik, R.,Feast, W.J., and Cacialli, F., *J. Appl. Phys.*, 84, 6859, 1998.
52. Karg, S., Scott, J.C., Salem, J.R., and Angelopoulos, M., *Synth. Meth.*, 80, 111, 1996.
53. Brown, T.M., Kim, J.S., Friend, R.H., Cacialli, F., Daik, R., and Feast, W.J., *Appl. Phys. Lett.*, 75, 1679, 1999.
54. Baldo, M.A., O'Brien, D.F., You, Y., Shoustikov, A., Sibley, S., Thompson, M.E., and Forrest, S.R., *Nature*, 395, 151,1998.

55. Jung, G.Y., Pearson, C., Horsburgh, L.E., Samuel, I.D.W., Monkman, A.P., and Petty, M.C., *J. Phys. D: Appl. Phys.*, 33, 1029, 2000.
56. Sterbacka, R., Juka, G., Arlauskas, K., Pal, A.J., Kllman, K.-M., and Stubb, H., *J. Appl. Phys.*, 84, 3359, 1998.
57. Cimrov, V., Remmers, M., Neher, D. and Wegner, G., *Adv. Mater.*, 8, 146, 1996.
58. Burns, S.E., Pfeffer, N., Grηner, J., Neher, D., and Friend, R.H., *Synth. Meth.*, 84, 887, 1997.
59. Janata, J., *Principles of Chemical Sensors*, Plenum Press, New York, 1989.
60. Gardner, J.W., *Microsensors*, John Wiley & Sons, Chichester, 1994.
61. Moseley, P.T. and Crocker, A.J., *Sensor Materials*, IOP Publishing, Bristol, 1996.
62. Snow, A.S. and Barger, W.R., in *Phthalocyanines: Properties and Applications*, Leznoff, C.C. and Lever, A.B.P., Eds., VCH Publishers, New York, 1989, p. 342.
63. Petty, M.C. and Casalini, R., *Eng. Sci. Ed. J.*, June, 99, 2001.
64. Barker, P.S., Chen, J.R., Agbor, N.E., Monkman, A.P., Mars, P., and Petty, M.C., *Sensors Actuators B*, 17, 143, 1994.
65. Gardner, J.W. and Bartlett, P.N., *Sensors Actuators B*, 18, 211, 1994.
66. Casalini, R., Wilde, J.N., Nagel, J., Oertel, U., and Petty, M.C., *Sensors Actuators B*, 18, 28, 1999.
67. Casalini, R., Nagel, J., Oertel, U., and Petty, M.C., *J. Phys. D: Appl. Phys.*, 31, 3146, 1998.
68. Jones, C.A., Petty, M.C., and Roberts, G.G., *IEEE Trans. Ultrasonics Ferroelec. Freq. Control*, 35, 736, 1988.
69. Burfoot J.C. and Taylor, G.W., *Polar Dielectrics and Their Applications*, Macmillan, London, 1979.
70. Richardson, T.H., in *Functional Organic and Polymeric Materials*, Richardson, T.H., Ed., John Wiley & Sons, Chichester, 2000, p. 181.
71. Petty, M., Tsibouklis, J., Petty, M.C., and Feast, W.J., *Thin Solid Films*, 210/211, 320, 1992.
72. Yano, K., Ishii, T., Sano, T., Mine, T., Murai, F., Hashimoto, T., Kobayashi, T., Kure, T., and Seki, K., *Proc. IEEE*, 87, 633, 1999.
73. Kouwenhoven, L.P. and McEuen, P.L., in *Nanotechnology*, Timp, G., Ed., Springer-Verlag, New York, 1999, p. 471.
74. Moore, R.G.C., Evans, S.D., Shen, T., and Hodson, C.E.C., *Physica E*, 9, 253, 2001.
75. Nabok, A.V., Richardson, T., McCartney, C., Cowlam, N., Davis, F., Stirling, C.J.M., Ray, A.K., Gacem, V., and Gibaud, A., *Thin Solid Films*, 327–329, 510, 1998.
76. Prasanth, N., Kumar, S., Narang, N., Major, S., Vitta, S., Talwar, S.S., Dubcek, P., Amenitsch, H., and Bernstorff, S., *Colloids Surfaces A*, 198–200, 59, 2002
77. Cousins, M.A. and Petty, M.C., Unpublished data (m.c.petty@durham.ac.uk).
78. Fink, H.-W. and Schneberger, C., *Nature*, 398, 407, 1999.
79. Dekker, C. and Ratner, M.A., *Physics World*, 14, 29, 2001.
80. Rakitin, A., Aich, P., Papadopoulos, C., Kobzar, Y., Vedeneev, A.S., Lee, J.S., and Xu, J.M., *Phys. Rev. Lett.*, 86, 3670, 2001.
81. Phadke, R.S., *Appl. Biochem Biotech.*, 96, 269, 2001.
82. Rao, C.N.R. and Cheetham, A.K., *J. Mater. Chem.*, 11, 2887, 2001.
83. Adleman, L., *Science*, 266, 1021, 1994.

appendix:
Summary of computer programs for analysis and design

Contents

What follows is list of computer software packages commonly used for the design and analysis of neuroprostheses and the processing of biological signals. This list is by no means exhaustive but is instead intended as a basic resource for assembling a suite of analytical and design tools. The list primarily consists of the more commonly used packages of which the editors are aware. Each software package is listed under the area where it is most commonly used.

Neuron modeling

1. *NEURON* (http://www.neuron.yale.edu/): Free, downloadable software for developing and exercising models of neurons and networks of neurons. The authors state that it is especially useful for problems where the cable properties of cells are important and where cell membrane properties are complex.
2. *GENESIS* (http://www.genesis-sim.org/GENESIS/): The GEneral NEural SImulation System is a general purpose simulation platform that supports simulations of complex models of single neurons through simulations of larger networks of more abstract neuronal components. The software is free and downloadable from the website.

Electronic circuit construction and simulation

1. *PSpice* (Cadence Design Systems, Inc.): One of the most basic circuit simulation products on the market and a common tool for electrical engineers.
2. *Electronics Workbench* (Electronics Workbench): Another of the most commonly used electronic design and automation products.
   (Note that many of the neuron models discussed in Chapter 3 have been modeled using programs such *PSpice* and *Electronics Workbench*.)
3. *Orcad Capture* (Cadence Design Systems, Inc): One of a family of tools for the design of complex analog and digital circuitry for VLSI systems.
4. *Micro Magic Tools* (Juniper Networks): A suite of design tools for the design of complex chips at the physical level. Often free for educational applications.
5. *Blast Software* (Magma Design Automation): An innovative suite of products that uses gain-based synthesis to synthesize millions of gates simultaneously. A well-received product among VLSI researchers.

Data capture and processing software

1. *LabVIEW* (National Instruments): One of the most commonly cited data capture programs. Part of a suite of programs that performs video and data capture and represents capture and processing functions as programmable function blocks.
2. *MATLAB* and *Simulink* (The MathWorks): Multifunctional programs for performing calculations, analyzing and visualizing data, and modeling and simulating complex dynamic systems; contains many specialized toolboxes for implementing common signal processing and communications system algorithms, including wavelets.
3. *IDL* (Research Systems, Inc.): Another widely used program for visualizing, processing and compiling data
4. *Mathematica* (Wolfram Research): An excellent data analysis and technical programming tool, with the benefit of direct document creation for technical presentations. Also available is the *Wavelet Explorer* package, which allows for a variety of wavelet-based processing functions
5. *RC Electronics*: A multichannel data collection program capable of multichannel oscilloscope and strip-chart recording and signal processing.
6. *MSLib* (www.igpm.rwth-aachen.de/urban/html/progs.html): Free wavelet processing software that can be downloaded with permission and a password from the author.

7. Wavelet software, general (www.wavelet.org/links.html#software): A website managed by *Wavelet Digest* with links and information on several software packages for wavelet analysis.
8. *THINKS* (Sigma Research): Highly recommended neural network development program with beginner and professional versions; considered to be very user-friendly.

Electromagnetic modeling

1. *FEMLAB Electromagnetics* (COMSOL AB): A component of the FEMLAB software for modeling and solving scientific and engineering problems. The electromagnetics module is capable of solving and displaying the results of several different electromagnetics problems.
2. *FEKO* (EMSS): A full-wave, method-of-moments-based program capable of analyzing the electromagnetics of dielectric bodies, planar stratified media, and similar problems commonly found in neuroprosthetic design.
3. *IE3D* (Zeland Software, Inc.): Also a method-of-moments-based electromagnetics design and analysis tool. Zeland also offers other products for specialized electromagnetics solutions.
4. *CAEME* (NSF/IEEE Center for Computer Applications in Electromagnetics Education): A popular electromagnetics design and analysis tool among academic institutions available from the National Science Foundation in the United States.

Index

B

B&B Medical Technologies, 328
Back-telemetry, 131, 139, 140–143
Backscatter load modulation, 140
Band model, 390
BarMed Pty., Ltd., 327
Baseline drift, 369
Battery technology, 337
Battle, Guy, 197
Behaving animals, chronic recording from,
    185
Bi-CMOS, approach to MEMS-based
    neuroprosthetic design, 357, 379
Bidomain model, 58
Binary wavelet packet tree, 206
Biochemical sensors, 128
Biocompatible materials, 129, 317
    in epiretinal neuroprostheses, 289
Bioengineering, effect of emerging
    technologies on, 4
Biofuel cells, 337
Biological biocompatibility, 280–281
Biomedical engineering, effect of emerging
    technologies on, 4–5
Biomedical signals
    features of, 195
    wavelets in processing of, 209–210
Biomimetic principles for construction of
    prosthetic limbs, 230
Biomimetic systems, 396
Biomolecular electronics, 12
BION system, 336
Bionic Glove, 324
Bionic Technologies, 168
Biopotential
    analysis, 83
    recording, 84
Biotelemetry, 130
Biotic hands, 11
Biphasic amplifier, 153
Biphasic stimuli
    for chronic stimulation, 120
    use of in CNS, 118–119
Bipolar cells of the retina, 268, 281
Bipolar cochlear electrode configurations,
    182, 254
Bipolar junction transistor (BJT), 358
Bipolar neurons, 100
Bipolar stimulation, 90
Bladder control, 232
    lower extremity motor prostheses and,
    331–332
    motor prostheses for control of, 327–328
Blindness, pathologies leading to, 271–273

Bone conduction implantable hearing
    devices, 241–242
Bone-anchored hearing apparatus, 241
Bootstrap reference, 153
Boron, use in fabrication of Michigan probe,
    173–174
Boston arm, 225
Brain injury, 310–312
Brain-computer interface, 278
Bray, Victor H. Jr., 238
Brindley, Giles, 262, 298
Brindley-Finetech Bladder Controller. *See*
    Finetech-Brindley Bladder
    Controller
Buffering, on-chip, 167
Buzaski, Gyorgy, 183

C

Ca channels. *See* calcium channels
CA^{2+}-ATPase pump, 24
Cable equation
    continuous form, 100–104
    discrete form, 104–106
    infinite, 111
Cable theory, 22, 100
Calcium channels, voltage gated, 22, 29,
    36–37, 79
Calcium-dependent gating, 44
Capacitive coupling, 351
Capacitive electrodes, 87
Capacitive modulation, 140
Capacitors, 362
Cardiac assist devices, 334
Cardiac pacemaker implants, 146, 262,
    318–319
Carrier mobility, 390–391, 401
Carrier velocity, 390
Case Western Reserve University, 329, 332,
    334
Cathodic stimulation, 54, 115
Cathodic threshold values, 42
CCD. *See* charge-coupled devices
Cell axon, 78
Cells
    communication with, 14–15
    damage, 86–88
    selective stimulation of, 88–90
Cellular excitability, 24
Cellular regeneration, 335
Central nervous system
    damage to, 308–313
    elements specific to prostheses of, 375
    geometric properties of neurons of, 100